HIDE
AND
SEEK

Books by Peter Hucthausen

Echoes of the Mekong

Hostile Waters (with Igor Kurdin and R. Alan White)

Frye Island: Maine's Newest Town, a History, 1748-1998

K19: The Widowmaker

October Fury

America's Splendid Little Wars: A Short History of U.S. Military Engagements, 1975–2000

Shadow Voyage: The Extraordinary Wartime Escape of the Legendary S.S. Bremen

Books by Alexandre Sheldon-Duplaix

Histoire des sous-marins des origins à nos jours (with Jean Marie Mathey)

Les sous-marins, fantômes des profonduers (with David Camus)

Histoire mondiale des porte-avions

HIDE
AND
SEEK

The Untold Story of Cold War Naval Espionage

**Peter A. Huchthausen
and
Alexandre Sheldon-Duplaix**

WILEY

John Wiley & Sons, Inc.

Published by John Wiley & Sons, Inc., Hoboken, New Jersey
Published simultaneously in Canada

Photo credits appear on page 399.

Library of Congress Cataloging-in-Publication Data:

Huchthausen, Peter A., date.
    Hide and seek : the untold story of Cold War naval espionage / Peter A. Huchthausen
and Alexandre Sheldon-Duplaix.
      p. cm.
    Includes bibliographical references and index.
    ISBN 978-0-471-78530-9 (cloth : alk. paper)
      1. Military intelligence—History—20th century. 2. United States. Navy—History—
20th century. 3. Soviet Union. Voenno-Morskoi Flot—History. 4. Naval history, Modern—
20th century. 5. Cold War. I. Sheldon-Duplaix, Alexandre. II. Title.
VB230.H84 2008
359.3'432097309045—dc22

                                  2008018997

Printed in the United States of America

10 9 8 7 6 5 4 3 2 1

To the many nameless combatants
who fought in the shadows to
keep the world free

PETER HUCHTHAUSEN
SEPTEMBER 25, 1939–JULY 11, 2008

Captain Peter Huchthausen, USN (retired), died in Normandy, France, on July 11, 2008. He graduated from the United States Naval Academy in 1962. Initially, he served as a line officer on the destroyer USS *Blandy* during the Cuban missile crisis, enforcing the naval blockade and verifying the removal of Soviet missiles from Cuba. Peter then served two tours of duty in Vietnam, first as a skipper of a patrol boat with the USN Riverine forces in the Mekong Delta. He returned to Vietnam on the destroyer USS *Orleck*, which provided naval gunfire support to army and marine forces operations along the Vietnam coast. Transferring to Naval Intelligence, he served on the NATO Naval Staff in London, England. He then returned to sea duty for a three-year tour as the USS *Enterprise* Battle Group Intelligence Officer. Selected for attaché duty, he was posted as the naval attaché to Yugoslavia and Romania and later to the Soviet Union. Following retirement in 1990, Peter authored seven books. Two of them were made into movies: the HBO movie *Hostile Waters* and *K19: The Widowmaker.* Peter began visiting Normandy in 1994 and was so taken by what the American airborne forces accomplished in Operation Neptune in June 1944 that in 2005 he decided to permanently retire near Sainte-Mère-Eglise.

Captain Huchthausen was awarded the Defense Meritorious Service Medal; the Navy Combat Medal with Combat "V"; two Joint Meritorious Service medals; the Vietnamese Cross of Gallantry with Bronze Star; two Presidential Unit Citations; the Vietnam Service Medal with five stars; the Armed Forces Expeditionary Medal; the Vietnamese Civic Action Medal; the Vietnamese Expeditionary Medal; the National Defense Medal; the Sea Service Deployment Ribbon; the Vietnamese Campaign Medal; and the USN/USMC Hostile Action Ribbon.

CONTENTS

ACKNOWLEDGMENTS

The authors express their sincere gratitude for the invaluable assistance rendered by Captain Uwe Mahrenholtz, Federal German Navy retired; Lieutenant Commander Desmond Robinson, Royal Navy retired; and Allan Shore, former Defense Intelligence Staff in Whitehall; and Ambassador Yoya Kawamura, Japan.

From Russia we appreciate the candid information from Fleet Admiral Vladimir N. Chernavin and Vice Admirals Valentin Selivanov and Yuri Kviatovskiy. We also thank Rear Admirals I. Ivanov and Lev D. Chernavin; Captains Lev Vtorygin, Sergei Aprelev, Nikolai Muru, Igor Kurdin, Yevgeniy Litvanov, Oleg Malov, and Valentin Shigin; Colonel Boris Grigoriev; and Colonel (Cosmic forces) Valeriy Pradishev (retired).

From Sweden we thank Rear Admiral Emil Svensson, Captain Erland Sonnerstedt, Captain Lars Wedin, producer Lars Borgnäs, and Professor Ola Tunander.

From the Naval Historical Center in the Navy Yard, Washington, D.C., we thank Dr. Edward Marolda, Kathy Lloyd, and Glen Helms.

From the Joint French Defense Historical Service (Service historique de la Défense), we thank Vice Admiral Louis de Contenson, Captain Serge Thébaut, Philippe Vial, Jean Martinant de Préneuf, and Cadet Mathieu Le Hunsec. From the former French Naval historical service (Service historique de la Marine), we thank Rear Admirals Alain Bellot, and Jean Kessler, retired.

Our gratitude also goes to Arthur D. Baker; Chris Carlson; Norman Friedman; Larrie Ferreiro; Colonel Werner Globke, federal German air force retired; Professor John Hattendorff, Naval War College, Newport, Rhode Island; Captain Claude Huan, French navy retired; Ambassador Yoya Kawamura, Japan; Patrice Loriot; Captain Uwe Mahrenholtz, federal German navy retired; Rear Admiral Jean-Marie Mathey, French navy retired; Norman Polmar; Bernard Prezelin; Lieutenant Commander Desmond Robinson, Royal Navy retired; Peter Swartz, USN (retired); and Allan Shore, former Defense Intelligence staff in Whitehall.

Special thanks are due to translator Inna Smirnova and our frequent Washington, D.C., hostess, Rona Feit.

NOTE TO THE READER

The following pages relate events explaining the critical role of naval intelligence from the end of World War II in 1945 until the collapse of the Soviet Union in 1991. Sensitive archives are usually kept secret until the date of their destruction, while more mundane intelligence papers are generally made available to the public after fifty or sixty years. Declassification initiatives sometimes shorten this delay, most notably in the United States. Personal recollections are often distorted by egotism, political or professional passions, close friendships, hatred, and vanishing memories. Bearing in mind these limitations, the authors present candid accounts depicting key events in the long behind-the-scenes Cold War naval intelligence scramble.

Most of the material from 1945 through 1962 was obtained from Western archives and recent Russian studies, books, and articles little known in the West at the time of this writing. For the period after 1962, the authors of this book have augmented these sources with interviews and their own personal experiences and firsthand knowledge.

Peter Huchthausen served at sea and ashore in antisubmarine warfare; was a senior U.S. naval attaché in Belgrade, Bucharest, and Moscow; and between assignments was part of the Defense Intelligence Agency attaché operations and human intelligence organization in Washington, D.C.

Alexandre Sheldon-Duplaix works in the French Joint Historical Service in Vincennes, France, and is a researcher and a lecturer on the history of the Soviet navy at the Defense Staff College in Paris. He was a naval analyst under contract with the French navy from 1987 to 1999.

Retired Soviet navy captain first rank Lev Vtorygin provided much of the material from the Soviet side to the authors over a period of fifteen years. Vtorygin had a successful twenty-five-year career in the Soviet navy. After his sea duty in destroyers in the Baltic Sea, he won coveted selection to attend the military diplomat academy in Moscow. He served with distinction abroad as an officer in military intelligence. His first assignment was as assistant naval attaché in Buenos Aires, Argentina, in the late 1950s, then in the Soviet Embassy in Washington, D.C., from 1960 to 1964. During the Cuban missile crisis in 1962, Vtorygin traveled along the East Coast of the United States and reported on the U.S. forces being readied to invade Cuba. In the 1970s and 1980s, he served in Fleet headquarters as an adviser in Vietnam and as a military analyst.

Throughout the book, the reader will encounter personal accounts by Peter Huchthausen and Lev Vtorygin that flesh out the other sources.

Peter Huchthausen gives his personal accounts as a former U.S. naval intellingence officer and attaché in Belgarde, Bucharest, and Moscow in chapters 5, 10, 15, 17. Alexandre Sheldon-Duplaix has used declassified U.S., British, French, and Swedish archives to write chapters 2, 4, 6, 9, 14, 16. The authors have shared the research and the writing of the remaining chapters.

INTRODUCTION

For many years, it was widely accepted that the greatest success of Allied naval intelligence in World War II was breaking the Japanese naval and diplomatic codes. This achievement led to the pivotal 1942 battle at Midway, in which the U.S. Navy defeated the Japanese fleet. Arguably, however, a lesser-known naval intelligence coup in 1945 may have given the United States the final missing elements it needed for the Manhattan Project to produce the first atomic bomb. The United States then used this bomb to end the war with Japan.

Although the media have publicized that Adolf Hitler had lost interest in a nuclear weapons program after the successful 1941 commando sabotage of the Norsk heavy-water plants in Ryukon, Norway—by the famed "Heroes of Telemark"—this may not have been true. A recent study has shown that the Nazis had perhaps accumulated enough fissionable material to manufacture an atomic bomb. According to Carter Hydrick, in his book *Critical Mass*, from June 1940 to the end of the war, Germany seized 3,500 tons of uranium compounds from Belgium and stored it in salt mines in Strassfurt, Germany. The amount of uranium held by the Nazis would have been then more than three times the amount possessed by the U.S. Manhattan Project.[1]

When the war turned against the Third Reich, and with certain defeat at hand, Berlin authorized the transfer of large quantities of uranium oxide to Tokyo. In 1944, three Japanese submarines carried German uranium to Japan but were sunk near Malaysia.[2] The Japanese submarine I-52 was sunk while carrying uranium from Germany to Japan in early 1945. Then, in January and February 1945, two German submarines attempted again to deliver uranium oxide. One was sunk en route and the other on her return voyage, after failing to reach Japan because of the Allied blockade. The German and Japanese submarines attempted to deliver their cargos of uranium oxide to the joint Japanese-German submarine base at Penang (Palau Pinang), for further shipment to Japan in fast warships. The Allied blockade of Japan was so successful that submarines were considered more vulnerable to attack than faster surface warships were. The German submarine presence in the Indian Ocean was called the Monsoon Group.[3] On February 9, 1945, HMS *Venture* torpedoed and sank U-864 off the coast of Norway while the submarine was on the way to Penang with a cargo similar to that of U-234's.

In March 1945, during the final months of Hitler's Reich, U.S. and Royal Navy intelligence learned from several sources that the Germans would make yet one more attempt to ship by submarine large quantities of uranium oxide taken from Belgium, Czechoslovakia, and Norway.[4] Code breakers confirmed this information while reading German and Japanese naval and diplomatic communications.

In late March, the Allies gained intelligence that the German submarine U-234 would be carrying uranium oxide and other defense treasures as a final effort to transfer vital technical material and knowledge from the dying Third Reich. The Nazis hoped that the shipment would allow Japan to fulfill its quest to construct a nuclear weapon. It is understandably not popular knowledge, especially in Japan today, that the Japanese army and navy were on the way to making their own atomic bombs in 1945. Indeed there are some controversial claims that the Japanese may have tested in Korea a small nuclear device in the weeks preceding Hiroshima. According to some scholars, Tokyo needed only additional weapons-grade enriched uranium to complete a weapon.[5] Had Japan succeeded in producing one or more weapons, it planned to drop them from airplanes launched from aircraft-carrying submarines, targeting a major city on the U.S. West Coast.[6]

Earlier in the war, Japanese submarines had already bombarded California. On the evening of February 23, 1942, the Japanese submarine I-17 had surfaced and fired sixteen shells at an oil refinery near Santa Barbara. Three-year-old Peter Huchthausen was among the bystanders evacuated from the beach while army planes were hurried to attack the submarine. This was Japan's first attack on the U.S. mainland, in what is remembered as the shelling of the suburb Ellwood. Twice the following night, more than a million residents of Los Angeles were awakened first by air-raid sirens and then by antiaircraft gunfire. For about half an hour, the 37th Coast Artillery Brigade expended 1,430 rounds of ammunition against "as many as fifteen airplanes . . . other than American Army or Navy," as reported by General George S. Marshall, the army chief of staff to President Franklin Roosevelt. The mysterious airplanes flew at various speeds, from what is reported as being "very slow" to as much as 200 miles per hour at altitudes from 9,000 to 18,000 feet.[7] No proper explanation was offered for this mysterious air raid that did not involve conventional aircrafts. But the West Coast now seemed within reach of Japan's submarine-based seaplanes. Three years later, the cargo of

U-234 could offer Tokyo a last chance to turn the tide of the war by striking the continental United States with a nuclear weapon.

The naval team at Bletchley Park, England, gained the intelligence on U-234's departure via decrypted messages. The intercept revealed that Germany, in a last-gasp effort to assist its ally, would dispatch a large, 1,763-ton displacement, type XB mine-laying submarine with 560 kilograms of precious enriched uranium oxide to sail to Japan.[8] Included in the submarine's strategic cargo was a German naval officer scientist, Korvettenkapitan (commander). Dr. Heinz Schlicke, a radio, radar, and fusing expert, was the inventor of the infrared proximity fuse, which both the Japanese and the Americans still needed to detonate an atomic bomb.

Japan, like the U.S. project Manhattan, lacked enough fissionable material and an adequate fuse to set off the fission chain reaction.[9] U-234 transported assorted types of technical equipment accompanied by two Japanese naval officers; several disassembled jet aircraft; and a famous Luftwaffe general, Ulrich Kessler, who was carrying plans for the *Amerika* rocket—the future missile with a range capable of reaching New York City from Germany.[10]

The submarine sailed from Norway in late April 1945. Then, shortly after Germany's surrender on May 8, U-234 was detected and chased by Allied antisubmarine warfare (ASW) forces. After several attempts by the Canadian navy to lure the submarine into Canadian waters, the commanding officer of U-234 decided to surrender to the U.S. Navy. Just after the submarine's commander informed the crew of his intent, the two Japanese officers committed suicide. The submarine surfaced off the coast of Nova Scotia, was captured by the destroyer escort USS *Sutton* (DE-771), and was escorted to Portsmouth Naval Base in New Hampshire. Immediately following U-234's mooring in Portsmouth, Dr. Schlicke and his infrared proximity fuse were escorted to Washington, D.C., while eighty gold-lined cylinders containing the uranium were spirited away to the Manhattan atomic bomb project.[11] Within weeks, the Manhattan Project had suddenly doubled production of enriched uranium and by the end of July had a sufficient quantity to complete a bomb.[12] The first U.S. bomb was shipped to the Pacific island Tinian aboard the cruiser USS *Indianapolis* (CA-35), loaded into a U.S. Army Air Force B-29, and dropped untested on Hiroshima. The rest is history.

The coveted prizes of the U-234, the uranium, and the jet aircraft with its invaluable secret passengers had escaped the notice of Soviet intelligence by only weeks, as the Red Army swept westward through the Baltic ports, scooping up all useful German technology in its path.

During nearly half a century of the Cold War, Western intelligence was pitted against top-notch Soviet and Eastern European intellects, who proved to be overwhelmingly successful, even though their countries imploded when the Berlin Wall was breached.

Hundreds of years of czarist authority and repressive Russian Orthodox Church followed by decades of Soviet Communism had thoroughly reduced personal initiative in the USSR's citizens. In this context, the most intelligent people naturally gravitated to fields that promised the greatest security and influence: the military and the intelligence services. With the overwhelming preponderance of military might and intelligence assets way out of proportion to what was necessary, the cream of the cream became entrenched in the top echelons of the intelligence services. There, they enjoyed perquisites that were not available to the average Ivan and Marina on the impoverished streets of Moscow or Leningrad.

In the West, by contrast, service in intelligence usually happened by chance. Few people studied with the goal of entering intelligence staffs and organizations. Yet World War II had created a need for talented and imaginative minds to form official groups that specialized in intelligence. Although Western democracies harbored an inherent distrust of those who worked in the shadows, victory brought much-deserved accolades and, on occasion, fame to these shadow warriors of the intelligence world.

Security was sometimes detrimental to efficiency. British intelligence was critical of the U.S. rigidity: "The American idea[s] on security . . . are really quite different from ours; every officer and man is ingrained from birth with a personal sense of responsibility toward his national security agencies, and is simply terrified of stepping outside the rules (even when they give some flexibility) for fear of risking his own neck."[13]

During forty-five years, naval espionage played a critical role in helping Washington maintain U.S. maritime supremacy over the expanding fleet of the Soviet Union and to assist Moscow in its asymmetrical effort to capture key technologies and tactical information necessary to make up for the shortcomings of its industry and military. Both sides harbored legitimate fears. The following pages will explain how naval minds had a very special part in building a clearer picture of the probable enemy's intentions and capabilities.

The effect on the course of the Cold War of known and less-known cases of naval espionage is discussed. Among the revelations or re-examinations are the following:

- Ian Fleming's special units and the Allied competition for Nazi naval technologies

- The Communist naval officer who attempted to take over De Gaulle's intelligence service
- The debriefing by the United States of returning Axis prisoners, defecting sailors, and persecuted prostitutes to collect intelligence across the iron curtain and in Korea
- Anglo-German penetration of Soviet waters in support of the anti-Communist resistance in the former Baltic states
- Stalin's naval thinking and war plans and their effect on Western intelligence
- The official Soviet investigation into the sinking of the battleship *Novorossiysk* and its lasting consequences
- A new look at Khrushchev's naval diplomacy and the meaning of the Crabb incident
- The manipulation of Soviet submarines reporting by the United States during the Suez crisis to influence Anglo-French planning
- The handling of a French air force general by a Soviet naval officer and GRU agent who provided essential information to the Kremlin before, during, and after the Cuban missile crisis
- The U.S. intelligence monitoring of the Soviet buildup and deception in Cuba
- The intelligence mistake that led to the U.S. escalation in Vietnam
- The good and bad fortunes of spy ships and an explanation for the attack on the USS *Liberty*
- The naval intelligence revolution of the 1980s and the competing U.S. and Soviet ocean surveillance systems
- The intrigues of a U.S. naval attaché in Belgrade, Bucharest, and Moscow
- The intelligence quagmire caused by the handling of nuclear naval reactors and radioactive waste after the collapse of the Soviet Union

This book also presents new facts about psychological warfare operations, the latest revelations on underwater incursions in Swedish waters, and how the Cold War navies tried to unravel strange encounters that apparently could not be attributed to the other side. Included are the Soviet naval intelligence instructions regarding "abnormal physical phenomena and objects."

1

VICTOR'S PLUNDER
1941–1945

Throughout the long years of World War II, it had become clear that most German military technology far exceeded the weaponry of the Allies. It was no secret that German tanks outperformed the West's mainline tanks and, until the advent of the Soviet Stalin T-34-85 battle tank, also outfought all the Red Army's tanks. The Germans used the first turbo jet engines in the Messerschmitt Me-262 fighter in combat in 1944. This paved the way for a series of modern German aircraft rolling off production lines in defiance of the heavy Allied strategic bombing of Germany. The scarcity of trained pilots, rather than a lack of material availability, prevented much use of these new airplanes. Even more ominous for the Allies were the growing successes of the German A-4 rocket, known more commonly as the V-2, and the promised threat of the A-6 and the A-9, called the *Amerika* rockets, with ranges stretching as far as North America.[1]

The use in combat of the Henschel Hs-293 glide bomb and its improved version with a radio guidance system made it the first naval cruise missile. This weapon proved devastating in the closing years of the war, causing the loss of one Allied battleship, half a dozen cruisers, and even more destroyers, as well as thousands of tons of merchant shipping. New German antiaircraft missile systems, *Wasserfall* and *Schmetterling*, had they become operational in greater numbers in 1944 and 1945, could have been ruinous to the Allied forces' heavy bombers. In May 1945, an advance naval technical exploitation team entered the Baltic port of Kiel ahead of the allied Twenty-First Army and found completed prototype models of the new German navy type XVIII, XVIV, and XXI submarines waiting at the piers to be manned with trained crews. A British member of the team said, "We didn't win the war any too soon."[2]

Why the large gap in military technology between the Allies and the German war machine? It was no myth that in Nazi Germany, scientists and engineers had been coddled and given immeasurable perks by politicians and military chiefs since the 1930s. In contrast, the U.S. and British military tended to shun scientists and modern innovations and to distrust new technology. Training manuals for the British tank corps still called for use of tanks in the role of cavalry as late as 1939. Compounding these obstacles, the subsequent Allied wartime blockade of minerals, petroleum products, and chemicals had compelled German scientists to develop an astounding array of synthetic fuels, materials, and even food products, which increased their technological lead in many areas over the Allies. The notion that because the Allies were winning the war their weapons were better was proved to be absurd. They were winning because they had more of everything and because the Nazi war machine was crumbling due to inept leadership, rather than from lack of modern military science. And as has been established in recent historical studies, the Germans were defeated on the ground by the Allied fighting men. Many of the much-touted German generals of the earlier blitzkrieg successes were outperformed not only by American and British generals, but, more important, by the soldier in the field. The Allied armies outfought the Germans on the ground primarily because of the rugged individualism and initiative of the individual GI and Tommy. As Secretary of the Navy James Forrestal had put it, "The guy with the rifle and machine gun is the man who wins the war in the last analysis and pays the penalty to preserve our liberty."[3]

Wartime Naval Intelligence Cooperation among the Soviet Union, Britain, and the United States

On June 22, 1941, on learning of the Nazi invasion of the Soviet Union, the British Foreign Office took the now unprecedented step of informing Moscow that the Finns, and therefore most likely their German partners, were reading Soviet ciphers.[4] Moscow was initially suspicious that there was a secret "capitalist" arrangement between Berlin and London that allowed Hitler to proceed with Barbarossa, the German invasion of the Soviet Union. The Foreign Office's gesture of giving the Soviets a heads-up, however, and its offer to send a tri-service mission to the USSR, accompanied by Churchill's supportive speech, were warmly received by Stalin. Within a month, a military cooperation plan was set up.

In July 1941, a Soviet military mission arrived in London, soon to be headed by Rear Admiral Ivan Kharlamov, while Britain dispatched a "30 Mission" to Moscow with its naval branch under the command of an admiral. By May 1942, naval intelligence exchanges had intensified. Every Tuesday the Admiralty representatives met with Kharlamov.[5] Intelligence on German naval weapons and on the Baltic theater, including Finnish shore batteries and ports, was passed on to the Soviet admiral, despite the British people's sympathy for Finland's war aims. In exchange, the Soviet navy furnished the Royal Navy with detailed characteristics of German battleships and heavy cruisers. The Royal Navy also presented the Soviet navy with intelligence on U-boat radio equipment and raider operations. Meanwhile, the Royal Navy had established a liaison office in the Black Sea under the command of a captain. His data on Romanian and Bulgarian Black Sea defenses were so useful to the Red Navy that he immediately got access to the Black Sea Fleet's daily intelligence summaries. The Black Sea Fleet later proved to be the most effective source of intelligence for the Axis forces' detailed orders of battle. Seeking fair trade, the Soviet navy used the British Black Sea naval mission as a back channel to request naval information.

By August 1942, Moscow had also authorized London to set up a communications intercept station called "Y Cottage" at Polyarny in the Kola Peninsula. This station fed the Admiralty information on German ships' movements in support of the Allied convoys to Murmansk and Archangelsk. After the losses experienced by convoy PQ 15 due to a lack of air cover, the Soviets stepped up their air operations in support of PQ 16. But when Churchill suspended the convoys after the PQ 17 disaster—this convoy was nearly completely lost when the British withdrew all escorts—an infuriated Stalin restricted the intelligence exchanges. Nevertheless, the joint Soviet-British code-breaking "Y Cottage" activities in Polyarny continued during 1943, despite incidents with the local Soviet commander. The Royal Navy's main concern was the harsh treatment of British sailors by local Soviet authorities, which could have endangered the Y station. As gestures of goodwill, Britain authorized two Soviet personnel to study British intercept procedures in England, and it presented a sample German Enigma machine to the Soviet naval mission in London.[6] In August, the Royal Air Force stationed a photo reconnaissance unit in the Kola Peninsula to assist a Soviet air force attack on the German battleship *Tirpitz*.[7]

Although cooperation was working in some areas, Admiral John Godfrey, the British director of naval intelligence, failed to obtain a naval mission in Vladivostok, after much effort. He also had to abandon his idea to send his brilliant and lively aide Lieutenant-Commander Ian Fleming to Moscow, where it was feared that Fleming would cause unwanted difficulties by playing tricks on the Soviets while acting as Godfrey's spy in the British military mission.[8] Indeed, collecting information on the host country certainly was almost as important as supporting the Soviet war effort. When Stalin asked Captain Jack Duncan over dinner how he would define his mission, the U.S. naval attaché candidly answered that he "was in Russia for the purpose of getting information." The Soviet leader called this statement "the most honest and straightforward" he had heard during the whole evening.[9]

On August 18, 1943, Admiral William Stanley, the U.S. ambassador, accompanied by Captain Duncan and his deputy, Commander Mike Allen, met with Admiral Nikolai Kuznetsov. The commander in chief of the Red Navy gave the U.S. officers presentations on the Soviet Northern and Pacific fleets, with the special request that this information should not be passed to the Royal Navy.[10]

While trying to play the Americans against the British, the Soviets were also acknowledging the leadership that the United States had taken in the course of the war. In effect, the U.S. Navy was allowed to have an assistant naval attaché in Vladivostok and could supplement Allied knowledge on the Soviet navy that was obtained in Murmansk and Archangelsk and in the Black Sea Fleet area by the other British and U.S. liaison officers. During 1942, the Red Navy had provided the U.S. Navy with general information on the German navy, along with data on raiders and U-boat refueling practices and secret information on armor and mines. In exchange, the United States provided the Soviets with its silhouette identification guide books on Japanese and Axis ships. Soviet intelligence on German, Japanese, and Soviet mines also proved very valuable to the Allies. Lieutenant G. B. Bassinger, USN, filed forty-four reports to the Office of Naval Intelligence (ONI). This cooperation ended in July 1944, however, when Admiral Ernest King, the U.S. Navy commander in chief, decided to withdraw the Western teams from Vladivostok due to lack of concrete results. Yet a senior Soviet meteorologist was stationed in the United States and remained there until the end of the war.[11]

As long as the Soviet Union had not joined the Allies in the war against Japan, exchanges of intelligence about this country were a more delicate matter. On February 2, 1944, Ambassador Averill Harriman raised the issue with Stalin. The Soviet leader gave his approval, and on February 28, a Soviet captain met with the new U.S. naval attaché, Admiral Olsen, and agreed on a procedure: each party would provide documentation on Japanese naval forces and a list of questions.

In early March, formal naval intelligence exchanges on this sensitive topic had at last been established. The Soviet navy provided the U.S. naval mission with information on convoys, installations on Sakhalin Island, ground tactics, naval training, and losses. The material was sent to ONI and on occasion to the White House. The U.S. side was instructed, however, "not to discuss or exchange navy order of battle intelligence, either surface or air." On November 22, 1944, Admiral King explained to the U.S. naval mission in Moscow that since the USSR was "still maintaining friendly relations with Japan, we cannot furnish information on Japanese Naval Order of Battle."[12]

Between February and May 1945, the Allies scaled up the quality of the information passed on to the Soviet Union by providing intelligence derived from decrypted German communications called Ultra Magic sources.[13] German forces, however, were becoming less capable of acting against the convoys heading for Archangelsk and Murmansk, and accordingly, the level of naval exchanges diminished considerably.[14]

Anglo-American cooperation concerning the Soviet Union was not perfect, either. For reasons that may have been related to doubts over U.S. cipher security or competition, the Royal Navy decided not to share with the United States the information obtained through the Soviet interrogation of captured Romanian naval personnel and passed on through the British liaison in the Black Sea.[15] But the most sensitive exclusive information that the Royal Navy sought from the Soviet navy was related to a devastating new weapon.

On July 30, 1944, the Soviet navy submarine chaser MO-103 was patrolling near the northern entry to the Bjorkosund in the Gulf of Finland. A squadron of Soviet minesweepers also operated in the vicinity. They sighted a periscope and signaled the chaser to drop depth charges on the intruder. The U-250, commanded by Werner Schmidt, was seriously damaged by the depth charges, and it briefly surfaced before sinking. This allowed six men, including the commanding officer, to escape and be captured. Soon afterward, the Finnish coastal artillery shelled the location of the sinking U-250, while German torpedo boats attempted twice to penetrate the area. This overreaction to the loss of U-250 appeared suspicious to the Soviet high command, which then made the decision to raise the U-boat.

Once the submarine was refloated and towed to Kronstadt, it was inspected. This turned up valuable documents, ciphers, and a coding machine. But its torpedoes were the real treasure, by far. Three were the new and yet unknown T-5 acoustic torpedoes. They were quickly taken to a navy facility and neutralized. Meanwhile, the Royal Navy was given ten hours to inspect the U-250 and found this examination profitable enough, despite the fact that the Soviets had removed the radio equipment and documents.[16] When Britain learned of the T-5 acoustic torpedoes, it negotiated the acquisition of the weapon at the highest level: on November 30, 1944, Churchill asked Stalin whether Britain could send an aircraft to take one of the torpedoes to England. Stalin answered with the following: "Unfortunately we cannot at the moment send one of them to Britain. The following alternative is possible: we can provide at once the military mission with drawings and descriptions of the torpedo; and when examination and tests are finished the torpedo itself can be handed over to the British Admiralty; or British experts can depart immediately for the Soviet Union to examine the torpedo in detail and make the required drawings."[17]

The Royal Navy accepted Stalin's second offer, and in January 1945, a group of British torpedo engineers, headed by Commander E. Conningwood, arrived in Leningrad to inspect the weapon.

Ian Fleming's Special Units, and the Seizure of Axis Naval Technology

As an Allied victory became more likely, the Allies accelerated plans to exploit Germany's astonishing technological superiority. As a result, at the Allied conference in Yalta in 1943, the Allied leaders, inter alia, clearly outlined plans for sharing the plunder of the Nazi war industry following Germany's unconditional surrender. In fact, plans to exploit German science and weapons technology were already well under way before Yalta. By spring 1942, imaginative minds in Britain had already seized on the idea of forming combat groups of engineers and technicians, guarded by Royal Marines, to grab German equipment, ciphers, and documents during raids against occupied Europe and then spearheading the major amphibious landings in Africa, Sicily, and Normandy.

The idea was first planned out by Ian Fleming, the special assistant to the director of naval intelligence (DNI). Fleming patterned his special force after German units that were initially employed during their invasion of Crete, when special commando forces accompanied the main assault force to capture British documents, codes, and other sensitive material. In a memorandum to DNI Admiral Godfrey on March 20, 1942,

Fleming described the German operation "as one of the most outstanding innovations in German intelligence." He suggested that British naval intelligence organize similar units for forthcoming operations against German-held territory.[18]

The first British teams of this type were placed under Royal Navy command and participated, initially with no success, in the abortive raid at Dieppe in August 1942. The special unit was then placed under the orders of Royal Navy commander Robert "Red" Ryder, a winner of the Victoria Cross, which was Britain's highest award for gallantry in action. The teams were given the temporary name of "Special Engineering Unit." Three such formations were created. One called Number 36 Troop was specially tailored to collect technical data and was manned specifically for the "grabbing of special equipment."[19] The troops' first operation was a success, when they landed in Algiers and captured and plundered an Italian headquarters a full two hours prior to the arrival of the main British assault forces.

During preparations for the 1944 Normandy landings, Fleming regained command of his special technical teams, by then called "30 Assault Unit" and nicknamed the "Red Indians." The teams had grown to more than three hundred men, a basic core of naval personnel protected by a larger force of Royal Marines. They had gained experience scouring the North Africa battlefields for enemy documents and communications equipment. After much success, the unit was recalled to Britain to train for further deployment into Germany. Following more victories during the liberation of Paris, the freelance units had gained a reputation for being wild marauders and were soon at odds with regular front-line forces. Their most notable accomplishments came during the capture of ports in North Germany, where they uncovered astonishing new German weapons and submarines.

Sensing the effectiveness of the British missions, the Americans were not to be outdone. In 1944, Admiral Ernest King, the U.S. Navy commander in chief, organized and deployed a Naval Technical Mission staffed by two hundred experts, including engineers, scientists, and weapons experts with their own aircraft, unlimited transport, and funds, to scour Europe for the U.S. Navy.[20] Meanwhile, during the liberation of Paris in August 1944, the British 30 Assault Unit, led by Royal Marines, as quoted from one of the unit's reports, "sped through empty Paris boulevards to scour dozens of German naval headquarters. At the Villa Rothschild, the principal German naval headquarters in Paris, the Marines fought briefly with defending Germans but found most buildings deserted, their sensitive material destroyed. Reluctant to believe the Germans could be so methodical, Granville (their leader) blamed French intelligence, claiming "They got here before us."[21] There was truth to his suspicion, because the individual Allies did not always share the results of their own searches.

Among the many technical teams scouring newly conquered areas in Germany was Alsos Team, the intelligence arm of the U.S. Manhattan Project in Oak Ridge, Georgia, which was then racing to produce an atomic bomb. (The word *alsos* is Greek for "grove" and was so named by the commander of the Manhattan Project, Brigadier General Leslie R. Groves.) This team was made up of navy, army, and other technical personnel. Its target was specifically to determine the real state of the German nuclear bomb project and to quickly recover as much uranium oxide as possible.[22]

In early 1945, as the eastern and western fronts converged toward Berlin, the Allies began to coordinate the collection of highly advanced technical and scientific data from the withering German armies. Not long after the Yalta meeting, at which the Allies pledged to cooperate in the plunder effort, it became patently obvious that the Soviets were cheating and obfuscating in their race to get to the prizes first. So the Allies did likewise. The geography of the rapidly moving front lines in 1945 gave some advantage to the western Allies, especially in the rocket and missile fields.

The Americans arrived first at the underground rocket construction and assembly area in the Hartz Mountains in a place called Nordhausen, which had been named the Mittewerk by the German armament planners. When elements of the U.S. Third Armored Division, named Task Force Welborn, broke through six fanatic SS companies outside of Nordhausen, the accompanying team of engineers and scientists found the elaborate long-range rocket and naval cruise missile assembly plant nearly intact. On April 11, 1945, as they sifted through the horrible remains of forty thousand slave laborers who had been worked to death at the plant, the incredulous technical team members found hundreds of intact V-2 rockets awaiting shipment and use and several Henschel 293 naval guided missiles, which had also been built and assembled there. They scooped up several hundred of the V-2s and the new *Wasserfall* antiaircraft missiles and quickly shipped them to Antwerp for further transport to U.S. army research bases in the States. The gruesome skeletal forms of more than seven hundred barely surviving laborers were cared for by army medical teams, while horrified war crimes investigators felt stupefied by their first grisly discoveries. They would soon move on to find more such camps on a much larger scale.

Nordhausen was the first of two underground sites constructed after the September 1943 Allied bombing of the primary German rocket site at Peenemünde, on the Baltic island of Usedom. Hitler quickly decided to relocate the plants to underground sites. In his befuddled brain, seeing this as the answer to impending defeat, he ordered five hundred thousand V-2 rockets to be built for his mythical final secret weapon onslaught, which would include the new unused antiaircraft missile systems and the first naval guided missile used in warfare at sea. The first batch of slave laborers had arrived at Nordhausen from Buchenwald to transform the underground site of a small ammonia mine into forty-six 220-yard tunnels, in which the secret rocket force would be constructed and assembled. It is estimated that about eighteen hundred workers died there each day from exhaustion and hunger or were murdered outright by the SS. When workers died while hand-digging the tunnels, the leading rocket scientist, Dr. Werhner von Braun, personally ordered replacements from other concentration camps, which the SS willingly provided. Braun, an SS major, was the man who would soon lead the U.S. space program.[23]

The second underground rocket site was under construction in Ebensee. This small town snuggled in the Austrian Alps, in a region called the Salzkammergut, provided the setting for much of the film *The Sound of Music*. At this site—one of forty-four satellite concentration camps of Mauthausen, Austria—slave laborers hewed tunnels similar to those at Nordhausen into the mountains and made ready to assemble more of the five hundred thousand V-2 rockets ordered by Hitler. The Ebensee camp was never fully completed, and during the last days of the war, it housed a petroleum refinery. More than eight thousand concentration camp inmates died in the hastily constructed effort there.

Just ahead of the U.S. Army's arrival at the underground sites, the German scientists and engineers hastily fled to Bavaria and holed up in the mountains awaiting their fate. Allied relations with the advancing Red Army appeared on the surface to be correct but were rapidly disintegrating. Nordhausen would soon be a part of the Soviet zone, as decided in Yalta, so Allied technicians stripped everything they found of value and blew up what they could of the rest. Allied technical teams were ordered by their local commanders to take anything that was useful, contrary to the supreme Allied commander's orders to leave it all in place, intact, until the agreed-on zone occupiers arrived. This same tactic was used with a great deal less finesse by the Soviets, where entire factories, design bureaus, and shipyards found in the British zones were hastily removed, lock, stock, and barrel, to the Soviet Union. Ebensee was liberated by the U.S. Second Armored Division and would remain in the U.S. zone in Austria. Although little of technical value was left in Ebensee, the plans for the rocket work at that site were recovered intact.

The Soviet army finally arrived to occupy the Mittewerk at Nordhausen two months after the Americans, on July 5. When the Soviets realized that this had been the main site of German long-range rocket construction, they sent Sergei Korolev—the future chief designer of the Soviet space program—to investigate. He then commenced a full study of all the German sites within the Soviet occupation zone, beginning with Peenemünde in the north, where much valuable material had been hidden after the Allied bombings. By this time, however, U.S. Army ordnance teams had found and retrieved fourteen tons of the rocket archives from a mine where they had been concealed and shipped them to the United States before the Soviets arrived.

Soon the Allies were advancing on the sacred grounds of the North German Baltic ports. With them were the British and U.S. technical intelligence-collection teams. The following intelligence report conveys the remarkable flavor of what transpired in the flickering last minutes of the Third Reich:

Wednesday 2nd May

Commander Hinds and Lieutenant Commander Blackler [of 30 Advance Unit] were sent into Luebeck in advance. . . . The *Hafenkommandant* had committed suicide before our entry. I directed Commander Hinds to proceed to Travenuende and Lieutenant Commander Blackler to join the armoured force about to occupy Neustadt. Shortly afterwards I proceeded by myself to Travemuende which had just been occupied by Commando Units and representatives of 30 Advance Unit. It was reported to me that a German flag officer across the river wished to surrender, so I commandeered a fishing craft and went over to the Priwall Air Station, but he failed to appear. . . . It was later reported that he was not a naval officer but a member of the Luftwaffe and his anxiety to surrender was due to his belief that he was in the Russian zone. At Priwall on the east side of the Trave River I found a large collection of motor transport of all types, filled with German officers, some of high rank, troops, mostly armed, women, luggage etc. I was approached by a Prussian colonel, mounted on a poor type of steed who requested that the party be allowed to cross the Trave by ferry. He also asked how far the Russians were off. I had much pleasure in informing him that the Russians were close on his heels, and that they were not to cross for the moment. There is no doubt of the great fear held by the Germans for the Russians at this time.

Friday 4th May

With a force of eight tanks in support, Lieutenant Commander Blackler occupied the barracks of the U-boat training establishment, which at the time was flying a Red Cross flag. While interviewing the Commandant, *Fregattenkapitaen* Schmidt, another officer attempted to shoot him, but was suitably dealt with by the escort. There was considerably difficulty with displaced persons, mostly Russians, who were out of hand, and in finding accommodations for refugees from the S.S. *Athen*, berthed alongside the submarine school. She was greatly overcrowded, with many of her 3,000 to 4,000 passengers suffering from starvation and in need of medical attention. Blackler went aboard this ship and found her in a most appalling condition of utter filth. He considered at the time that the only thing to do was to take her out of harbour and sink her, but early the next morning at 0400, she caught fire. Blackler took charge in clearing her and eventually succeeded in having her towed out into the roads. Then Blackler showed me the bodies of a number of refugees, about fifty, who had been taken out of the S.S. *Athen* and shot by the SS on the morning of his arrival. From evidence obtained in the *Athen* it appears these victims had been selected at random by SS from the refugees in *Athen*, and had been executed by them at the harbor entrance within a cable of this prison ship. They were probably Poles or Russians. I viewed most of the bodies, and each one had been killed by a burst of machine-gun or Tommy gun fire in the head whereby they were terribly mutilated. No heart shots were apparent. It's hard to imagine what could be the object of such sadism at such a stage of the war. Subsequently I heard that a large barge was found on the rocks close to the harbour entrance. The barge contained a large number of dead bodies (600–1,000) of victims who had been killed by Tommy-guns and being axed in the head. Evidence showed they had spent seven days in the hold without being allowed out, and this barge with others had arrived in tow from the eastward.

The senior military officer, Brigadier Mills-Roberts, ordered civilians of Neustadt to clear the barge of the bodies, and bury the victims, but not before Field Marshal Milch [Erhard Milch, the state secretary of Herman Goering's Luftwaffe], who had arrived to surrender, had been compelled to view both the bodies and the execution place on the barge. A subsequent remark of his to the effect these were only

Poles or Russians, so infuriated the Brigadier, that he seized the Field Marshal's baton out of his hands, and beat him over the shoulders with it, breaking it to pieces. Some of the SS thought to be concerned in this slaughter were arrested later.[24]

The Submarine That Could Win World War III

Just four days prior to the German capitulation on May 8, the British 30 Advance Unit was converging on the prized naval ports from the west. U.S. Navy captain Albert Mumma accompanied the unit that was headed for the main German naval headquarters in Kiel. This city was the lair of the celebrated Admiral Karl Doenitz, the father of the submarine fleet and briefly the successor to Adolf Hitler. On the way, however, there were many adventures. The advancing team found that the Germans had been systematically destroying all of their laboratories, had vacated most concentration camps, and had burned secret documents. In Luebeck, the team found a hydrofoil capable of speeds up to 50 knots, human torpedoes, and a two-man midget submarine. The biggest prize was the Walterwerke, the home of the high-speed, hydrogen peroxide–fueled torpedoes and submarines. Production had just ceased the day before, on May 3, and the teams found two sabotaged hydrogen-peroxide submarines, U-1408 and U-1410, smashed at the piers.

Thousands of naval personnel surrendered quietly to the small advance unit, including the plant owner, Helmuth Walter.[25] Initially, Walter was uncooperative. A hard-line Nazi, he was determined to destroy all of his valuable new work. Later, however, upon receipt of written orders from former German submarine force commander Admiral Doenitz, Walter began to cooperate. Royal Navy commander Aylen, accompanying one of the units, wrote,

> On May 7, Walter began his revelations, starting with the admission that prior to the incineration, all the documents had been microfilmed and the cans hidden in the coal cellars. For the first two weeks we found new weapons at the rate of two a day. [Combustion] chambers were hauled up from flooded bomb craters, key torpedo data dug up from underground, a miniature twenty-five knot U-boat salvaged from the bottom of a lake, parts of [a] Messerschmitt jet engine [were taken] from a train on the Danish border, and there were prototypes of new and ingenious weapons: long-range guns, mine sweeping devices and jet powered grenades.[26]

One of Walter's most coveted realizations was the Type XXI Electroboote. The Type XXI was the first real submarine: a streamlined diesel-electric 1,600-ton boat with better batteries than its predecessors, allowing for submerged speeds as high as 18 knots. These were superior at the time to all western submarines in endurance, submerged speed, passive sonar capabilities, and sonar (anechoic) hull shielding.

When U.S. and British advance units found the German type XVIII, XIX, and XXI submarines intact, they snatched all they could and informed the Soviet advance teams that the missing submarines had been scuttled or were damaged beyond repair. Tom Bower, the author of *The Paperclip Conspiracy: The Hunt for Nazi Scientists*, wrote,

> While the Walterwerke and the submarines were rapidly stripped of equipment, the Admiralty embarked on a deliberate deception. A top-secret cable from London gave instructions that if the Russians inquired about the survival of equipment, Allied officers should issue bland rebuttals. Three prototype U-boat hulls were on no account to be sunk, but if the Russians [asked] questions about the vessels, they should be informed that they were scuttled and their machinery sabotaged. . . . Anything likely to emphasize the importance of these vessels should be avoided. Pleased with the Machiavellian guile of their "denial policy," the Admiralty informed Washington: "No important naval unit has fallen undamaged into Russian hands and all surviving U-boats and important surface ships have been captured."[27]

The cheating worked both ways. The Soviets had begun to deceive their allies soon after the Red Army captured a German torpedo research center in Gdynia, Poland. After a high-level exchange of correspondence between Churchill and Stalin, an Anglo-American technical team set out to visit Gdynia. In an expert act of obfuscation by the Soviets, however, the team was sent via Sweden, Romania, and Iran, and it never arrived in Poland. In another such case, the Soviets greeted an Anglo-American team in an occupied part of Prussia near Koenigsburg to view captured German naval equipment, only to then tell the team that the area was not yet in Soviet hands. The Allies later discovered, however, that Soviet recovery teams had already captured the plans of all of the important naval designs in Gdynia and had squirreled them east to Leningrad, where they would form the backbone of the Soviet postwar naval buildup.

Splitting the German Fleet

According to plans agreed on in Potsdam in July 1945 and implemented by the Allied Control Commission in Berlin, the remaining units of the German fleet were to be divided among the Allies. The commander in chief of the Soviet navy, Admiral of the Fleet Nikolai Kuznetsov, recorded in his autobiography the tensions that surfaced during the negotiations:

In mid-June 1945 I heard from General of the Army A. I. Antonov, Chief of the General Headquarters, that I was to leave for Potsdam to attend a conference of the Allies.

At dawn on July 14 our plane took off from the Central Airport and steered a course for Berlin.

On July 16 on the newly-built platform of the railway we—Marshal Zhukov, General of the Army Antonov, Vyshinsky, who then was our Deputy Foreign Minister, and the author [Kuznetsov]—were to meet the Soviet delegation, led by Stalin. Exactly at the scheduled time a steam locomotive with several cars pulled up to the platform. Stalin stepped off from one of the cars. He wore his usual grey service coat (though he already had the title of Generalissimo). Warmly greeting us, without lingering at the station, he got into a car. Together with Molotov and Zhukov, he went to Babelsberg—the residence of the delegations.

Though in the Far East, the war was still continuing, all members of the delegations were in a victorious frame of mind. However, the heads of government of the USSR, USA and Britain had other serious and difficult questions facing them.

I for one was concerned with the question of division of the captured Nazi fleet.

Despite the victory and the outwardly excellent relations between the Allies, unlike at the Crimean conference, here, in Potsdam, many wide-ranging political matters caused debates. I distinctly remember an angry exchange between Stalin and Churchill over the division of the German fleet. The British stubbornly declined an equal division of this fleet while Stalin insisted on it, motivating his stand by the role the Soviet armies and fleets had played in defeating Germany.

Quite often, difficult questions would be put off "until better days" and the delegations would pass on to other questions. That was what happened this time.

But when only two or three days remained until the end of the conference I grew anxious and reminded Stalin about the captured fleet. The three commanders-in-chief—of the USSR, USA and Britain—were assigned to meet with foreign ministry representatives and draft a proposal.

Admirals King and Cunningham and the author met on the upper floor of the Zezilienhof Castle. I was lucky to preside over this conference and decided to insist at all costs on a solution satisfactory to the Soviet Union. I had been so ordered by the Supreme Commander-in-Chief. We argued till we finally hit upon an unorthodox decision to divide the surrendered fleet into three "approximately equal parts" and draw lots. I feared that Stalin might be displeased by such a solution but everything went off well. One way or another, the Allies divided among themselves more than 500 military vessels and 1,329 auxiliary craft. We received 155 combat ships.[28]

The Soviet navy would gain significantly from the decision to divide up the German fleet. Soviet naval reparations from former Axis navies after the war, which were granted at the 1945 Allied Berlin Conference, helped compensate for Soviet wartime losses and a period of stalled shipbuilding. From the Germans, the Soviets received one damaged and incomplete 20,000-ton aircraft carrier, *Graf Zeppelin*; the 13,000-ton battleship *Schleswig-Holstein*; the 6,000-ton light cruiser *Nuremberg*; ten destroyers; and ten operational U-boats, including four type XXI Electroboote.[29] From the Italian navy, the Soviets later received the 24,000-ton battleship *Giulio Caesare*, the light cruiser *Emanuele Filberto Duca d'Aosta*, four destroyers, fourteen torpedo boats, and two submarines. From Japan came six destroyers and numerous small combatants.[30]

In addition, in 1945, conquering Red Army forces acquired vast amounts of naval plunder from the Soviet zone in northern Germany, primarily in the ports on the Baltic. There, they found intact unfinished ships, entire submarine sections and propulsion machinery, whole factories, and large quantities of scientific and technical data. Soviet forces seized undamaged and partially completed German U-boats, notably the modern diesel-electric type XXI. Three had been allocated to the Soviet navy as war reparations, while some of the twenty unfinished boats in Danzig were moved to Leningrad but were never completed and were eventually scuttled or scrapped.[31]

The Soviets were quick to recognize the value of the industrial assets found in the Baltic ports. Soviet commanders set up a headquarters in occupied Berlin specifically to sift through the vast amounts of confiscated technology. Priceless to the development of future Soviet long-range diesel attack submarines were the German-designed Kreislauf closed-cycle turbine propulsion systems for sustained high-speed underwater endurance without air intake through a snorkel. The most valuable long-term asset was the central German submarine design bureau at Blankenburg, seized in 1945, from which Soviet builders adopted the German modular, prefabricated system of double-hulled submarine construction. This is the highly efficient method of building entire submarine cross-sections at widely dispersed locations, transporting the sections by barge over protected inland waterways, and assembling them inside a construction hall, a procedure that is still used today in building modern nuclear submarines. Soviet forces also gained an early naval cruise missile advantage over the West by using German rocket engineers and V-2 underwater missile canisters captured at Peenemünde, to become the world's first navy to employ antiship cruise missiles on a wide basis.

Tracking the Father of the Naval Cruise Missile

Of most intense interest to Allied naval technical teams, which were already roaming the ports in northern Germany, was the Viennese rocket engineer and inventor Dr. Herbert Wagner, who had worked at the Henschel Aircraft plant at Himmelberg outside Berlin. He had designed and built the world's first cruise missile to be used in combat and a sophisticated antiair missile defense system called the *Schmetterling* (butterfly). Wagner was also the father of the ingenious He-293, a glide bomb that turned into a cruise missile, which was mentioned earlier in this book's introduction. Along with his cronies, Wagner scurried south to Hitler's redoubt and was scooped up and arrested in a quaint mountain cottage near Oberammergau.

There, several American Alsos Teams had been sent out by General Groves. Members of one Alsos Team, headed by the U.S. Navy commander Henry Schade, found to their great surprise not only Dr. Wagner, but his associate Werhner von Braun and half a dozen leading long-range rocket scientists from the Nordhausen underground works.[32]

The American team initially hid the group in the small town of Bad Sachsa, south of Hanover, then quickly sent the scientists to Paris for safety. This was a prudent move, as it is now known that a team of Soviet army special forces was roaming the U.S. zone, trying to kidnap these very Germans from the Americans.[33] Swift action on the part of the Americans saved these most-wanted experts from being squirreled away to the darkest regions of the Soviet Union. Many other experts were indeed later captured by the Soviets.[34]

Dr. Herbert Wagner's Henschel Hs-293D naval cruise missile, which inspired both U.S. and Soviet developments.

Fortunately for the Americans, Dr. Wagner proved cooperative and had the foresight to realize that his brightest future lay with the Americans, in a country where he could best employ his expertise for personal gain. His He-293 air-to-surface cruise missile had been incredibly successful. Originally called a glide bomb, this missile had been designed and tested by Dr. Wagner at the Henschel Aircraft works. His deputy, the talented Hans Muehlbacher, also from Vienna, devised the guidance system for the missile. Muehlbacher is a most colorful character who, at the time of this writing, was ninety-one years old and still played the concert violin. He was the inventor of stereo phonics. The aeronautical engineers Reinhard Lahde, Otto Pohlmann, and Wilfried Hell had developed the missile further for launch by aircraft.[35] It was built as an aircraft fuselage with stubby wings, weighed 550 kilograms, and carried a 295-kilogram warhead adapted from a German SC500 aerial mine. Wagner had intended to use the missile configured as a BV-143 aircraft body to drop a missile into the sea, which would submerge and become a torpedo. (Wagner's eventual work at the U.S. naval missile test center at Point Magu, California, led to the development of Asroc/Subroc antisubmarine missiles.)

The Hs-293 had been first tested by the Viking Fighter Squadron 100 at Peenemünde-West in early 1942. The test squadron then deployed to Athens, where the missile was initially launched from a Heinkel-111 and then a Dornier-217E against Allied ships in Bari, Italy, with devastating results. Later in 1942, it was used in combat in tactical support of the German Sixth Army in Stalingrad. Equipped with a new guidance system, it was later used against Allied ships during the invasion of Sicily and then against ships of the Normandy invasion fleet. Its total maritime damage was as follows:

Sunk

Italian battleship *Roma*, 41,000 tons

2 cruisers

10 destroyers (one of these was the U.S. Sumner class destroyer the USS *Meredith*)

DD 726, sunk off Utah Beach during the June 6, 1944, Normandy invasion (and whose engineering plant has been partially recovered and can be seen today outside Bayeux)

10 merchant ships

1 flak boat

2 LTC

1 LST

Heavily damaged

4 battleships

6 cruisers

12 destroyers

29 merchant ships totaling 215,000 tons [36]

Enrolling Nazi Scientists and Weapons Specialists

The cooperation of Nazi scientists was indispensable in training the Allies to use the technology they recovered as war booty. Without the German scientists' compliance, would some of the major postwar achievements have been possible or so rapid? The most significant weapon was the ballistic missile V-2, which had already demonstrated its devastating effects on British and Belgian cities in the closing days of the war. The V-2 had been a German army project, but it was being modified to be launched from canisters towed at sea by submarines.[37] Even though this naval application was not the first development that the Soviet and Western Allies had in mind, the V-2 served as a basis for the Soviet R-11 rocket, which on September 16, 1955, became the first ballistic missile to be launched from a submarine. This was five years before the firing of the first U.S. *Polaris* missile in June 1960.

Back in July 1945, the Soviets had settled down in their occupation zone around the missile facilities of Peenemünde and Nordhausen and established a clearinghouse for captured Nazi scientists. The headquarters chosen by leading Soviet scientific intelligence teams, headed by Major Boris Chertok, was the elegant house called Villa Frank, in Bleicherode, where Werhner von Braun had lived since the bombing of Peenemünde. There, the Soviet intelligence team set up a safe house, hoping to lure German scientists who either had not made up their minds which side to support or were captured by Chertok's engineering snatch teams. Wrote Chertok in his memoir *Rockets and People*,

Our headquarters was called RABE, an acronym that stood for *Raketenbau und Entwiklung* [missile construction and development]. Our "cover" had emerged—we established a place where German specialists scattered by the war could take refuge. This was clearly a guerilla operation on our part that could lead to diplomatic complications with the Allies, especially since the border was only sixteen kilometers away, and immediately beyond the border was a town where, according to our intelligence, the Americans command had assembled several hundred German specialists.

But we still needed authentic Peenemunde missile specialists. For this I set up a secret second program, which I entrusted to Vasily Kharchev. His task was to establish a network of agents, and if necessary, personally penetrate into the American zone to intercept specialists before they were sent to the United States. [Kharchev] assigned this program the code name of *"Operation Ost"* (East). Semyon Chizhikov was instructed to supply Kharchev with cognac, butter, and various delicacies "on account" for Operation *Ost*. The division chief of staff agreed to open and close the border between our zone and the American zone at Kharchev's request. Pilyugin undertook a special mission . . . and brought back many dozens of wristwatches to be used as souvenirs and "bribes" for the American border guards. Vasily Kharchev could barely sleep because of his intensive German and English studies.

The first success of operation *Ost* was to win over and bring to the RABE staff an authentic specialist on the combat firing of V-2 missiles, Fritz Viebach. The Americans unexpectedly gave our operation *Ost* a boost. Early one morning, I was awakened by a telephone call from the town commandant. He reported that his patrol had stopped two Jeeps with Americans who had apparently burst into the town and were trying to abduct German women. The latter raised such a ruckus that our patrol had arrived. The arrested Americans were raising Cain over at the commandant's office. They explained that these women were the wives of German specialists who were supposed to be sent to America. I asked the Commandant to serve the Americans tea and offer them some Kazbek cigarettes and promised to be there soon.

I woke up Chizhikov and Kharkov and ordered them to find some cognac, some good snacks, and to set the table at once. When I appeared at the commandant's office, the din was terrible. The four American officers, each trying to out yell the others, were communicating with the commandant through two interpreters—a German interpreted from English to German and a Russian lieutenant from German to Russian and vice versa.

I introduced myself as the Soviet representative for [the] German missile specialists. I asked our American friends to calm themselves and take a break from their tiring work by joining us for refreshments at the Villa Franka. They responded with an "Okay," and the cortege set out for our villa. Chizhikov had not let me down. When the Americans looked at the table their eyes lit up. All four young Yankees broke into smiles and exclamations of approval followed. . . . We found out that in September and October all of the German specialists that the Americans had named as war criminals would be sent to Witzenhausen via France to the United States. But several of their wives or mistresses had remained in the Soviet zone, in particular Bleicherode, and the Germans categorically refused to go without them. On behalf of the command, the Americans requested that the Soviets help them return these women to them. . . . [A] week later we received a report through our new female network of "agents" that Frau Groettrup, the wife of a German specialist, wanted to meet with us. She said her husband Helmut Groettrup was von Braun's deputy for missile radio-control and for electrical systems as a whole.[38]

The Groettrups settled in a separate villa and were offered a very high salary and expensive food rations, compared with those of the other Germans. Groettrup stayed and headed a contingent of German missile experts that worked for the Soviets. Later, the same group tried unsuccessfully to raid the U.S. camp to kidnap von Braun and Wagner.

The Soviets immediately began to exploit the material they had found in their zone of occupied Germany. According to the British intelligence digest, the Soviets succeeded in assembling a small batch of A-4s and began flight tests from Peenemünde and Gdynia.

Spiriting Nazi Scientists to the United States

Operation Paperclip was in full swing. This was the joint U.S. Army and Navy program to capture the top German scientists—especially in the fields of rockets and weapons—and move them quickly to the United States, disregarding the simultaneous hunt for Nazi war criminals. This controversial program won the ire of those most intent on capturing key Nazi war criminals and eventually trying them in Nuremberg. After being captured, figures such as Drs. Werhner von Braun and Herbert Wagner were spirited to the United States, while army and navy intelligence officers falsified their records to satisfy U.S. Immigration authorities and the State Department requirements for de-Nazification.

The United States lost no time in jumping into the race with the USSR. In February 1945, the U.S. Army Ordnance Department established the White Sands Proving Ground in southern New Mexico for the purpose of testing new rockets, including the captured German V-2. Of the four hundred German rocket scientists who surrendered to the United States, one hundred were recruited under the Paperclip operation. In November 1945, they arrived in the United States onboard the liner SS *Argentina* and were sent to Fort Bliss near White Sands in January 1946 to assist with V-2 launchings.[39] After the failure of the first firing on April 16, a successful flight was completed the following month, which marked the beginning of six years of V-2 tests. The Naval Research Laboratory was associated with the V-2 early on to make measurements in the upper atmosphere. The navy also proceeded with its own trials on board the aircraft carrier USS *Midway* (September 1947) and from White Sands. The V-2 provided the United States and the Soviet Union with their first opportunity to fire large missiles.

The presence of the German scientists working for the U.S. Army and Navy in the immediate postwar period was not popular. Many Americans were against giving jobs to these men, who, despite their personal loyalties, had only months earlier been working for the Nazi regime. The Federation of American Scientists protested in a letter to the U.S. government, stating that it was an affront to the people of all of the countries that had fought beside the Allies. To that protest, the U.S. Navy offered the following response, which was published in the 1946 *ONI Review*:

> To those petitioning American scientists, we offer the following proposition: If they will first amend their protest to declare that Russia's importation of hundreds of German scientists to work in their laboratories was an "affront" to the American people, we'll gladly give

the protest a second reading. If they will then convince Russia that she should forthwith deport their captive scientists back to Germany, we'll go further and give the protest a second thought. Until those two conditions have been met, we must continue to believe that America's defense and security are better served by having German scientists—with their knowledge of V-1 and V-2 bombs, atomic energy, cosmic rays, and all other awful forces of destruction—working in our laboratories rather than in Russia's.[40]

Marshal Stalin's Amorous Naval Connection

One intelligence report from the period was unique for defining one of Stalin's lesser-known connections with his navy. On February 1, 1946, an obscure intelligence report was written by a Royal Navy intelligence officer assigned to the captured German port of Wilhelmshaven, Germany. It contained bizarre personal information about the Soviet leader Generalissimo Joseph Stalin. The information had been gleaned from a Soviet naval officer, who was about to return to the USSR from occupied Germany. Captain Third Rank Rodion Kirkevitch was the chief naval engineer of the Soviet Naval Mission in Wilhelmshaven, which was being dissolved in February 1946. In a rare one-on-one meeting with the British officer—usually, the Soviets were accompanied by political commissars, as minders—Kirkevitch expressed some reluctance to return to Moscow and to see his family. According to the sad Russian, in 1942, Stalin had married Kirkevitch's then twenty-year-old sister, who bore Stalin two children. One was a son, Alexander, born out of wedlock; the second, a daughter, Nadia, was two years younger. Besides these two children, Stalin had two sons—Yasha, who had been killed in a German POW camp, and a second son, Vasily—as well as a daughter, Svetlana.

Stalin's affection for Kirkevitch's young sister had cooled considerably by 1946 because he began an affair with the daughter of Soviet marshal Semen Timoshenko. (Timoshenko, a two-time winner of the highest award, Hero of the Soviet Union, was the famed leader of the Red Army's South-West Front, which included Stalingrad during the 1942–1943 victory over the German Sixth Army.) Kirkevitch stated that he would not be surprised if Stalin, ignoring all laws, divorced his sister and married the new girlfriend. "Stalin is all powerful and does what he pleases," said the poor captain. "Of course my sister will be well pensioned off, but, nevertheless will put up a fight." Talking about his future, Kirkevitch said, "Stalin had suggested I leave the navy and promised a high official post in the civil service. But I am afraid that Stalin's benevolent attitude toward his brother-in-law would certainly change, when Stalin decides to get rid of his present wife." Kirkevitch foresaw that "I will in all probability fall in disgrace, lose my post and find myself in a very precarious position."[41]

Although the war was over and the plunder of Nazi technology well underway, the uncertainty of the geopolitical European boundaries would give both sides opportunities to penetrate the other's innermost sanctums.

2

PENETRATIONS
1945–1952

Both East and West began full-blown naval espionage and clandestine operations during the two years following Germany's collapse. Yet at the same time, information was still being exchanged between the U.S. and Soviet navies. Scientists and officers were even permitted to visit defense factories in the continental United States to research and shop.[1] With the weakening of the Grand Alliance, however, the U.S. Navy was closing its doors.

In January 1946, the U.S. government instructed the U.S. embassy in Moscow and the naval attaché to make sure that all U.S. technical experts in certain key areas—those dealing with vacuum tubes, radio, radar, navigational aid, television, infrared and ultraviolet equipment, medical research, and physical or chemical research—were leaving the USSR. Washington also asked the embassy to provide lists of U.S. companies that were still doing business in the USSR.[2]

Growing fears of active Soviet intelligence in January 1946 caused the U.S. Navy to caution its commands abroad not to use ordinary mail to transmit classified reports through areas "under Russian influence."[3] But the first precise revelation of Soviet espionage in North America came from a GRU (Main Intelligence Directorate) code clerk working at the USSR Embassy in Ottawa. Igor Gouzenko's confessions to the Royal Canadian Mounted Police and to the Federal Bureau of Investigation (FBI) were a wake-up call to U.S. Navy security. According to the defector Gouzenko, a Dr. Richard Steinberg from the U.S. Naval Research Laboratory had passed on to a Soviet agent in Canada the secrets of the advanced proximity fuse, which many believe was the invention that won World War II. The proximity fuse was a radar device developed by the U.S. Naval Research Laboratory, which made antiaircraft gunfire twenty times more accurate and which, in the advanced infrared version, was used to detonate the atomic bomb. Dr. Steinberg's work on naval operations was related to the proximity fuse, but it was uncertain whether he had the full technical knowledge to compromise this secret. The atomic scientist and spymaster Julius Rosenberg was also accused of having passed a replica of the proximity fuse to the Soviet Union and may have been the main source.[4] Ironically, a few days before these revelations, on November 1, 1945, Secretary of the Navy James Forrestal had authorized Naval Intelligence personnel "to conduct investigations of naval personnel and civilians under naval control in cases of actual or potential espionage, sabotage, or subversive activities, and in those cases which relate to the security of classified naval information."

Following Forrestal's authorization, the Sabotage, Espionage, and Counter-Subversion Section in the Office of Naval Intelligence (ONI) began detailed background investigations focusing on organizations that solicited naval personnel for membership.

Venona

Meanwhile, the full reach of Soviet subversive activities was becoming clearer. British public opinion seemed to accept the theory of Soviet sabotage to explain how twelve ships, including the liner *Queen Elizabeth*, had mysteriously caught fire during the first months of the year.[5] British naval intelligence reported that former SS and active Nazis were being extensively used as spies by the Soviet MVD (Ministry of Internal Affairs). One report stated that German mothers were seeking aid from the Red Cross in Berlin to trace "thousands" of boys ages thirteen to seventeen who had disappeared and were reported to be at training camps for nefarious activities in accordance with the new ideology.[6] The ramifications of Soviet espionage seemed global. From decrypted Japanese codes, it became clear that Tokyo had obtained very sensitive U.S. documents about the war effort in General Douglas MacArthur's South West Pacific Area command. This was achieved by tapping into a Soviet espionage net that was then active against British secret programs for atomic and biological weapons, as well as guided missiles, to be tested in Australia's deserts. Security was poor in Australia, especially in laboratories where detailed research was being conducted.[7]

But the worst fears were confirmed during July 1946, when U.S. Army Signals Intelligence cryptanalyst Meredith Gardner began to analytically reconstruct an NKVD (Soviet Secret Police) codebook. The information recovered was code named Venona. Between August 1946 and May 1947 the same army signal intelligence unit deciphered Soviet traffic linking

NKVD activities in Latin America directly to the Manhattan Project and to the War Department. Several of these decoded signals exposed Soviet moles in these two organizations. Hundreds of cover names were found and decrypted hiding the identities of agents, organizations, people, and places. Cover name Kapitan was found to be President Roosevelt; Babylon was San Francisco; Arsenal, the U.S. War Department; Enormoz, the Manhattan Project; and Anton was Leonid Kvasnikov, the head of Soviet NKVD espionage efforts against the Manhattan Project. Anton was also in charge of all of the intelligence collection against the U.S. Army Air Force and Navy jet aircraft, radar, and missile programs. A series of messages unveilled the NKVD's efforts to track down Soviet sailors who had deserted their merchant ships in various U.S. ports. More decoded messages revealed how U.S. citizens who were vocally communists were being recruited for espionage work. These disclosures contradicted earlier views that the Soviet services would not target Communists out of fear of being detected too easily. Unfortunately none of the Venona messages were deciphered in real time, thus leaving to conjecture the actual extent of communist infiltration.[8]

On August 9, 1946, ONI requested its naval attaché offices around the world "to collect all possible data pertaining to the political convictions of prominent scientists in the countries to which they are accredited."[9] ONI explained that "scientists of communist inclinations keep Moscow abreast, not only of their own developments, but of any information that becomes available to them."[10] ONI was especially concerned with the fact that scientists tended to share their ideas with their colleagues, regardless of the political loyalties, and that, as such, they represented an ideal funnel of information to Moscow.

Taking further measures to protect against Soviet penetration, Secretary Forrestal refocused the naval security effort in two directions: investigating subversive activity and conducting background investigations to determine the true loyalty of all naval employees.[11] Naval security services and the FBI placed informants on U.S. merchant ships to identify crew members suspected of subversive activities. In October of that year, the government implemented a federal program to further weed out those suspected of disloyal behavior. When evidence of disloyalty was found, the navy then passed the further investigation to the FBI. The United States needed to employ qualified scientists free of obscure backgrounds. The need for such experts was intensified by the urgency to develop new and increasingly complex weapons. Thus, these slow and detailed investigations severely hindered recruitment for important programs. In July,

ONI first introduced the polygraph lie detector tests. This mechanical and electrical device is used to detect and record certain physiological changes that are induced in an individual by his emotional response to certain questions, this being involuntary for the most part.

ONI, however, did not feel that using the polygraph for security clearance was adequate as the sole investigative technique, and this was used merely as a valuable adjunct. ONI recognized that the lie detector's effectiveness could be thwarted by a properly trained individual. Thus security organs could not use the polygraph as legal evidence.[12]

The polygraph was not widely in use when Joseph Barr joined Sperry Gyroscope, an important navy and army air force subcontractor involved in top secret research programs on missile guidance and radar. In 1946, Barr was a thirty-six-year-old engineer in military electronics. He was also a talented musician and an ardent communist who was being run operationally by Julius Rosenberg. Rosenberg later indicated that Barr had passed on to him and to the Soviet Union an exhaustive description of U.S. missile guidance research. For more than a year, Barr was allowed access to top secret information while his background remained unchecked. In June 1947, however, Sperry Gyroscope asked the FBI to perform a security check, which revealed Barr's communist connections. Sperry Gyroscope terminated Barr's contract, and in January 1948, the engineer sought safer ground in Scandinavia and France, with the ultimate goal of moving to the Soviet Union.[13]

Meanwhile, Barr indulged in his passion for music, cultivating a friendship with the French composer Olivier Messian. But the arrest of Julius Rosenberg's brother-in-law on June 16 prompted Barr to go East for good. In Moscow he was given a new identity as Joe Berg from South Africa and was assigned to Prague, where he started to work in a state-owned electronics and telecommunications conglomerate. Berg alias Barr was later recalled to Moscow, where he developed the Soviet navy's most modern combat systems under the direction of navy chief Sergei Gorshkov.[14] For decades, U.S. intelligence remained completely in the dark as to the whereabouts of Barr. He had been reported missing from his French address in Neuilly-sur-Seine when the legal attaché of the U.S. embassy came to look for him one month after he had fled.

One FBI case concerned a U.S. Navy officer, whose activities eventually became well publicized. Lieutenant Andrew Roth, a naval reserve officer, enjoyed numerous contacts in the navy and a convenient profession as a journalist for cover. He was found to be using his many newspaper connections to gather information on the latest naval developments.

In February 1947, the Navy Department issued an alert seeking Roth. He was at the time reportedly making an around-the-world tour as a representative of the *New York Nation*, the *Montreal Star Weekly*, *Telepress*, and *Yomiuri-Hochi* newspapers. Roth was reported to be "anti-British" and "strongly suspected by ONI of working for the Communist Party." Naval attachés were requested to discreetly inform the U.S. ambassadors and ministers of the countries in their travel itineraries, "in order that they may be cognizant of the subject."[15] Roth later fled a grand jury indictment issued in August 1948 after he was caught by the FBI passing classified documents to Philip Jaffe, a Soviet spy. Jaffe edited the New York–based magazine *Amerasia*. Roth escaped imprisonment because the FBI had used illicit surveillance methods. He managed to escape to England, where he became a bylined columnist for the *Guardian*.[16]

Realizing the extent of Soviet penetration, the West was faced with a major counterintelligence challenge. Thousands of investigations and security checks overtaxed the FBI's and ONI's resources. As of February 1948, ONI had a total of 15,000 pending investigations, which were increasing at the rate of 850 per month, and most of them remained inconclusive.[17]

Among the most bizarre cases of the time was that of an American seaman discharged from the navy as a certifiable psychopath. After treatment, he managed to join the merchant marine with the intended purpose of acting as an agent for the ONI. He tried unsuccessfully to be recruited by contacting the U.S. attachés in Copenhagen and Helsinki during his port calls. His conduct resulted in a general warning to all of the U.S. attachés around the Baltic to beware of the eager agent wannabe.[18]

Mariners and Fishermen

Soviet and Eastern European merchant and fishing vessels were convenient platforms for launching espionage operations. They were often used to insert or recover agents around the world. These ships could also conduct reconnaissance near military bases and could report by radio any encounter with U.S., British, or other Western warships on the high seas. The merchant ship and the trawler potential for engaging in clandestine operations became quite obvious to Western intelligence services during this period. With the increase in reports that Eastern agents were infiltrating the United States, in early 1947 the ONI mobilized its naval attaché network to track down Soviet agents and revolutionaries being smuggled into Western Europe, the United States, and Latin America on board Soviet and Eastern European merchant ships. On September 16, 1947, Washington informed its naval attachés in Moscow, Athens, Rome, Paris, and Mexico City that in late February, the SS *Rossya* had brought 13 NKVD agents to New York; later the *Rossya* had taken another 12 agents into Mexico; and in July, a group of 17 operatives—hidden among 500 Jews from Haifa—had allegedly entered the United States. In total, the *Rossya* was said to have carried approximately 190 agents to various Western countries.[19] The U.S. assistant naval attachés in Odessa forwarded the names of suspects living in the United States who had made or were attempting to establish contact with visiting Soviet merchant ships coming from Odessa.

On March 7, 1947, the attachés provided the names of Vladimir Ruhl and Gregori Keresman, two suspected communists who had met with the chief mate of the merchant ship *Sukhona* at the Claremont terminal in New York City. According to the naval attaché's information, Keresman's sister was an MVD major. Three other names of alleged Soviet agents already in the United States—Kokianye, Potorzhinski, and Baranovsky—had also been provided by the Odessa attaché officer. The first worked in an unspecified defense plant and the second was said to be a spy of major importance infiltrating New York theatrical circles. The third disappeared underground.[20]

Numerous Polish defectors provided a unique window on Soviet-sponsored espionage conducted from Polish merchant ships. One defector explained to the interrogators from the intelligence section of the U.S. naval command in Germany that all marine radio communications and electronics in the Polish merchant marine were controlled by Morska Oblusga Radiowa—"Mors"—which had its head office at 5–7 Ul. Zygmunta Augusta, Gdynia, Poland:

> All ships have to have a Polish radio operator, even those traveling to China with an all Chinese crew. In general only proven Communists are used for that job, for apart from their actual work as radio operator they are used for intelligence work. These radio operators are given training in intelligence work. They are equipped to submit all intelligence information they collect to the intelligence branches of the Polish Air Force and the Polish Navy and to the counter-intelligence branch. In addition they are charged with collecting information on technical developments and experiences gained abroad. These radio operators are of great value to the Polish intelligence agencies and are said to have performed valuable services. All radio repairs on foreign vessels

staying in Polish ports are carried [out] by "Mors" employees. These repairmen are instructed to examine the radar units, the radio stations, the sonic altimeter and the direction finding equipment, to look for any new developments and note these down.[21]

Another Polish defector recalled how his ship the *Hugo Kollontaj*, bound for South America, got a mysterious substitute for its political officer. This newcomer proved to be an agent who could conveniently vanish ashore for the duration of the port visits:

All of a sudden [on August 28, 1951, our political officer] packed up and left the ship, saying that this voyage will be served by another K.O. [political officer]. The new man arrived shortly before the departure. He was a medium built man of about 35 years, with black hair, thin oval face. His name was Miklass. It seems he did not care a damn about his functions as the K.O. All the time, he was seen writing something or sunning, or just loafing around. When [the] ship called at Dakar, Rio de Janeiro, Santos, Montevideo, he used to leave the ship and did not return until shortly before the departure. From Santos he traveled by rail to Sao Paulo, where there was a large Polish community. The source, (a seaman) and many other seamen believe that this man was a special agent. Once source picked up a chatter between the chief steward Sentowski who was an undercover UB [counterintelligence] man and some other man. They referred to Miklass as a major. . . . A friend of the source serving on the Polish ship *Olstyn* recalled three young men attached to the crew as apprentices and who knew nothing about navigation. In Greece, they went ashore and did not return until shortly before departure bringing along one man who was put in a special cabin. He did not leave his cabin until the ship passed the Kiel canal.[22]

In another case, an East German paramilitary unit based in Sassnitz, consisting of young men posing as fishermen, visited southern Sweden and Denmark aboard their trawlers. On November 26, 1951, they arrived in Copenhagen en route to the Kattegatt. The suspicious circumstance was that they entered Copenhagen through a very narrow and little-used channel between the islands of Zealand and Amager, instead of taking the commercial channel to the east of Amager. They claimed to have lost their way, but it was immediately suspected that they had chosen this deserted passage to land people secretly. The trawlers were placed under police guard at Copenhagen, and personnel were not allowed ashore. They finally departed on November 28.[23]

During the fall of 1951, Soviet trawlers also extended their fishing activities to unusual grounds—namely, the east and south coasts of Iceland—raising suspicions as to the real motives of their presence within Icelandic territorial waters, a few miles away from the U.S. air base at Keflavik. Local excitement over the Icelandic Coast Guard's arrest of one Soviet trawler was heightened by the coincidental appearance of two U.S. destroyers, which happened to anchor on either side of the Soviet fishing vessel. Local gossip commented on the speed with which the U.S. Navy appeared on the scene; however, the destroyers did not know the trawler's identity. The press also reported that the crew members were mostly of the blond Scandinavian type and the captain was a widely traveled man who spoke good English and was believed to have fished in Icelandic waters previously because of his apparent knowledge of local navigation. From various reports, ONI noted that the trawlers seemed to be engaged only in fishing herring, as they encroached upon Icelandic fishing grounds with their larger nets, violated territorial waters, and attempted on occasion to escape the coast guard. Yet the nearness of the fishing vessels to the shore on both sides of the Reykjanes peninsula led ONI to suggest that they might have also attempted to chart the coastline beaches and water depths for possible future amphibious landings.[24]

One of the vessels that engaged most actively in intelligence activities was the Polish liner *Batory*. Under the title "A Floating Spy-Centre in the Baltic," the Swedish newspaper *Stockholm Tidningen* published an interview with the seaman Stanislav Kreft, who had leaped overboard from *Batory* while in Copenhagen harbor to escape the socialist paradise and expose the ship's real purpose:

> The crew on board wonder why it is that the *Batory* is still permitted to visit so many [Western] ports, and why it is that so many capitalist passengers continue to use this vessel, since more often than not they are under the watchful eyes of the Security Police. These agents even drift around amongst passengers. Maybe the reason why so many Englishmen travel on the *Batory* from India to England is because the food on board is superb. But none of the passengers has any suspicion that on board there are some mysterious individuals, all well-trained in foreign languages and in the service of the Military Intelligence Bureau. This applies for instance to the two photographers who work for the bureau—one quite exclusively, who roams around every port with his German Leica camera, and has excellent opportunities of using a vantage point on the vessel itself for photos of coastal fortification, harbour entrances, fjords, oil-tanks, etc. . . . Once he had the job of

photographing Hammersfest, and this was arranged by means of "engine trouble," on which occasion this photographer had the opportunity of rowing ashore.[25]

With its many vessels and numerous seamen, the Greek merchant fleet harbored agents from the banned Greek Communist Party OENO, which was fighting a bloody civil war at home. ONI updated the Greek Ministry of Merchant Marine–compiled lists of OENO members serving on East European ships. These lists were passed to the U.S. Coast Guard for coordination with ONI.

OENO party representatives and operation centers were identified in most Mediterranean and European ports.[26] Greek captains with suspected communist sympathies and vessels penetrated by OENO members were also monitored and their itineraries studied carefully.[27]

A Mole in Rue Royale

The U.S. Navy was not alone in combating the apparent influx of communist intelligence agents. In December 1951, the ONI stated that communist penetration of the French navy was "small in comparison with that in the other French services."[28] In 1953, there were eleven or more suspected "valuable KGB agents" working in France. The British, who had not yet taken the full measure of Soviet infiltration into their own security apparatus, shared this view: "The [French] Navy is estimated most secure, the Army less secure and the Air Force, being most heavily infiltrated, least secure." Besides the "extensive penetration of the [French] political system by Communists," a "natural garrulous tendency in the French character" and a "French lack of security consciousness" were to blame.[29] But according to the former KGB historian Andrew Mikroshin and the British scholar Christopher Andrews, even the conservative French navy, "La Royale," had been infiltrated by Soviet intelligence:

> The [KGB] Paris residency ran more agents—usually at least fifty—than any other KGB station in Western Europe. Its most remarkable achievement during the fourth republic (1946–1958) was the penetration of the French intelligence community especially the SDECE [Service de Documentation Extérieure et de Contre-Espionnage], the foreign intelligence agency. An incomplete list in KGB files of the residency's particularly "valuable agents" in 1953 included four officials in the SDECE (codenamed NOSENKO, SGIROKOV, KORABLEV and DUBRAVIN) and one each in the domestic security service DST (GORYACHEV), the

Renseignements Généraux (GIZ), the foreign ministry (IZVEKOV), the defence ministry (LAVROV), the naval ministry (PIZHO) . . . and the press (ZHIGALOV).[30]

On March 2, 1942, General De Gaulle had summoned Vice Admiral Emile Muselier to his office in London. Muselier, having retired in 1939, rallied De Gaulle in England and accepted his authority despite being his senior. The admiral had been tasked to command the Free French naval and air forces and had just fulfilled De Gaulle's orders to occupy St. Pierre et Miquelon, Vichy's outpost off Canada. But De Gaulle was furious. "I know all about the deeds of Labarthe and [Captain] Moret," he said, referring to Muselier's deputies. "I also know that in due course you will resign from the National Committee and express your opposition to my policy. But I won't let them [Labarthe and Moret] have it their way. Your telegrams and coded messages are in my possession." Muselier did not lose his countenance. "If you had wished to see Captain Moret's telegrams I would have been pleased to show them to you," he retorted. "But even if you had them in your possession, I very much doubt that you could decipher them." The admiral then reminded De Gaulle of a previous injustice when the British had arrested the admiral as a consequence of forged documents establishing connivance between Vichy and Muselier. De Gaulle replied by quoting word for word the content of Moret's messages. Muselier went pale and immediately vowed to investigate this breach of security. He had been betrayed by one of his men. The following day, Muselier took a public stand against De Gaulle before the national committee and offered his resignation, which was accepted.[31] But his apparent attempt to depose De Gaulle with the help of Labarthe and Moret had failed.

Emile Muselier was a controversial figure. He had served in the Black Sea with the French squadron during the Russian civil war. His patrol boat had been severely damaged by the Bolsheviks in obscure circumstances. How he had managed to have his wounded crewmen treated by the Bolsheviks and subsequently return from Red-controlled territory long puzzled many in the French Navy.[32] Muselier was known for his liberal sympathies, which helped boost his career in 1924 when the left won the elections. It was later rumoured that he had Russian friends, and Peter Wright, the former head of British counterintelligence, believed that he was a Communist.[33]

Muselier may have been unjustly accused, but André Labarthe was later betrayed by the Venona intercepts as being a Soviet spy.[34] He had first worked for the GRU reporting on the meager developments of the Free French military scientific research before being reassigned to the NKVD after his dismissal from the Free French technical branch.[35] One of Labarthe's associates in London was Captain Raymond Moullec, alias Moret—Muselier's assistant in London and a close friend in retirement. As a lieutenant he had been an outstanding naval attaché in Spain during its civil war, where the Soviet naval attaché "Commodore Nicola," two years his junior, was none other than Nikolai Kuznetsov, the future commander in chief of the Soviet navy.[36] In London, the "Red tandem"— Muselier-Moret, supported by Labarthe—attempted to take control of the Free French intelligence organization headed by the famed but inexperienced Commandant Passy. Historian Thierry Wolthon was told that Labarthe later confessed to the French counterespionage that he had introduced Moret to Veniamin Beletski, the NKVD resident in London, and that Moret-Moullec was passing military and strategic intelligence to the Soviets.[37] Labarthe's testimony is the only evidence indicting Moullec, and it is not conclusive. But it appears that through Labarthe, Moscow had tried to place a man sympathetic to the Communist cause at the head of the Free French secret service.

After being made a rear admiral, Moret-Moullec retired in 1946 and became an adviser on military and naval matters to Maurice Thorez, the French Communist Party leader.[38] Moullec publicly took a stand against France's colonial wars and supported Petty Officer Henry Martin, who was being tried for sabotaging the French escort aircraft carrier *Dixmude*. He and Muselier were often seen walking together near Toulon, where the latter had retired.[39] By then, Moret-Moullec was no longer in the navy and had been barred from any future reactivation for his public statements against the war in Indochina. He could not have been PIZHO, the Soviet mole within the Naval Ministry. To this day, PIZHO's identity remains a mystery.

Supporting Liberation Movements

The Eastern Bloc was not the only side probing and then inserting its intelligence personnel into forbidden areas, nor was it alone in running covert operations to support its national interests. The Prague communist coup in March 1948 and then the Berlin Crisis through the summer of 1948 convinced London and Washington that their intelligence approach to the Cold War was perhaps too soft. In September, the U.K. chief of staff told Cabinet ministers that Britain should seek "to weaken the Russian hold over the areas that were then dominated by the Soviet

Union" and that "all possible means short of war must be taken."[40] In Washington, Secretary of Defense James Forrestal held a meeting with Admiral Roscoe Henry Hillenkoetter, the first director of the Central Intelligence Agency (CIA), about expanding U.S. covert activity across the Iron Curtain. Funding grew from $2 million in 1948 to nearly $200 million by 1952.[41] Covert operations, which had been at that point subordinated to George Kennan's policy-planning staff at the State Department, were re-subordinated to the CIA in October 1950, when Walter Bedell Smith took over from Hillenkoetter as director of central intelligence (DCI).[42] For Kennan, the individual success of particular covert actions was relatively unimportant and what mattered was to again maintain the pressure on the communists. Others in the military wanted to practice and prepare for larger-scale covert operations that would be carried out during a future and inevitable war with the Soviet Union.[43]

Despite the creation of the CIA in 1948, ONI retained it own human intelligence organization. Through the interrogation of defectors and merchant seamen and through its attaché network, ONI also routinely collected names and information that might prove essential for future covert operations. Increasing reports from agents working primarily for U.S. and British services behind the Iron Curtain indicated that morale in the Soviet Union was poor and only 3 percent of the population were Communist Party members in 1948.[44] The wartime breach in the Stalinist society was comparable to the intellectual ferment produced among Soviet officers by their contacts with Europe during the Napoleonic Wars. Therefore, according to ONI analysis, many in the Western intelligence community expected that another Decembrist uprising could occur and threaten the Stalinist regime:[45]

> Resistance to Sovietization has been very strong in the three Baltic countries, Estonia, Latvia, and Lithuania, and the Soviets have ruthlessly deported so many of the natives that there appears to be some danger of their losing their national identities. . . . Most Ukrainians were ready to accept the Germans as liberators from the Soviets, and would be even more eager to welcome liberation by America. . . . The people of Belorussia do not seem to have a national entity as well defined as some of the others, but they are presently united in their hatred of Sovietization. The passion for national independence also burns brightly in the three small but strategically important Georgia, Armenia and Azerbaydzhan. Perhaps these people do not hate the Soviets with the same bitterness as the Ukrainians because the Soviets

have not bled them to the same extent for the Five [Y]ear plans, possibly reflecting the fact that Stalin and so many of the Party leaders were from these countries. But they are painfully conscious that their own individual civilizations flourished while the Russians were still primitive nomads and look scornfully on Russian culture even today. Naturally they deeply resent absolute Soviet Russian control and hope for the day of deliverance.[46]

But in January 1948, the State Department and the military and naval attachés in Moscow shared the impression that, as one attaché described it, many Soviet citizens are "persons who look upon war as the only possible release from the new despotism." Such dispirited citizens were found not only in newly acquired territories and satellite countries but also in large sections of the original Soviet Union in unknown but appreciable percentages. Dreher, the assistant naval attaché in Odessa, had reported that in spite of the destruction and the hardships caused by the Axis intervention, the percentage of people holding this view appeared to be very high in Ukraine. An American engineer who had returned to the United States after having spent sixteen years in the Novosibirsk-Omsk region had stated in no uncertain terms that "the people in those regions look towards America as the only hope for release from their present situation," Obviously, people working in the northern and Siberian regions had been brought there against their will, but the assistant naval attaché in Odessa thought that this feeling was widespread across the Soviet Union:[47]

Within the first weeks in Odessa, I began to hear tales of dissatisfaction, disaffection, and even bitter hatred towards the regime. People were sure that war was just around the corner and wanted to know what I thought about the situation. Under questioning they admitted that it was war with America they were thinking of, and that most of them looked forward to it as a means of liberation from the "*Sovietskaya Vlast*," the Soviet rule. At first I regarded people who told me such things as counter-agents trying to get me to declare myself in sympathy with them and draw me out, so I treated them with extreme caution. . . . But as the year went on and I heard this dissatisfaction and hatred voiced with increasing frequency and under circumstances in which the contact could hardly be other than genuine and purely incidental, I realized that it must be accepted that these sentiments were indeed strong and widespread in Odessa. . . . One night I dropped in to the bar at one of the main hotels and took [a stool] next to a pair of Soviet Lieutenant-colonels. Soon we were exchanging joking remarks. What is your nationality was the inevitable question. I always watched very closely when I told anyone that I was an American in order to detect, if possible, any instantaneous reaction, but in this case it was unnecessary. He screwed his face up into an exaggerated look of alarm and glanced comically all around. I smiled and said: "Yes, I know that it's pretty bad to be talking to Americans, so we'll just stop the conversation." He glanced all around again. The stools just beyond his friend were empty, and the barmaid was about 10 feet away. "No. Listen to me, Americans. You'll never see me again. If I should meet you I won't even recognize you. You know." And he held up his hands with the index and second finger of the one crossed over the index and second finger of the other in front of one eye—the symbol which I had seen so often in Russia—"behind the bars!" "You understand. But while you're here I want to talk to you. I want to tell you a few things. We're all for you, Americans. Ninety-five percent of us are for you. But do it now, while it's easy. Don't wait while the Soviets build more planes and tanks and guns. Do it now." As we talked on for another 15 minutes, I inquired as to his background. In such a case, I always wanted to know whether this person was another disgruntled Ukrainian Nationalist, the son of a "Kulak," or what special reason he might have for hating the regime. But this man was a native born Muscovite, a product of the regime and as a military officer, subjected to the most intense political indoctrination of all. Yet that was his answer. I am reminded of a letter which two girls in one of the Soviet Black Sea ports handed to a seaman on a foreign ship there in 1949, asking that he send it to some American newspaper for publication. Written in schoolgirl English, it contained the following direct sentence: "Every time we read in our papers of some new American preparation for war, we become gladder and gladder."[48]

This feeling was also widespread among other Eastern European nations. Returning from Constanta, Romania, the master of a merchant vessel reported to the Seattle CIA office that the Romanian underground was in control of the pilot organization. The source had noticed that the pilots always found pretexts to delay the entry of Soviet vessels but expedited the three-hour inspection of the entering and departing Western cargo ships in only fifteen minutes, confirming the influence of the underground movement. The source felt that it would be easy to insert or extract agents in or out of Romania using cargo ships and felt certain of the people's complete cooperation should an opportunity present itself to oust the communists.[49]

In Poland, the following joke was often told: "'What is Poland?' a student was asked at an examination. 'Poland is my mother,' replied the student. 'Excellent! And wh o is Stalin?' 'Great Stalin is my father,' retorted the student. 'Excellent! Examination is passed.' 'One second please,' said the student. 'May I make a remark that I would be happy to see my parents separated?' "[50]

But a Polish informant to a naval intelligence interrogator was more cautious. He stated that in his opinion, the Poles would offer resistance to the communists and to the Soviets only if they were assured of success beforehand. To his knowledge, no effective underground resistance organizations existed, and while some minor industrial sabotage was taking place, it was of no great consequence. The Poles, he said, had been deeply disillusioned by how the agreement at Yalta affected Poland, and while they remained basically pro-Western, particularly pro-American, they were not inclined to risk their lives. The informant was certain that in a time of crisis, the Catholic Church would provide a ready-made organization for resistance. In his words, the West did not need to worry about building up an internal resistance organization as long as the church continued to exist.[51]

Mariners provided ONI with invaluable HUMINT (human intelligence) information on dissidents who would be susceptible to offering some help behind the Iron Curtain. Among them were prostitutes who were often persecuted by the authorities for taking hard currency inside the country. Through mariners' interviews, ONI kept track of the the prostitutes and the bars where citizens gathered to express their negative attitudes toward communism. In Gdynia, Poland, "business girls" tended to be anticommunist because they were actively chased by the police. Most of the Scandinavian, British, and U.S. seamen on shore leave ended up at the Grand Hotel, a drinking and dancing establishment. The women were very clever at outwitting the police, being careful not to be seen leaving the hotel in company with seamen. Some of these girls were well-known by seamen who called regularly at Gdynia and were experienced and astute money changers who readily exchanged currencies with sailors whom they recognized. Again, the police were eager to apprehend any women who possessed foreign money. One of the most popular prostitutes was a twenty-five-year-old known as Jeannette who spoke French, German, and Polish, as well as broken English and Scandinavian. Another was the thirty-five-year-old Lucia, who was much in demand by seamen. Both operated from the Grand Hotel and were considered genuine anticommunists. On the opposite side, the dancing establishment next door to the Danish consulate was a favorite night spot for Communist Party members. The prostitutes there were not considered reliable.[52]

In Weimar, East Germany, the most popular brothel was a restaurant named Zum Wassertor, owned by a fifty-five-year-old widow. Seven or eight women operated from the place, which was frequented chiefly by fishermen and merchant seamen and only occasionally by members of the maritime and people's police. The acting madam of the place was Erna M. She was about thirty years of age, attractive, and married to an old man who offered no interference. Frau M. was described as reliable, absolutely pro-West, and anticommunist. The source said that she would give refuge to people from the communist regime who were seeking aid and that one could gain her confidence by giving her the regards of Christian B.[53]

Inserting agents across the Iron Curtain was no easy task, though. While a ship was at anchor, parties of security guards and customs officials armed with revolvers proceeded to search the vessel. During this time, the crew was mustered into one room and often given a physical examination. Cameras, binoculars, the ship's wireless, and firearms were locked and sealed. All currency was checked on arrival and departure. Armed guards were posted while the ship was alongside. The same procedures were followed in China, often under the supervision or in the presence of a Soviet official. A small boy was frequently used to search unusually small compartments of the ship that were not accessible to adults.[54]

Ferret Flights and Electronic Intelligence

It was a HUMINT event that prompted ONI to refocus its energies on electronic intelligence, a facet of the trade that was at the frontier of scientific, operational, and strategic intelligence. From the experience of World War II, it seemed likely that collection and analysis of enemy radar signals would become major priorities for naval intelligence. But in 1949, a defector from the Soviet Ministry of Shipbuilding opened a window to Soviet developments, which prodded the U.S. Navy to seek better electronic countermeasures (ECM) and improved electronic intelligence capabilities to defeat these future Soviet systems. Naval analysts studied lists of Lend-Lease Western radar equipment purchased by the USSR on the open market. Western attachés were asked to photograph Soviet electronic antennae and equipment.

German developments in direction finding (DF), intercept receivers, and tape recording guided the Naval Research Laboratory in producing sensitive collection sensors and electronic countermeasures equipment.[55] The Navy Research Lab's Countermeasures Section recognized the vulnerability of U.S. radar and conducted research in signal intercept, analysis, and location equipment in order to produce electronic counter-countermeasures (ECCM) that were designed to defeat Soviet electronic countermeasure systems. Under Howard Lorenzen, the lab also developed intercept equipment for shipboard and airborne reconnaissance that covered the entire frequency spectrum, from a few kilohertz to a few megahertz.[56]

To gather electronic data on the Soviet radar networks, electronic intelligence-collection "Ferret" airborne missions were flown directly toward and along the Soviet borders. Navy P2V Neptune and PB4Y Mercator "Ferret" aircraft and surface ships such as the heavy cruiser USS *Columbus* were fitted with intercept systems developed by the lab. A. L. Dwyer, a naval aviator attached to the Intelligence Office of the Alaska Naval District, initiated electronic intelligence flights along the Kamchatka–Bering Sea littoral of the Soviet Union from 1949 to 1951. The signals he detected were principally metric radar and E/F band coastal surveillance radar (metric radar at 72 megahertz; E/F band radar at 3,000 megahertz) but they provided unique data on Soviet defensive efforts.

One of the first incidents of these hazardous missions was the April 1950 disappearance of a U.S. Navy PB4Y *Privateer*. This large aircraft, which carried a crew of ten, six of whom were electronic technicians, took off from Wiesbaden in West Germany on April 8, 1950, en route to Copenhaguen. At 2:40 p.m., it made its last radio transmission over Bremerhaven in West Germany. According to the USSR, an aircraft, which they identified as a bomber, was detected on radar at a distance of about 350 miles from Copenhagen, over Leyeya Cama—7 miles within Soviet territory. It was promptly intercepted by a patrol of Soviet fighters and ordered to land. The Soviets maintained that the intruder had opened fire on the fighters, which had then shot it down. All evidence led to the assumption that the bomber was the missing *Privateer*.[57] Its equipment was later reportedly recovered by Soviet divers, which suggested that it had gone down above the Baltic.

In the Mediterranean, the navy's early surface electronic intelligence–collection operations targeted Albania, Bulgaria, and Yugoslavia. Five destroyers carrying naval communications units, designed to intercept and record all electromagnetic radiation, initiated this collection using trained operators.[58] Meanwhile, the navy established a panel of the Joint Communication and Electronics Committee to share the information being received by the three services about Soviet electronic warfare. Under an operation code-named "Dragon Return," German and Austrian scientists who had worked in the USSR as prisoners of war were interviewed. These men revealed that the Soviets were duplicating the German high-frequency direction finder Wullenweber antenna array used for communication intelligence and antisubmarine warfare. Annual reports were prepared by the Joint Communication and Electronics Committee to update Soviet radar order of battle. These also promulgated lists of shipborne electronic suits and IFF (identification friend or foe) equipment to differentiate unfriendly from friendly signals.[59] The CIA's directorate of plans (later known as the directorate of operations) also expanded its collection of electronic intelligence.[60] As a result of this joint effort by intelligence collectors and analysts, it was soon discovered that Soviet radar was far better than anticipated by skeptics who believed that the USSR was far behind the United States in electronics.

In one incident related to electronics and communications intelligence, the U.S. Seventh Fleet confronted Communist Chinese jamming against one of its communication ships. According to clauses contained in the 1949 treaties, the United States still enjoyed access to Chinese seaports. A U.S. ship specially equipped for communications had been stationed in Tsingtao to ensure that radio communications could be made between U.S. ships in Chinese ports and the Pacific Fleet naval command. One day, however, radio communications were interrupted by an unusual interference. The U.S. Navy, suspecting that the radio signals were being jammed, dispatched a small ship equipped with direction finders to locate and direct a party of marines to destroy the Chinese transmitter that was causing the interference. This, the marines did.[61]

Surface and Submarine Reconnaissance

Not long after World War II ended, naval intelligence began to ask operating naval forces and merchant ships, including foreign-flag vessels, to collect reconnaissance at sea by observing and reporting on Soviet and Chinese warships. These ships were also trained to observe and report specific activity in ports. ONI provided cameras to select merchant ships and organized an elaborate system of debriefing by naval attachés abroad or district intelligence officers in the continental United States. A special code designation protected the identities of merchant ships' observers. This reporting proved invaluable at the time and was one of the few sources available for reporting Soviet ship movements.

The Navy Photographic Interpretation Center on the Potomac River near Washington analyzed and filed important photography, which later became useful during the 1962 Cuban missile crisis, when a massive Soviet merchant sealift took place. Besides the electronic intelligence-collection missions performed by specially equipped destroyers in the Mediterranean during 1949 to 1951, submarines joined the reconnaissance effort by patrolling along Soviet coasts. In May 1948, in a first mission, USS *Sea Dog* (SS 401) patrolled the Siberian coast with an intelligence officer from the Alaskan Joint Staff aboard. He provided current intelligence on Soviet radio calls and frequencies and estimated locations of Soviet air bases for making communication intercepts.[62]

In August 1949, USS *Cochino* and USS *Tusk* conducted initial operations along the Kola Peninsula, searching for evidence that the USSR had detonated an atomic bomb. *Cochino*'s crew members certainly had no desire to make the headlines of the Soviet armed forces daily *Krasnaya Zvezda*, but during the mission, a battery explosion sank the *Cochino*. Her crew was rescued by her patrol partner USS *Tusk*. That month, the Soviet Union began what would be a many-year-long string of press releases denouncing the "suspicious training" activities of U.S. submarines near the USSR's coasts.[63]

During the same tense postwar period of jockeying for information, British intelligence initiated a campaign of reconnaissance along the Soviet shoreline.[64] Commander J. G. Brookes, a senior Royal Navy intelligence officer, helped set up Operation Hornbeam—the British fishing fleet's watch on the Soviet Northern Fleet. With an alleged £10,000 reward per trip for the crews of trawlers from Hull and Grimsby, fishing intelligence on Soviet naval operations in the Barents Sea was becoming a business of its own. Launched in the summer of 1949, the trawler *Lancer* made a total of forty-five spying missions in the next five years. During her first voyage to the Arctic Circle, she recovered the fuselage and intact cameras and film of a crashed American reconnaissance aircraft. In June 1950, she captured a live twenty-four-foot Soviet torpedo and brought it back.

These trawlers recorded Soviet marine and aviation radio traffic and photographed Soviet Northern Fleet warships, often at close range. On occasion, certain trawlers disembarked British naval intelligence teams, which paddled ashore in kayaks from the edge of the 3-mile territorial waters limit to hide radio receivers. These radio receivers were placed on the Soviet Kola Peninsula from west of the Murmansk Fjord to the mouth of the White Sea. They would record local signal activity on tapes that would later be recovered and analyzed at an intelligence center in Pitreavie, Scotland. In October 1950, Soviet Captain Second Rank Bakhmutov, the head of the intelligence staff of the Soviet Northern Fleet, gave the following description of British activities:

> During October 1950, 65 English fishing trawlers were observed in the waters of the Barents Sea, periodically approaching close to shore and on more than one occasion violating our territorial waters. Thus, for instance, on 26/9 the trawler H-42 "*Swanella*" was arrested in our territorial waters. The vessel's captain, who presumably spoke Russian, was systematically noting all Soviet civilian and naval ships entering or leaving the Kola Sound.

Conclusions:

1. Voyages by English naval vessels to the Barents Sea and trips along the Murmansk Coast by fishing vessels are being carried out mainly with the intention of studying the operational area, sea conditions and the meteorological environment.
2. Systematic visits to the Barents Sea by English naval vessels, trips along the Murmansk Coast and frequent violations of our territorial waters by English fishing vessels testify to active intelligence-gathering in this theatre of operations.
3. Presumably on certain English trawlers there are Royal Navy intelligence officers.[65]

Nazi Pirates and Freedom Fighters

In the Baltic, British naval intelligence also ran agents using former wartime German Kriegsmarine units under the cover of the British Baltic Fishery Protection Service. Between 1945 and 1956, German navy Kapitanleutnant (lieutenant) Hans-Helmut Klose was employed by the British in mine-clearing operations. After clearing mines in the Skaggerak and the North Sea, he began to work for British intelligence, initially to nurture contacts he had in Poland; in the Soviet Baltic republics of Estonia, Lithuania, and Latvia; and in the former Soviet occupation zone, renamed the German Democratic Republic.

In October 1966, a series of newspaper articles released in the German Democratic Republic first revealed the history of Britain's operations using former German fast attack (PT) boats to insert agents and carry out intelligence work in the Eastern Bloc countries.[66] The articles, which were aimed at compromising the operations, claimed that the men used in the operations were former German sailors under the command of a man called the "Red Fox." These operations consisted of inserting agents and performing other clandestine work in the Baltic countries. According to the East German press, the agents were trained by Western agencies: the British Secret Intelligence Service (SIS, also known as MI6), the U.S. Central Intelligence Agency, and the postwar German Organization Gehlen (OG). These secret operations were actively opposed in a behind-the-scenes war with the Soviet intelligence services. Many of the operations were alleged to have originated out of a Hamburg yacht marina. The East German articles claimed that the main perpetrator of these pirate operations was the ex-Nazi naval officer Hans-Helmut Klose, who was motivated by his personal desire to continue the war against the USSR.

British fast patrol boats initially towed the rubber craft, but the trip from Gosport, England, to the eastern Baltic target coasts took around sixty harrowing hours in rough seas. Then, in 1951, a small tanker called the *Dievenow*, flying the flag of Frontier Control, towed a number of the fast boats to Cuxhaven from where they commenced their duties. The operation—cover name Klose—initially worked under British MI6 control and in close coordination with the CIA, the OG, and some Scandinavian intelligence services. The men operating the fast patrol boats were experienced sailors who knew those waters from having worked during the closing years of World War II as part of a Kriegsmarine flotilla based in Ventspils in the eastern Baltic. The boats had rescued high-ranking German officers and intelligence teams from the advancing Soviet army.

From 1946 to 1949, during the Soviet blockade of Berlin, the Western Allies used former German sailors to land agents and saboteurs in the Baltic States. Their activities were not viewed as violations of international law, because the West did not recognize the Soviet occupation of Estonia, Lithuania, and Latvia.[67]

The Baltic operations increased in tempo when the Western Allies feared a Soviet invasion of Berlin. Henry L. Carr was the original postwar coordinator of the North European Directorate of MI6. Under his command was a Baltic Section led by Alexander McKibbin, who had three subordinate groups. The Estonian section was led by former Nazi SS colonel Alfonse Rebane. Former Luftwaffe officer Rudolph Siljaris ran the Latvian Section, and the history professor Stasys Zmantas managed the Lithuanian Section. In 1946, the three sections kicked off an operation called Jungle, in which they inserted indigenous agents into each of the countries to link up with anti-Soviet elements. In Estonia, for example, their native collaborators were called "forest brothers." The MI6 sections organized the anti-Soviet citizens in each country to observe, collect intelligence, and perform specific sabotage acts against the communist nomenclatura. Recent estimates of the results state that the casualties of this secret warfare totaled 75,000 local civilians; 30,000 underground fighters; and about 80,000 Soviet troops, primarily in the Baltic States, East Germany, and Poland.[68]

Even neutral and nonaligned Sweden was actively collecting information. Since the end of the war, the Swedish intelligence service branch, called C-Bureau, began to conduct independent secret operations in the same Baltic waters, primarily as intelligence collectors against the Soviet armed forces. They did this while the CIA carried out Red Fox operations, inserting agents and provocateurs by parachute. On the other hand, the British services preferred to use high-speed PT boats and fast, semirigid Zodiac boats for their nefarious actions.[69]

3

ALLIES TO ANTAGONISTS
1945–1952

Five years prior to the outbreak of World War II, Soviet dictator Joseph Stalin began to take major strides to give his coastal defense navy a new blue-water dimension. In Stalin's mind, the political alliances of the 1930s had created a new threat to the USSR.[1] These alignments exposed the Soviet Union's real strategic impotence when it proved unable to send a fleet to support the Spanish Republic, while fascist governments were quick to align themselves with Franco's forces. On December 24, 1935, the Party organ *Pravda* published an editorial warning "the enemies of the proletarian State" that "the temporary weakness of the Soviet Navy would soon be overcome. Stalin then charged his most ruthless industrial assistant, Andrei Zhdanov—already a key figure on the Soviet Supreme Navy Council—to lead a crash expansion of naval shipbuilding and to acquire badly needed Western naval technology.[2]

Stalin purged V. M. Orlov, his navy commander, who had refused to send a squadron to the Mediterranean to challenge Hitler and Mussolini and support the Spanish Republic. To succeed him, the Generalissimo picked his Pacific Fleet commander, the young energetic Nikolai G. Kuznetsov, who would become an outstanding war leader. By 1939, naval spending made up 18 percent of the Soviet defense budget. The war, however, put an end to the planned "Big Fleet" of 699 combatants envisioned in the ten-year naval program of 1938.[3] With the defeat of Germany and Anglo-Saxon dominance on the oceans encircling the Soviet Union, Stalin once more contemplated building a powerful navy to confront his former allies and maritime superpowers, the United States and Britain.

Stalin's Blue-Water Dreams

During Winston Churchill's visit to Moscow in October 1944, Stalin remarked that Hitler's most serious mistake was to have attempted to conquer Europe while its fleet remained inferior to the combined naval forces of its adversaries.[4] That same month, Stalin gave the order to prepare a new ten-year plan for military shipbuilding along the lines of the plan that had originally been established in 1938.[5]

By a fierce emphasis on scientific and technological progress, albeit at heavy cost to an already devastated populace, the USSR was rebuilding its industrial base with the objective of becoming a world-class navy. The result was a rapidly expanding core of well-equipped shipyards, which incorporated newly acquired material and techniques that had been gained as spoils of war. Meanwhile, institutes and design bureaus were hard at work on creating new weapons, including missiles that would be used to conduct the main strike on the adversary. The study of Lend-Leased and captured technology had revealed the weaknesses of Soviet naval and military research. Stalin himself clearly set a priority on missiles. On May 19, 1946, the Soviet leader signed a decree of the Soviet Council of Ministers creating a Special Committee on Rocket Technology.[6] The wording of the text showed the USSR's total dependence on German technologies and the tribute to intelligence:

> reproduce as a primary task the FAU-2 [long-range guided missiles] and *Wasserfal* [antiaircraft guided missile] using national resources . . . consider the following work on missile technology to be the primary tasks: (a) recreate the technical documentation for the FAU- 2, *Wasserfal*, *Rheintochter*, *Schmetterling*; (b) restore the laboratories and test facilities with all of the equipment and instruments needed to conduct research and trials . . . ; (c) train Soviet specialists who have mastered the construction of FAU-2 missiles, anti-aircraft guided missiles, testing methods, to produce parts for assembling rockets . . . prepare facilities to house the German design bureaus and specialists . . . allow the Special Committee for Rocket Technology and the ministries to order different equipment and systems from scientific research institutes laboratories and State rocket weapon firing ranges in Germany as war reparations . . . order the Special Committee to present the Soviet Council of Ministers proposals for sending a purchasing commission to the United States to procure equipment and instruments for scientific research institutes laboratories on rocket technology . . . establish as a State priority rocket technology development and direct all the ministries to carry out the tasks related to rocket technology as their primary tasks.[7]

In 1947, Stalin dismissed his navy chief because the officer had opposed some of Stalin's reforms. Demoted to vice admiral, Kuznetsov was put on trial a year later for having shared the German T-5 acoustic torpedo technology with the British Admiralty. This was an obvious subterfuge, because Stalin himself had signed the letter authorizing British technicians to examine the weapon. But the Generalissimo liked Kuznetsov well enough that by February 1950, the vice admiral had been entrusted with the 5th Fleet in the Pacific, while his former deputies were left in jail or in disgrace.[8]

Stalin, however, was dissatisfied with his own reforms. As he had feared, the army was treating the navy as a auxiliary force, which was not the direction that the Soviet leader wanted for his sea service. To establish the standing of the navy vis-à-vis the army, Stalin created a Ministry of Naval Forces and a Main Naval General Staff on February 26, 1950. In July 1951, he forgave Kuznetsov, who was immediately reappointed navy minister.[9] Stalin's choice reflected both his liking of Kuznetsov's integrity and his ongoing desire for an independent-minded, seagoing navy centered around battle cruisers and cruisers.

The Generalissimo was planning for a fleet that would achieve dominance in the waters surrounding the Soviet Union in cooperation with the air force and would serve as a training base for the officers of a future blue-water navy. "I advise you at this stage to have a few more light cruisers and destroyers," Stalin had told N. G. Kuznetsov. "Things are going poorly for you with the cadres. Cruisers and destroyers would allow you to train good cadres." But, as historian Natalya Yegorova remarked, despite these extraordinary efforts, the realization of the ten-year shipbuilding program moved forward with great difficulty, and by the end of 1949, the Ministry of Shipbuilding had not fulfilled the plan.[10] If war had broken out, Stalin's navy would again have been restricted to supporting ground forces in the coastal areas.[11] In that context, the three battle cruisers that were laid down in 1951 were to have been the forerunners of his projected 1956 second ten-year shipbuilding plan and would have been instruments of prestige for the Soviet Union, proving to the world its superpower status.[12]

But Stalin and Kuznetsov also fathered the Soviet variant of the future "capital ship" of the Cold War: an entirely new platform that might have won a war with a single shot, the nuclear submarine.

Worrying intelligence reports from the United States had revealed the construction of an underwater platform not dependent on outside air: a true submarine that ran on nuclear energy and was capable of producing oxygen and water. Following a September 9, 1952, decree from the Central Committee of the Communist Party, the USSR launched a crash-course program to build an atomic submarine. Project 627 was not merely a Soviet replica of the American attack submarine *Nautilus*. It was initially designed to be a strategic platform intended to fire a single 27-meter-long nuclear torpedo against the probable enemy's main city port of New York.[13]

Assessing Stalin's Intentions

Such insights into the Kremlin's naval plans would have been priceless to the foreign attachés in Moscow. But it seemed that for the time being, there was no traitor in the Soviet navy with such knowledge, and Western intelligence services were left to make the best of the bits and pieces that they could collect in a closed society. Contrary to the British, the U.S. Office of Naval Intelligence (ONI) did not underestimate Stalin's blue-water plans. By early 1946, it viewed the Soviet Union as a potential competitor on the high seas:

> The strength of the Soviet Russian Fleet must take into account Russia's naval position in the past, the will to maritime power displayed by Russia's ruler and the future naval potential of the Soviet Union. All available evidence indicates that the Soviet Navy of today has inherited the "big Navy" traditions of its Imperial Russian predecessor. The geographical position, industrial strength and prestige of the Soviet Union appear to assume a substantial basis for future sea power. Hence the present composition of the Soviet Fleet should be regarded as a transitory stage in the reconstitution of Russian naval strength.[14]

To monitor Soviet developments, the Western Allies relied on teams of specialists, interrogators, and human intelligence collectors. The United States and Britain had captured the former Wehrmacht intelligence organization operating on the Eastern Front nearly intact. This was the efficient network run by German general Reinhart Gehlen. With the ideological split widening between East and West, the Gehlen operation proved invaluable, and it was soon resuscitated. It became an active intelligence network run initially out of the CIA and eventually, following the peace treaty, by the new Federal German Republic (West Germany) from headquarters outside Munich. It would become known as the Bundesnachrichtendienst, or BND. But most of the information was collected by the Western military and naval attachés posted in Moscow. These spies in uniform were employed within each country's diplomatic missions, embassies, and consulates throughout the world. It was an established method used by most countries since the nineteenth century.

In addition to their diplomatic and liaison responsibilities, military and naval attachés are the ears and the eyes of their intelligence services. The work of the attaché, like that of a journalist, is reporting. The first U.S. naval attaché was sent abroad in 1882 when the ONI was created and Secretary of the Navy William E. Chandler directed, "In order to collect information, a corps of correspondents, in the persons of Naval Attachés to our foreign legations . . . will be organized."

Attachés work in the field of human intelligence and are protected by diplomatic accreditation. The methods they choose to collect intelligence is directly driven by the availability of vital information in the host country. In an open society, attachés glean their knowledge by reading published literature or asking their host armed forces for the desired data. In closed societies, they resort to the theft of information, by either discreet photography, eliciting it from local officers, simply paying informers, or more modern technical means, such as placing remotely controlled black boxes to gather desired information.

The age-old Russian tradition of cloaking all military information in secrecy forced military and naval attachés in the USSR into extreme roles as discreet intelligence collectors. When information was thus denied, an attaché was forced to resort to more imaginative techniques to accomplish his reporting mission. The attaché traveled to acquire information and, when necessary, gained access to restricted vantage points to observe naval forces during construction or repair. The attaché's mission was to detect change and report new military developments: for example, the prototype for a new class of submarine concealed under camouflage. The pressure on the attaché to seize such opportunities was intense, since his actions were driven by the close competition in antisubmarine research and development. A new device spotted on a Soviet submarine during construction or launch might portend an innovative weapon system that would require a different countermeasure on a Western ship under construction at the same moment, so close was the rivalry.

During the era of Soviet naval expansion, the naval attaché's focus was reporting on the Soviet fleet. Opportunities for success varied widely by location and political setting, with the Moscow-based attachés being virtual prisoners in the controlled information environment of the USSR.

Moscow Posting

In 1945, the largest naval attaché office in Moscow was that of the United States, with an admiral, a captain or a colonel, and six junior officers.

Britain, France, Canada, Turkey, Sweden, and Norway had fewer people posted in that city. The United States also enjoyed the most privileged Allied observation posts in the Soviet Union, with assistant naval attachés being resident in Archangelsk, Odessa, Novorossiyisk, and Vladivostok. The rationale for this larger presence was to assist the Military Mission to the USSR and oversee the massive Lend-Lease program. Their mission was to coordinate American cargo ships and aircraft bringing Lend-Lease aid to the Soviet Union. But the rising tensions profoundly changed the nature of the attachés' work. For Rear Admiral Leslie G. Stevens, the American naval attaché in Moscow who succeeded Houston L. Maples in late 1947, the only justification for his office was to write intelligence reports: "This office can continue to exist only through the direct results that it produces."[15]

The attaché effort was guided by listings of requirements that were sent to the attachés abroad. The most pressing priorities for both the army and naval attachés at that time was to uncover evidence of Soviet progress in atomic energy for the development of a nuclear bomb. Soviet improvements in guided missile and submarine construction were also among the highest intelligence priorities. In a September 1948 "Guide for the Collection of Information on Submarines," ONI underlined the necessity "to determine submarine construction and the extent of the program without actually observing the construction at the yards, for example, by noting manufacture, transportation, and shipment of component parts." Attachés were encouraged to pay special attention to alterations to tunnels and bridges, which might indicate plans for the mass production of submarines and an intention to transport the sections overland for assembly at some distant point.[16]

For Rear Admiral Stevens, the Soviet press furnished "the only sources from which this office can expect to provide a regular output of information in the fields it is supposed to cover that are freely and openly available to the Russian public."[17] Past experience had taught Maples that "a patient accumulation of minor details eventually forms a pattern which becomes apparent to the sensitive intelligence officer, and the mass of such material publicly available in Russia provides many such details." Attaché Stevens further considered that the volume of this open-source material was "so great" and "individuals with sufficient background and knowledge of the Russian language to scan the material in Washington" were "so few," that his office had to digest the press information and not simply collect it.[18]

The publication of the "Stalin Prizes for outstanding work in the field of arts and science" was sometimes a good indicator on the progress of Soviet research.[19] For a trained observer, three awards were related to

Soviet efforts to develop an atomic bomb: two senior scientific workers at the Radium Institute had been distinguished "for their work related to the self induced fission of uranium," and another award referred to "super quick photography and its use in investigation of the processes of explosion and impact."[20]

Interaction with the Soviet Navy

Parades, official visits, and private parties always produced some kind of intelligence, and for that reason the various nations tended to barter these opportunities for their respective military and naval attachés. From the Soviet side, these invitations were kept to a minimum. In 1947, American naval attachés in Moscow complained that they had not been invited to any social functions since Victory Day (May 1945), while their Soviet counterparts in Washington had been asked to the New York Navy Day Fleet Parade (October 1945), an amphibious exercise in the Caribbean (May 1946), and numerous other events. Relations with attachés were handled by the Foreign Relations Office, and foreign attachés could not go directly to the Red Navy, which made socializing with Soviet naval officers nearly impossible. Attachés were never invited to private get-togethers, and Maples had never heard "of a single case of a higher political or naval person having foreigners in to a private party."

The lack of invitations to social activities being issued to the American naval attachés in Moscow and their difficulties in obtaining Soviet navy publications finally led Washington to retaliate. On May 12, 1947, Chief of Naval Intelligence Thomas B. Inglis informed Maples in Moscow on his new stand toward the Soviet navy: "For your information while it is customary for ONI to supply the various Naval attachés in Washington with OPI news releases, copies of All Hands and various personnel lists, these are not now given to Naval attachés of the USSR. Neither are the latter included in inspection trips such as the US Naval Academy or various other Naval stations or ships. The CNI has omitted the names of the attachés of the USSR from private social functions such as cocktail parties and dinners at his residence. . . . He does not intend to include them in such invitations until informed by you that such invitations are extended to our representatives in Russia, or until you recommend more cordial treatment for some other reason." Relations between the two superpowers were not broken, though. Inglis was ready to tell the newly appointed Soviet naval attaché, Rear Admiral Evgeni Glinkov, that any time he had a proposal for reciprocity, Inglis would be glad to study it. Inglis had been willing to receive Rear Admiral Glinkov on May 5 and would continue the delivery of hydrographic material to the USSR if the Conference of the International Hydrographic Bureau held in Monaco were to convey a Soviet willingness to cooperate.[21]

Noting that the Soviet Union had issued new sanctions for espionage, Admiral Maples was rather pessimistic about the future: "In view of the publication of the decrees on disclosure of state secrets, it seems highly unlikely that there will be any relaxation of the current restrictions on visits and travels by the Naval Attaché or members of his staff."[22] This situation prompted the State Department to ask General Bedell Smith, the American ambassador in Moscow, to raise the question. Smith did not feel that the United States could continue "to overlook this complete lack of reciprocity." As an apparent result, in September 1947, the chief of naval intelligence was happy to learn that the naval attaché in Moscow had been invited to the Red Air Fleet Show, the "first [invitation] of its kind in many months." Inglis hoped that "other invitations productive of intelligence" might be forthcoming. As a consequence, he informed Stevens of his intention "to extend Soviet naval personnel [in Washington] an invitation to a comparable US naval flying exhibition."[23]

Marius Peltier, the French naval attaché, was in a similar situation. On one occasion in front of his hotel, he met a former acquaintance from his days in the Allied commission in Berlin. The Soviet naval officer looked embarrassed, and Peltier's questions got nowhere.

"What are you doing now?"

"I don't know, I am waiting for orders."

"In the meantime, where are you? Here, in Moscow?"

"Yes."

"I would like to chat with you. Can we have lunch?"

"Naturally, but I am very busy, and I can't tell you when."

"Call me at my hotel."

"Certainly, but excuse me, I now have an urgent meeting; I shall ring you."

He fled and never called.[24]

Meeting the Top Brass

For months, Peltier had asked in vain for an audience with the head of the navy. One day, Peltier's phone rang: "Lieutenant Commander N., from the Foreign Relations Office speaking; the commander in chief of the Soviet navy shall receive you later in the day. Do you accept?" Caught by surprise, Peltier readied himself and waited for his escort. International tensions were building up, and Peltier wondered what Admiral Kuznetsov would say to him.

After recalling his good memories of an earlier visit to Paris, the admiral briefly explained his concept of international relations: "You are an officer, like me; we are bound by duty; if I am asked to strike an adversary I shall strike him with all my strength, and strike him again, whoever he may be; do you understand? Now, as long as I don't have to strike, we are friends." And the admiral concluded the meeting with a hearty handshake.[25]

Shortly after that meeting, Peltier replaced the departing French military attaché and took over his position as head of the military mission. The latter had complained to the Soviet authorities about the lack of opportunities to meet with Soviet officers other than those of the foreign relations. Almost immediately, the departing military attaché and Peltier got an invitation to the Metropole Hotel, where they were met by two Soviet generals and a navy captain. The senior general was straightforward, sharp, and witty. Peltier took notice that truth in earnest is often the best weapon and that the general was a brilliant philosopher. "I am a materialist and therefore I am free," proposed the general. "You don't prove anything," said Peltier. "History will prove that we were right," shot back the general. From this exchange, Peltier concluded that the USSR's military school provided unity of thought not only on strategic and tactical subjects but also on philosophical topics.

Attaché Performance

The real performance of the naval attachés was measured by the quality and timing of their intelligence reports. American, British, and Swedish correspondence was generally more detailed than that of the French, because their overall knowledge of the Soviet navy was deeper. On the atomic bomb program, American sources indicated that the Soviet Union was making progress. In December 1946, Harshaw, the assistant U.S. attaché in Odessa, had noted that shipments of food products and machinery to Bulgaria were probably in exchange for a recent load of radium ore from that country.[26]

Then a Russian source revealed that her father's cousin, a former ammunitions minister, later identified as Vanikov, had been recently decorated. This led to the speculation that he was being rewarded for significant progresses in the atomic bomb program.[27]

Even if the U.S. naval attaché office concluded that Soviet naval technology was markedly inferior, it did not disregard Soviet technological abilities. In particular, Stevens praised Soviet engineers for being able to reverse engineer and produce the B-29 bomber: "The manufacturing difficulties that were encountered in the United States in initially fabricating the B-29, the difficulties that were apparent in our Navy in studying the conversion of the standard B-29 to an anti-submarine airplane, and our own experience with alterations in fuselage and tail contour, all combine to give striking evidence that Russian design and manufacturing ability as shown by their bringing out a number of airplanes in this class, should not be underestimated."[28]

American attachés considered the Soviet naval officers' esprit de corps to be strong and the quality of their education on the rise.[29] In July 1948, a German source told the American naval attaché that the Soviet officers' promotion system was again linked with political qualification. From 1940 until the last quarter of 1947, Soviet officers had been promoted only in accordance with their military ability, following three criteria: personal qualification, technical qualification, and tactical qualification. Then, a fourth column on political qualification was added. The source further explained that in 1947, only 15 percent of the military officers were members of the Communist Party. An officer who was up for promotion but could not be promoted due to a deficiency in one of these four qualifications received a maximum probation period of two years, which could be extended in case of illness. During that time period, he was expected to rectify his shortcomings. Another failure meant retirement.[30]

The American assistant attachés were able to determine that political qualification also became paramount in the port administration. Port officials of Vladivostok and Odessa were increasingly criticized for their poor showing. These criticisms were accompanied by a mounting pressure for customs officials to join the party or give up their jobs. "Midnight visits of NKVD plainclothesmen" to the homes of several officials and their subsequent disappearance had become common.[31]

Back in 1945, the U.S. naval attaché offices that had been established in Soviet ports were as good as the British offices at reporting on the Soviet navy order of battle. But with the new political context, ONI instructed them to do better. Sometimes photographs could be made from a hotel window or from a visiting American Liberty Ship. In one case, the assistant attaché in Odessa captured a new landing ship with his camera but noted that "it was deemed inadvisable to try to take any later shots . . . because of a spectator in the park across the street who discovered my presence at the window and kept a weather eye peeled in this direction."[32] In 1948, with the naval attaché in Vladivostok gone, the U.S. Navy started to rely on its consulate to monitor the activities of the Pacific Fleet. On July 20, 1948, James Risk, the vice consul, reported to Rear Admiral Stevens, the new U.S. naval attaché in Moscow, on considerable local activity around Vladivostok, despite little change in the status of the ships present. The observations were made from a back window on the second floor of the American Consulate. Sometimes the consulate windows were shaken by gunfire. As the diplomat noted, "The frequency of the firing practice obviously argues a large volume of training," something that he related to the fact "that the majority of the Military personnel seen about the streets appear to be raw recruits."[33]

Falling in Love

U.S. food aid came regularly into Odessa and Vladivostok, creating mixed feelings of gratitude and suspicion among the local officials, the security services, and the population. American crewmen spent their free time in bars chasing girls, which disappointed the Soviet Intourist organization because it was trying hard to get them interested in cultural activities. Incidents were not rare, and racial tensions onboard ships were reflected ashore with fights between black and white sailors. But Harshaw thought that "Considering the number [of American sailors] ashore and the potent vodka it could be worse." Health issues were also a concern. As Harshaw explained, "The situation of venereal disease is bad. Almost all who go ashore get [a] disease."[34] In Vladivostok, Soviet authorities had placed more restrictions on shore distractions, to the satisfaction of the U.S. assistant attaché, who remarked, "I am quite sure all are much better Americans after their short visit here . . . no one wanted to stay or to take their Party membership. The NKVD kept the girls away from them so that there wasn't much for them to do except walk around and the weather was bitterly cold which sort of made going ashore very unpopular after their first taste of the beach."[35]

The Soviets' attitude toward the Americans residing in their ports was not always predictable. Ronald Day, an American civilian in charge of the assistance program in Odessa, ran over and killed a man while leaving the home of a Soviet official where he had had dinner. The employee seemed sober when the accident happened but was heavily intoxicated when the militia came to his hotel for a deposition. To the naval attaché's complete surprise, the local authorities had not tried to take advantage of the American's helpless situation and had solved his difficulties, although it later emerged that the victim was an agent provocateur who might have deliberately provoked the accident.[36] The incident reflected internal strife between the local authorities, who were grateful for U.S. aid, and the security services. But Day was not as lucky in his love life. Upon learning of Day's affair with his Russian secretary, the MVD had given the girl thirty days to get out of town, and in fact she never returned from a professional visit to Moscow. The American protested, to no avail, and upon leaving the Soviet Union, he was informed that both his girlfriend and a port official, who had tried to help them get married, had been deported.[37]

Western Ship Visits

Compared to the continual flow of American cargo ships delivering food and equipment such as steam engines and tractors to the ports of Vladivostok and Odessa, Western warships' port calls in the Soviet Union were rare. There were three notable exceptions, however: one was the British carrier HMS *Triumph* in Leningrad during July 1946; the second was the British cruiser *Aurora* in Sevastopol; and the third, the USS *Starr* (AK-67) in Vladivostok in January of that same year. All three port calls presented valuable opportunities to collect intelligence on the Soviet navy and on Soviet society through numerous interactions between the crews and the population.

HMS *Triumph*, wearing the flag of Admiral Lord Fraser, G.C.B., K.B.E., sailed from Portsmouth on July 21 with HMS *Rapid* in company and reached Kronstadt on the twenty-sixth. The *Triumph* fired a salute of twenty-one guns to the USSR, which was returned by a shore battery, and then one of seventeen guns to the flag of the Soviet commander in chief of the Baltic Fleet, Admiral Tributs. The salute was returned by his flagship, the *Oktiabraskaia Revolutia*, a reconstructed battleship dating from 1911. After the two admirals had ceremoniously exchanged full-dress visits, while Russian press photographers "copiously photographed the *Triumph*'s masthead aerial array," there was a formal luncheon party of fourteen courses aboard the Russian flagship. Lord Fraser left in the British destroyer for Leningrad and then departed for Moscow, where he attended the Red Navy Day celebrations as the guest of Admiral Kuznetsov. The British crew members seemed to have been impressed with the friendliness of the people and the spontaneous welcome they received. In general, the crew members' experiences tended to make them dislike the Soviet way of life. Some sailors who had previously been attracted to Communism found less to admire in its practice, and a typical remark was: "I didn't think that was the way Communism worked—if it is I have had it."[38]

The *Starr*'s port call in Vladivostok was not as pompous. It was marked by an incident that the U.S. assistant naval attaché described as follows: "The Soviets had tried to put one of their English speaking NKVD boys aboard dressed up in Navy Uniform. . . . I had met and worked with the guy . . . back in the 'Cloak and Dagger' days [World War II], and spotted him much to his surprise. When it came to the Captain's attention, that the bird was asking too much questions, we kicked him off the ship and permitted only the two regular liaison officers aboard." These actions by the NKVD guards reflected the end of the grand alliance. In Vladivostok, Ryan noted "the general tightening up by the Soviets around us . . . especially noticeable by the way we are watched." The telegraph company had also tried to cut off their communications with Moscow. Anticipating the possibility of his office's closure, Ryan expressed his anger at the U.S. Navy's passivity: "You undoubtedly received that dispatch from the department giving the number of Soviets and offices they maintain on our West Coast. . . . Why the hell . . . pardon the strong language . . . can't we do something about their spreading all over the US like a band of locusts and we are limited to 19 kilometers outside of the city. . . . It has been my experience to learn they think a hell of a lot more of you if you treat them on that basis. As it looks to me, even in the States, the 'hand kissing' period is definitely over. . . . Does our Navy know it?"[39]

Moving Around

Traveling around the Soviet Union was a priority if attachés wanted to gain firsthand intelligence. This usually consisted of information about the morale of the people and their living conditions and sometimes details of those military objectives that could be observed. As Ryan had pointed out with some rudeness, however, foreigners' movements were severely limited. The usual means that the Soviets employed to restrict travel was by setting up roadblocks and making it impossible for foreigners to obtain gasoline or accommodations. Military and naval attachés were required to obtain permission to go any farther than 62 miles from the center of Moscow. Often, permission was not granted; attachés were told that the request was "being considered." By comparison, Soviet naval attachés in

Washington were asked to notify the chief of naval intelligence three days in advance regarding trips from Washington. And the protocol and liaison sections complained that the Soviet naval attaché's office was "casual" in carrying out these provisions, often giving notification just prior to departure and sometimes subsequently. Since civilian employees were not required to ask for permission, it was believed that "frequent unreported trips" were made, and that some of the "civilian employees" were officers in the Red Navy. The instruction concluded, "Whereas in Moscow, the U. S. Naval attaché is denied permission to travel, there are no restrictions on the travel of the Soviet Naval attaché in this country [the USA]."[40]

Usually, attachés circulated U.S. publications among fellow travellers to create a bond and study people's reactions. Once, on a trip to Odessa, attaché Maples tried to socialize by showing American magazines to random train companions. As he reported, "The magazine they were really interested in was *Life* magazine, their favorite." When they saw a photograph of Franklin D. Roosevelt, they immediately put their fingers on it and said that Russian people loved Roosevelt. Maples's impressions of Ukraine were that very little reconstruction had taken place, although food products were not lacking. In Odessa, Maples was impressed by the manner in which German prisoners were treated: "They looked in excellent shape, healthy, well fed. . . . There is evidence that they are better treated than Russians themselves."[41]

Another keen observer was Captain S. B. Frankel, who was lucky enough to have traveled twice during 1947. In March, he arranged to visit Riga and returned to Moscow via Tallinn and Leningrad. He noted the presence of two hundred passengers standing outside without tickets and showing signs of malnutrition; they were obviously in search of food. Having lived in Riga from 1936 to 1938, Frankel remarked that the main change was the population transfer. Now, 60 to 70 percent of the population was Russian. Forming a very small part of the city's population, the Latvians had no hope for the future except for a "gentle war of liberation." A new class of Soviet citizens looked down on the Latvians for not accepting the "much more advanced Soviet culture." Frankel's visit was cut short by the NKVD. As he explained, "I was unable to detect the presence of a recording device in the room, but I twice caught the maid listening at the door." Four times, the two visitors were photographed by a camera concealed in a package. A chance encounter with an Intourist interpreter they had met earlier gave advance warning to Frankel that their application for further travel to Tallinn and Leningrad had been denied. For two days, the two visitors took various pictures of the city and of the war damages with the approval of the NKVD, until their film was confiscated by a militiaman and was never returned, despite the Soviets giving them a formal apology and promising to do so.[42]

That same year, Frankel left the Soviet Union. Instead of sailing directly to the United States, he chose to go through the Soviet-Turkish border and visit the Soviet Republic of Armenia in the process. There, he observed two groups of German and Japanese prisoners near the Turkish border and also noted that "both these groups appeared to be well fed and in remarkably good humor." Frankel then crossed into Turkey. He reported that the Soviet border was a high barbed-wire fence extending in every direction to the horizon. Between this fence and the similar fence on the Turkish border, the no-man's-land was freshly plowed to detect border crossers.[43]

4

THE KOREAN WAR
1950–1953

Communist North Korea's invasion of South Korea on Sunday morning, June 25, 1950, came as a tactical and strategic surprise. It was only one year after the test of the first Soviet atomic bomb. The subsequent war in Korea would dramatically change the East-West dynamic, and many feared that it might escalate into World War III.

The sudden attack was a blow to the first CIA director, Rear Admiral Roscoe Hillenkoetter—himself a survivor of another Sunday's surprise attack, at Pearl Harbor nine years earlier. He was soon dismissed, having been made a scapegoat for top-level misjudgment. In fact, the CIA's agents in North Korea had reported increased troop movements and an armored buildup, which allowed the agency to circulate warnings to President Harry Truman's staff five days prior to the offensive.[1] But the U.S. intelligence focus in Asia had been primarily against the Soviet Union, and the warnings went unheeded.

This lack of trust in the fledgling CIA's reporting was due, in part, to the arrogance of Pacific Supreme Allied Commander General Douglas MacArthur and his staff based in Japan, who relied solely on their own intelligence assets. U.S. electronic and communication intelligence collection against North Korea had ceased with the departure of U.S. forces in 1949. The CIA subsequently relied mostly on former Office of Strategic Services (OSS) agents who had been left in place following the defeat of the Nationalist Chinese. At the time, MacArthur was primarily concerned with the Japanese communists and had been unwilling to extend support to the CIA, which ran the remnants of the OSS stations that continued to report in China. These units were, however, still supported by the navy's Seventh Fleet, despite MacArthur's disapproval.[2]

The South Korean capital, Seoul, fell quickly, and Washington naturally assumed that Moscow was behind the attack. Since the end of World War II, Stalin had been intimidating the Allies in Greece, Turkey,

Iran, Czechoslovakia, and Berlin. In 1950, President Truman regarded the North Korean attack southward as a military aggression similar to Adolf Hitler's moves prior to World War II. Persuaded that this invasion could be the opening campaign of World War III, Truman committed U.S. ground and naval forces to defend South Korea and Taiwan. With full UN support, the Seventh Fleet bombarded North Korean units moving along coastal or inland roads and assisted the quickly retreating UN troops to hold the perimeter around the southern port of Pusan. The quick reaction by Truman and the Seventh Fleet no doubt deterred the Chinese Communists from launching a long-threatened amphibious assault on Chiang Kai-shek's refuge in Taiwan.[3]

"Blackbeard" on the Road to Inchon

The Seventh Fleet's unchallenged mobility in the Pacific enabled General MacArthur to stem the North Korean advance outside of Pusan. In mid-September 1950, he launched a daring amphibious assault on the North Korean west flank at the port of Inchon. The British contributed Royal Navy volunteers and Royal Marine Commandos to MacArthur's huge intelligence and special operations organization, which was called Combined Command for Reconnaissance Activities in Korea. There, they were joined by 320 volunteers from the U.S. Army and the U.S. Navy, along with 300 South Koreans, to form the Special Activities Group. Transported onboard the American submarine USS *Perch* (SS-313) and the British frigate HMS *Whitesand Bay*, they raided Kusan and diverted attention from the planned landings at Inchon.[4] Meanwhile, continuous photographic and electronic reconnaissance flights pinpointed the radar stations near Inchon and quickly knocked them out of action with air and naval bombardement.

But by far the most crucial naval intelligence collection was performed on the ground by U.S. Navy lieutenant Eugene F. Clark, a volunteer from General MacArthur's intelligence staff. He would soon be remembered as "Blackbeard" for his expedition with 150 South Korean troops on the enemy shore. U.S. Navy senior historian Edward Marolda describes Clark's pivotal role during this decisive amphibious operation:

> As this armada approached the narrow channel leading to Inchon in the early morning hours of 15 September, a beacon suddenly shined from the top of a lighthouse that had been out of operation for some time. Inside the lighthouse was Lieutenant Eugene F. Clark, who had been executing a daring intelligence mission behind enemy lines since

the beginning of the month. The brave and resourceful naval officer had been landed on a nearby island, Yonghung Do, with a second American and a small party of South Koreans to learn about local tides, currents, and other information that would be valuable to allied amphibious planners. Clark and his men gathered their intelligence, fought a small naval action with the Communists in which the enemy lost two boats to accurate machine-gun fire, and repaired the light. The enemy overran Yonghung Do, caught and executed 50 villagers who had helped the Americans and South Koreans. But the "Blackbeard of Yonghung Do," as Clark would soon be called, avenged them by accomplishing his very important mission.[5]

After several days of bloody fighting, the Allies quickly took Inchon, moved inland, and recaptured Seoul. The Inchon assault had dismembered most of the North Korean People's Army, liberated South Korea, and enabled UN forces to advance into North Korea, where they occupied the capital, Pyongyang. In that city, several items of U.S. Navy equipment—radio and electrical instruments, as well as a crated jet engine for an F-80 fighter—were found in the Soviet Embassy and at the airfield, wrapped and awaiting shipment to the USSR. Soviet intelligence had been caught completely off guard by the Allied flanking maneuver.[6] A number of North Korean officials were also captured there, including employees of the Korean-Soviet shipping company Mortrans, who had visited Dairen and Vladivostok several times and were able to provide useful information should the war be extended to Soviet-controlled territory.[7]

General MacArthur planned another amphibious assault, at Wonsan on the Sea of Japan. But the North Koreans had successfully placed between two thousand and four thousand Soviet-made magnetic and contact mines in the approaches to the harbor, which sank a number of Allied minesweepers. Finally, on October 25, 1950, the 1st Marine Division disembarked and moved toward the Yalu River. Within two weeks, it was met by Chinese so-called volunteer forces, which forced the overextended army, marine, and South Korean units to retreat to the coast and re-embark under the protection of Seventh Fleet air cover. By Christmas Eve 1950, the U.S. Navy had completed the withdrawal from Hungnam of 105,000 troops; 91,000 civilian refugees; 350,000 tons of cargo; and 17,500 military vehicles.[8] From then on, the units withdrawn from the North Korean east coast would fight for another two years to restore their lines near the original demilitarized zone and to preserve South Korea's independence. During this time, naval intelligence played a critical role in three areas: air and surface reconnaissance, mine and coastal warfare, and antisubmarine defense.[9]

Targeting, Damage Assessment, and Pilots' Morale

Although U.S. forces had occupied South Korea for several years following the Japanese capitulation in 1945 and had carried out extensive mapping, according to ONI their knowledge of the terrain at the end of 1950 was inadequate:

> From the time of the Japanese surrender until our occupation troops were withdrawn and the power of the government handed over to the Koreans, a matter of some forty-five months, the area was visited by thousands of Americans in numbers probably greater than the total of western travelers in previous recorded history. With unexcelled opportunities for observation, reporting and photography, few beyond the handful of intelligence officers attached to the various staffs and commands made even perfunctory attempts to increase our usable knowledge of the area. The result was that when the North Koreans swept across the 38th parallel, the resisting forces found themselves dependent on old issues of the *National Geographic* for terrain studies, snapshots borrowed from prewar visitors for target identification, and nineteenth century pilot books for coastal information.[10]

Because of the drawdown of qualified manpower after 1945, the U.S. Navy lacked sufficient air intelligence officers. It was forced to cooperate closely with the U.S. Air Force to obtain intelligence support for bombing railroads, highways, tunnels, and bridges. The U.S. Air Force flew photographic reconnaissance missions using RB-45C aircraft based in Japan. Its four General Electric J-47A jet engines placed the airplane in the 550-mile-per-hour class, enough to avoid the Soviet-built and sometimes-piloted MIG-15 jet interceptors. A navy liaison officer recalled a reconnaissance mission above North Korea onboard an Air Force J-47A on November 26, 1950:

> She lifted off lightly as a gull on the first lap of our photographic reconnaissance mission along the entire length of the Yalu River. . . . We continued to climb at 2,000 feet per minute . . . and over the west coast of Japan we leveled off at 30,000 feet, cruising at an economical speed of 493 miles per hour. . . . The pilot announced that tip tanks were dry and sure enough the wings were no longer bowed! Behind us for miles swirled

telltale vapor trails, deadly fingers pointing us to enemy fighter and flak battery alike. . . . Now we entered high nebulous clouds which obscured the Yellow Sea. We would have to drop a mere 20,000 feet if the field at Antung, our first objective[,] was to be photographed. Apparently we gave little warning for the first MIG-15 could be seen taking off as we went by. We were not alarmed. . . . He would never catch us, and lucky would be the MIG pilot who sighted us by chance from a good attack position. Fire control had failed too, for there was no [flak] visible. Northeast of the Yalu we turned and twisted under the accurate coaching of the radar operator, keeping always to our own side of the tortuous stream, yet making near vertical shots of international bridges and Manchurian airfields [close at hand]. . . . It was now 1310, exactly 3 hours since takeoff. A fuel plot . . . indicated a shortage near home if we did not take immediate action. . . . We zoomed to 40,000 feet, leveled off and slowed down to 475 miles per hour. . . . Below lay Kanazawa, 150 miles west of Tokyo, and we started the letdown at 2,000 feet per minute. . . . Thirteen-thousand-foot Fuji was a white molehill far below as we streaked down silently at nearly 600 m.p.h. . . . Time was 1420. We had flown a distance of 600 nautical miles from Chongin, Korea, in 70 minutes, the last 150 miles requiring only 15 minutes. A . . . normal approach at 140 m.p.h., and routine touchdown completed our 1,630 nautical mile mission in 4½ hours, over-all. . . . We were quite ready to believe the pilot's claim that his argus-eyed machine was capable of photographing anything within [a 1,000-mile] range despite the best interference the enemy could muster.[11]

During the second year of the war, the fast carrier force (Task Force 77) began to rely mainly on its own aerial photography for intelligence. Commander Task Force 77's photographic reconnaissance plan of May 2, 1951, provided for most important bridge and airfield coverage every fourth day, city coverage once a week, and damage assessment on the day of the attack, whenever possible. The introduction of an excellent K-25 camera package on four propeller-powered AD Skyraider aircraft of each carrier air group resulted in better reconnaissance and damage assessment. Detailed panoramas showing enemy shore guns and small boat activity helped the fire-support ships in their counterbattery operations. Soon, the entire North Korean east coast rail line from Wonsan to the Chinese border had been mapped, while route study booklets enabled the pilots to locate bridges and avoid antiaircraft batteries. Photographic intelligence also provided a means for verifying agent reports on targets in enemy territory. Last but not least, photographic intelligence increased the effectiveness of aircraft attack tactics and improved pilot morale.[12]

Covert, Clandestine, and Related Naval Activities in Korea

In the early months of the Korean War, little coordination existed between UN naval forces and the covert groups that were operating behind enemy lines along the coasts of Korea. On several occasions, UN agents or commandos were fired upon by friendly naval forces. Progressively, recognition procedures and small-craft movement schedules were established to support two guerrilla groups, the Yo Do Island Intelligence Collection Center and the Nando Island Army Task Force Kirkland, operating along the Korean east coast. Their agent networks provided valuable local target information for the fleet, often acting as shore fire controllers. But one of the most successful SIGINT (signal intelligence) coups was carried out by a covert CIA group on the west coast. In May 1951, a converted armed junk commanded by a former air force master sergeant was used to cut the communications cable that ran from North Korea to China. The Chinese were now forced to use radio communications, which provided a rich source of intelligence for the rest of the conflict.[13] In early 1952, the other clandestine groups operating on the west coast were integrated under Commander Covert Clandestine and Related Activities, Korea (CCRAK), which had its headquarters in Seoul.

Captured North Korean documents revealed the morale effects of the Seventh Fleet's shore bombardments and the North Korean emphasis on mine warfare to counter Allied naval forces:

> Day after day, the Bestial American imperialists are violating the territorial waters of the republic with their well prepared fleet and are brazenly terrorizing peaceful cities and ports with naval bombardments. The naval units of our sacred People's armed forces are engaged in bitter fighting, unparalleled in world history, on the vast surface of waters and in numerous instances displaying heroism. The writer penned this brief article, with the intention of giving his readers a general knowledge of the mines, which are now being used extensively by our naval units in dealing with enemy naval units and which have played havoc with enemy vessels.
>
> —"Problems Concerning Sea Mines," Kim Won Mu, Military Knowledge, June 1951, Military Publications Department, National Defense Ministry, North Korea.[14]

Prisoner-of-war interrogations provided occasional information of tactical value about the North Korean navy. But the army-dominated interrogation process was unsatisfactory from the navy's perspective. By May 1951, the two hundred POW interrogation reports that had then been completed mentioned nothing of naval interest, although some of the prisoners had lived in or passed through the ports of the peninsula. In October, ONI activated an interrogation desk on the ground in Korea, and the Commander Naval Forces Far East (COMNAVFE) began to get concrete results. By then, the North Koreans had laid more than a thousand Soviet-made mines in South Korean waters, using junks fitted with racks. The minefields were composed mainly of Soviet-moored contact mines, types M-26 and M-KB. Prisoner-of-war interrogations revealed that mines used in the campaign were shipped from Vladivostok to North Korean ports in junks capable of carrying ten mines each.[15]

Nevertheless, most POW interrogation reports were still lacking in substance. Such was the case in the interrogation of First Lieutenant Han Ho, the chief navigator of the North Korean navy First Flotilla, 599th Military Unit. Ho was in fact a medical officer and appeared or pretended to have very little knowledge of his unit, which was composed of ex-Soviet sub-chasers MO class.[16] Captured documents from that same unit were more instructive. They gave complete descriptions and performances of the G-5 class Soviet-built torpedo boats and of the losses sustained during earlier engagements. They also provided intelligence on their past operations out of Wonsan, and even provided a Soviet manual on torpedo boat tactics.[17] Army interrogators missed a key opportunity to warn of mines when the location of a minefield that was obtained during the interrogation of prisoners was not communicated to operational naval forces.

The Episode of Lieutenant Commander Kim Chung Uk

At long last, in the summer of 1952, the U.S. Navy got its sought-after defector. Lieutenant Commander Kim Chung Uk may have been the most valuable North Korean navy officer to volunteer information to the UN forces. His interrogation gave advanced warning on the buildup of the North Korean torpedo boat flotilla on the east coast. Kim Chung Uk had attended the naval academy at Wonsan from July 1947 to October 1949 and upon graduation had received the rank of lieutenant. From October 1949 to April 1950, he had been assigned to the Wonsan naval base as a gunnery officer on three minesweepers.

Promoted to lieutenant commander in April 1950, he had become the chief of staff of the Chinnampo defense unit. In November 1950, he deactivated his unit, fearing a UN invasion, and subsequently was apprehended by Korean naval security agents and taken to Antung in Soviet-controlled Manchuria for questioning. Having been cleared by his superiors, he was reassigned to the navy special training school at Sang Sam Bong, where he studied until January 1951. He was then made chief of staff of the Tap Dong's naval headquarters combat-training branch before taking command of the Konwon navy defense unit's independent battalion. From August 1951 to January 1952, he helped the 17th Mechanized Division make defensive preparations against a possible UN parachute assault. During this period, Kim decided to surrender with his entire battalion to UN forces. The plan was discovered, and he was arrested but managed to escape and flee to Pyongyang. There he obtained a master sergeant's uniform, and as Master Sergeant Kim Chol, with a fictitious unit, reported to the North Korean army. From February to June 1952, he was assigned to prisoner-of-war exchange reception at Pyongyang, working as a supply clerk and a secretary to the political commissar. On July 28, he crossed into UN territory, where he offered his services. The source spoke with confidence. Kim's apparent sharp memory, combined with his willingness to talk, yielded a wealth of information.

Prisoners had earlier said that North Korean torpedoes were older types, non-homing and fueled with kerosene and expoding only on contact. North Korean torpedo boats based in Wonsan belonged to the Soviet G-5 type and were well maintained.[18] But Kim reported a 60-knot magnetic homing torpedo allegedly of German design being manufactured in Soviet Russia, which would be provided for the North Korean torpedo boats that were about to be commissioned.

Kim Chung Uk was one of thirteen North Korean navy officers who had been sent to the Soviet Union to study. On their return, they were to be assigned as patrol torpedo (PT) boat flotilla commanders, using Soviet PT boats and the new magnetic homing torpedoes. One hundred more North Korean navy officers were being trained at Yujin, and at the end of their training, they would be given PT boats and magnetic homing torpedoes to operate against the UN naval forces. Kim said that the graduation would be in the fall of 1952 and that he thought there would be some new type of action after that time. Kim's information clearly exposed the Soviet involvement in the conflict, despite Moscow's denial.

It was learned from interrogations of Kim that the Soviet torpedoes had been shipped from Khabarovsk via Vladivostok. Before the invasion, in mid-October 1949, about thirty fifth-year students from the Leningrad Naval Academy had come to Korea as far south as Wonsan to practice with magnetic torpedoes. The Soviet PT boats were very low, flat, longer, and painted a darker gray than the North Korean navy PT boats. Mines were stored "in hill back of coal loader" and in the village of "Panggum-Ni."[19] Mines were brought from Sang Sam Bong as required for the mission. Kim's report of motor torpedo boat training was further supported by agent reports of PT boats undergoing exercises in the vicinity of Unggi Bay. The development of an effective PT unit on the east coast of Korea offered a serious threat to UN naval vessels operating in this area and seemed like a logical development for the North Korean navy. Kim also provided information on the joint North Korean and Soviet naval intelligence effort. He described a hydrographic office at Unggi with about twenty men, plus Soviet advisers whose mission was to study UN ships, the various types of vessels, and weapons.[20]

Monitoring Soviet Shipping and Naval Activities

Soviet support to North Korea and this potential naval threat forced the COMNAVFE to keep the La Perouse Strait and the vicinity of Vladivostok under constant surveillance. In December 1950, the USS *Besugo* (SS 321) was assigned to the La Perouse Strait patrol area, but weather conditions were so bad that reconnaissance efforts were ineffective, and the submarine patrols were discontinued for the duration of the winter months.

From April 4 to December 6, 1951, submarine reconnaissance was conducted continuously in the La Perouse Strait area. Both visual and photographic collection provided much-needed intelligence information on Soviet ships and shipping trends.

Submarine patrols of La Perouse Strait were resumed on March 1, 1952, and continued until December 10; visual and photographic surveillance was conducted of shipping east, north, and west of Hokkaido. The submarines also made reports to COMNAVFE on Soviet and Chinese Communist sea and airborne activity. In December 1952, a reconnaissance patrol was conducted by *Scabbardfish* (SS 397) off the South China coast. On January 22, 1953, patrols of the La Perouse Strait were again resumed in order to maintain continuity of shipping surveillance and to provide submarine crews with experience in cold weather operations. USS *Remora* (SS 487) was one participant. Limited amphibious landing and raiding operations from submarines were also carried out from February through July 1953. The cessation of Korean War hostilities on July 27, 1953, caused no change in the submarine reconnaissance operations.

A periscope photo of a Soviet cargo ship taken on May 7, 1953, by the USS *Remora* (SS-487) off Vladivostok. U.S. submarines monitored weapons shipments to North Korea.

Pomfret (SS 391), on station at the time, remained on patrol until relieved in August by *Ronquil* (SS 396).[21]

Chasing the Shadow Submarines

At the beginning of hostilities in 1950, the most urgent menace to the U.S. Seventh Fleet and Allied navies would have been from the Soviet submarine forces in the Pacific. At the time of the invasion, the commander in chief of the Pacific Fleet estimated that a total of seventy-eight Soviet submarines were operating in the Far East and must be considered a major threat to the UN forces.[22] Consequently, the Military Intelligence Section of General Headquarters Far East Command began evaluating the Soviet navy's submarine facilities, summarizing information obtained from interviews with repatriated Japanese prisoners of war. Astonishingly, these Japanese repatriates contributed 99 percent of the information obtained on the Soviet Pacific Fleet submarines' facilities. The repatriates, while aboard the ships that were returning them to Japan from the Soviet Union, filled out personal data forms, which were then scanned by interrogation teams. Likely informants with technical qualifications were next carefully interrogated. Results of port interrogations were forwarded to the Allied Translator and Interpreter Section, Tokyo, for analysis. Informants who appeared to possess valuable information were called for a second detailed interrogation lasting approximately eight hours.[23]

In the late 1940s, there were 300 Japanese prisoners of war working at the Soviet submarine base of Uliss Bay in Vladivostok. In addition, 350 were confined at the Metallist Steel Works plant northwest of Vladivostok; 200 in the Dalzavod's repair yard, including 50 highly skilled Japanese employed in positions of responsibility in the machine shops; and 200 in the Daltroy Dry Dock and shipyard, both located in Vladivostok's Golden Horn Bay. No less than thirty-four POW interrogations were used to prepare a detailed report on these last two yards. Three prisoners had been able to locate the four concrete underground warehouses where torpedoes were stored approximately 5 miles northeast of Vladivostok. Three more repatriates had observed a magnetic mine storage tunnel at the Uliss Bay submarine base and another underground mine depot nearby. Similar reports were obtained on the other Soviet Pacific Fleet submarine facilities, such as Sovietskaya Gavan, where 750 Japanese prisoners were working back in 1947. Twenty-nine of them had provided a description of the base and of its five tunnels running into the side of the hill to accommodate submarines during the ice-free months.

Magadan Harbor could also be used as a temporary submarine base, and, interestingly enough, a Japanese repatriate had reported that the Soviet navy feared sabotage by the local inhabitants, who were disenchanted with Communism. Based on such sharp Japanese observations, the final report, dated August 10, 1951, and covering submarine facilities across the Soviet Union and submarine sightings in the Far East, stated:

> Currently available information identifies forty-four submarine facilities in the Far Eastern USSR, including twelve submarine bases or naval installations, eleven shipyards, nine underwater ordnance or submarine parts manufacturing installations, a floating dry dock and a submarine supply pier. . . . Observations through July 1950, covered submarines sighted at or near forty-four locations in the Soviet Far East, including Sakhalin, the Kuril islands, Kwantung leased territory, Korea and China.[24]

When the Korean War broke out, the rules of engagement for antisubmarine defense issued by the Seventh Fleet were the following:

Unidentified submarines may be attacked and driven off by any means available in self defense or when offensive action against our forces is indicated. In interpretation of the foregoing, continued submergence of an unidentified submarine in position to attack our forces operating against North Korea or in support of the assigned mission with regard to Formosa is considered to indicate offensive action against our forces.[25]

During the first ten months of the Korean War, there had been ninety-six submarine contacts outside the Korean operating areas while over the next eight months there were eighty-eight more.[26] Veteran U.S. Navy sonar operators Donald C. McElfresh and Allison E. March claimed that they conducted a twenty-nine-hour attack on a submarine sonar contact in the Yellow Sea on July 28–29, 1951. At that time, Soviet submarines were based at Port Arthur, Manchuria, where 275 Chinese sailors were being trained, but this specific incident was never acknowledged by the U.S. Navy. The hostilities in Korea officially ended on July 27, 1953, with the signing of the Armistice at Panmunjon.

During the thirty-seven months of the conflict, there had been no submarine incident to actively test the UN's antisubmarine warfare capabilities. The lessons of the Korean War antisubmarine defense experience were summarized in part by the final Pacific Fleet Evaluation Report. It was recognized that "[Escort of heavy units and logistic groups] would have been entirely inadequate in the face of opposition and that [electronic countermeasures and radar intercepts] continued to be the most effective means of initial detection of submarines."[27]

The Opening Campaign of World War III?

Stalin had made his Georgian compatriot Lavrenti Beria a deputy prime minister. Beria had also been tasked by the Generalissimo to make atomic weapons, and the former NKVD chief retained a partial control over the Soviet State Security apparatus through his atomic espionage spy ring. His son Sergo even claims that Beria had delayed the explosion of the hydrogen bomb until after Stalin's death because he feared that the paranoid Soviet leader would feel encouraged to initiate a world war. As Sergo remembers, Stalin had ordered the North Korean invasion as a first step toward the inevitable Armageddon which would do away with capitalism:

[Marshall A.M.] Vassilievskiy told me that he himself drew up the plan for aggression against South Korea. After the initial successes, the General staff and our intelligence services warned Stalin that conquest of all Korea should be avoided as otherwise a Western counter-attack would not be long in coming. . . . He did not yet feel ready for a world war and wanted to gain time for putting the final touches to his preparation. Therefore it seemed to him preferable to make use of the Chinese. At this time I was working on missiles for use against warships at a distance of 100–150 kilometers. These missiles could be given a conventional or a nuclear charge. . . . In 1951, Stalin decided to lend a hand to the Chinese troops who were in difficulty after the American push in the spring. He convened the Politburo and ordered that preparations be made to send part of our air force to China, equipped with these missiles. . . . He wanted to use them against the Americans' aircraft carriers and warships. . . . While I was rejoicing to see that weapons made by me were going to be put to use, my father spoke up to say that by acting like this we were going to start the Third World War. . . . My father emphasized his point, saying that we could not allow ourselves to break with the Western World and that a new world war would mean the end of Europe. . . . None of the members of the Politburo, not even Khrushchev, I must honestly acknowledge, wanted war with the West. . . . After 1949 nobody believed any more in the likelihood of American aggression. . . . It was then that the Americans organized a leak, and document describing their plans fell into our hands. It was a warning, intended to let us know that any attempt at aggression on our part would unleash a reaction from them. My father thought, however, that the Americans had not conveyed their message vigorously enough. More than that was needed to cool Stalin's ardor. . . . In 1952, the whole country was on a war footing. The objectives of the Central Committee were quite clear: we were preparing for the third world war and it would be a nuclear war. All the country's resources were mobilized. . . . As a general rule we began to mass-produce weapons before we had tested them. Stalin took the risk in order to gain time. . . . What for Stalin was essential was to take Germany, the only country in Europe he thought of as formidable. . . . Though still disarmed, Germany had an economic potential so dangerous that he preferred to destroy it. He summoned me on several occasions around May 1952 to ask if our missiles would be able to demolish the bridges over the Rhine or wipe off the map this or that industrial centre in Germany. . . . He wanted to know if we could destroy the dykes of the river Ruhr. . . . At this time our armament industries were working as in wartime. Our preparations clearly showed that what we had in mind was an offensive war.[28]

Sergo Beria's insights into Stalin's plans reveal that the aging generalissimo wanted to make full use of his future nuclear weapons to end the East-West stalemate. This cataclysmic outcome was also being contemplated by British leaders. They were already very much afraid of losing Honk Kong in a feud with China and felt that Western Europe and the Middle East were not well defended to stop a Soviet invasion. Furthermore, historian Robert Aldrich has found out that Vice Admiral Eric Longley-Cook, the director of British naval intelligence, feared that the United States might launch nuclear strikes against the USSR from bases in the United Kingdom without first consulting with Downing Street.[29] Longley-Cook had warned Prime Minister Winston Churchill that the United States was seriously weighing a nuclear strike against Russia before it could hit the United States:

> Many people in America have made up their minds that war with Russia is inevitable and there is a strong tendency in military circles to fix "the zero date for war." It is doubtful whether, in a year's time, the U.S. will be able to control the Frankenstein monster which they are creating. There is a definite risk of the U.S. becoming involved in a preventive war against Russia, however firmly their [North American Treaty Organization] allies object. "Let's Use It Now." [Americans are saying that] "we have the bomb, let's use it now while the balance is in our favor. Since war with Russia is inevitable, let's get it over with now."

Longley-Cook then quoted one U.S. general who pointed out that the West could not afford to wait for Europe or America to be devastated by a Soviet nuclear strike: "We have a moral obligation to stop Russia's aggression by force, if necessary, rather than face the consequences of delay." Then the British Director of Naval Intelligence quoted another U.S. general as saying America was already at war with the Soviet Union and that "whether we call it a Cold War or apply any other term we are not winning the only way we can be certain of winning"—that is, "to take the offensive as soon as possible and hit Russia hard enough to at least prevent her from taking over Europe."[30]

5

REQUIEM FOR A BATTLESHIP
1955

The explosion, the capsizing, and the sinking of the Soviet battleship *Novorossiysk*, with the loss of 609 men in Sevastopol on October 29, 1955, was not only the world's largest peacetime naval disaster, but the start of major changes in the Soviet navy. Apparently, the loss of *Novorossiysk* led to the final sacking and demotion of navy commander in chief Nikolai Kuznetsov. He was replaced by Sergei Gorshkov, who would remain in the top command position for a quarter century and would catapult the Soviet navy into the forefront of the world's modern nuclear navies.

Novorossiysk was the former *Giulio Caesare*, which had launched in 1911 and been transferred to the Soviet navy in 1949 as part of the Allied dismembering of World War II Axis navies. After a major overhaul, the vessel became the flagship of the Soviet Black Sea Fleet. On the fateful night of its demise, two large explosions occurred beneath the hull, and then the battleship's main forward magazines blew up, causing her to capsize and trap hundreds of crewmen below decks. The ship settled in the muddy bottom of the 17-meter-deep waters of the harbor, just several hundred meters from the downtown Sevastopol fleet landing. Despite scores of trapped crewmen tapping from the hull's interior, only nine were removed four days later by rescue teams that cut though the hull.

News of the battleship sinking was hidden from the public for thirty-three years. The entire country, with the exception of those who lost family members, was kept in the dark about the loss of the ship. Soviet control of information kept the event so secret that even Western naval intelligence was aware of only fragments. In 1958 ONI issued a report of the sinking based partly on the testimony of defector Nikolai Artamonov (detailed in chapter 6). The ONI summary stated that although there was conflicting evidence, the final evaluation was that the battleship did capsize and sink on October 30, 1955:

> In the early morning of 30 October 1955, as the boat carrying flag officers ashore approached its landing, an explosion was heard from the direction of *Novorossiysk*. The disaster alarm was sounded at the naval base, and vice-admiral Parkhomenko returned to the ship. . . . A hole in the hull was patched and the flooding stopped. The decision was made to beach the ship. . . . This motion caused the patch to rupture. Flooding resumed, and within a very short time the ship capsized. There was apparently heavy loss of life; one report stated that out of about 1,300 on the ship only about 300 topside personnel were saved. . . . Additional confirmation has been received from a Soviet naval officer who defected in 1957.[1]

And in the 1957–1958 edition of the vaunted *Jane's Fighting Ships* the following appeared: "Battleship *Novorossiysk* (formerly the Italian *Guilio Caesare*) is reported to have been sunk by drifting mines in the Black Sea in October 1955, the casualty list running into the hundreds." Details of the ship's sinking and the covered-up investigation began to seep out to the press on May 14, 1988, the day after the death of Gorshkov, who had been replaced as commander of the Black Sea Fleet just weeks before the battleship blew up. Following public exposure of the disaster, long suppressed evidence came to light that pointed to causes for the ship's loss that were contrary to the findings of the official 1956 inquiry. The events surrounding the explosion of the battleship are still subject to debate and investigative research unlike any other occurrence in Soviet naval history.

The ship's keel was originally laid in Genoa on October 24, 1910. She was launched on April 2, 1914, the first of three of the class: *Giulio Caesare*, *Leonardo Da Vinci*, and *Count De Cavour*. Two additional ships were built, practically the same, which were called *Andrea Doria* and *Caio Duilio*. They were the most powerfully armed combatants of their time, designed to counter the most likely enemy of the era—Austria-Hungary. The entire class of these battleships had a tragic fate—all were lost.

The battleship *Giulio Caesare* had been transferred to the Soviet navy in February 1949, as war reparations. A small working party of former Italian sailors and yard workers, which remained onboard until the turnover, acted sullen and seemed to loathe the Soviets. The petulant Italians resented the transfer of their fair battleship to the Red Navy and were often heard muttering under their breath *"Marina Rossa,"* after a long string of undecipherable Italian obscenities. Rumors spread among the befuddled Soviet sailors that embittered Italians had sequestered explosives in voids throughout the ship, which were primed to explode when the main engines reached critical rpm vibrations. Some Soviets said that lube oil seals on the main engine reduction gears were sabotaged to slowly drain the gear oil to the bilges, causing the massive bull gears to overheat and melt together.

On February 6, 1949, the special team from Sevastopol raised the Soviet naval ensign, and the former *Giulio Caesare* joined the rolls of the Soviet navy as battleship *Novorossiysk*. The ship was received by a special command headed by Admiral G. I. Levchenko. During the turnover, as reported in the communist jargon of a report dated 1949, "The Soviet sailors, displaying a high degree of professional competence and courage, overcame major technical difficulties inherent in the unfamiliar ship and quickly prepared it for the voyage to Sevastopol. The primary burden of preparation lay on the shoulders of the personnel of the engineering department, under the leadership of Captain First Rank L. A. Rulyev, who was decorated for the operation with the Order of the Red Banner and promoted to the rank of Engineer Rear Admiral."

The special Soviet crew believed the Italian rumors and attempted to check all of them. Finally, after a long delay, the skeleton crew guided the great battleship through the eastern Mediterranean, the Sea of Marmara, the Dardanelles, and the Turkish Straits. They steamed cautiously on to Sevastopol, home of the Black Sea Fleet. The gleaming white city, a sparkling jewel of the Crimea, had a population that was more than 85 percent Russian, surrounded by a sea of Ukrainians and Tatars. In the middle of May 1949, the ship was docked in the Sevastopol northern dry dock. Specialists were astonished not only by the elegant outlines of the underwater hull, but by the exceptionally heavy underwater sea-growth on the hull—the result of many years spent in warm Mediterranean waters without dry-docking. Her degaussing coils were corroded and in very poor condition.

Captain Nikolai Muru's Account

A firsthand account of the rescue after the explosion was provided by a doctor of technical science, former navy captain first rank Nikolai Petrovich Muru, who was a survivor of the rescue operation aboard the battleship and a participant in the complex salvage operation in 1956. Muru had served on a score of surface ships in his younger days as an officer. He became well-known as a squadron engineer and was frequently called on to assist other engineers in resolving problems with engineering plants on ships throughout the fleet. His uncanny ability to provide innovative solutions for severe engine casualties, especially in the field of underwater salvage and damage control, soon won him a reputation throughout the fleet and a position at Leningrad's Dzerzhinsky Order of Lenin Higher Naval Engineering School. Muru had participated in many renowned salvage projects, such as the recovery of the lost S-80 diesel submarine in 1969 and the attempted recovery and the final demolition of the missile destroyer *Otvazhny*, which exploded and sank in the Black Sea in October 1974. As a captain, he taught ship stability and damage control to young naval engineers for the last eight years of his active career and for ten more as a civilian, following his mandatory retirement at age sixty-five.

The battleship crew had been awakened at 1:30 a.m. by a powerful jolt caused by a severe explosion in the bow. It shook the entire ship from stem to stern. Crewmen aboard cruisers moored nearby noted that the explosion was accompanied by a flash of light and the ejection of thick black smoke. It was later determined that the flash was from gasoline igniting in the commanding officer's gig, and the thick smoke was actually large amounts of silt ejected upward from the sea bottom.

At first, only the sailors who happened to be in the explosion area understood the true seriousness of the incident. Those who remained alive after the initial blast, having been thrown undressed from their berths, desperately attempted to escape their rapidly flooding, silt-encrusted, smoke-and-gas-filled quarters. They frantically tried to make their way topside.

The powerful underwater explosion occurred near the ship's bow, about 35 to 40 meters away from the stem and slightly to the right from the centerline. It led to the destruction of all longitudinal deck plates to a height of 16.9 meters, from the keel to the first deck of the superstructure (in total, 136 millimeters). There was a gaping hole (14 by 4 meters in size, with torn metal extending 2 to 3 meters around the perimeter) in the main deck forward of the first tower of the main battery.

During the explosion, a great mass of bottom silt was thrown onto the raised foc'sle and decks located below. It was established later, after inspection by divers, and was confirmed after the raising of the ship that the explosion tore a huge hole of 150 square meters in the hull casing on the ship's starboard side. There were dents and cracks in the hull all around the hole.

Practically all of the bow compartments and voids (from frames 23 to 50) of the ship were flooded immediately after the explosion. Seawater poured through the holes in the hull into the weapon-handling rooms and the main armament magazine. After taking on about 3,500 cubic meters of water, the ship developed a down trim by the bow that gradually grew as a result of water spreading through deck hatches and partition doors toward the remaining bow compartments.

The explosion instantly knocked the main electrical generating plant out of order. The emergency diesel generator situated in the bow was also immediately destroyed, which plunged the battleship into complete darkness. Emergency lights went on only in the ship's stern section. Thanks to the fast actions of the engineering department personnel, power was restored in the after section of the ship in about six to eight minutes, which turned on the lighting in every compartment except the bow. Emergency parties from a number of surrounding ships arrived right after the explosion to help the crew of *Novorossiysk*. Rescue parties arrived from the cruisers *Mikhail Kutuzov, Kuibyshev, Frunze, Molotov, Slava,* and *Kerch,* along with their engineering officers. The commanders did not initially realize how dangerous the situation was; however, several officers attempted to organize a tow for the damaged battleship, stern first, to the nearest shoal.

The combination of factors—the flooding of the higher compartments, with no ballast water in many compartments below the armored deck—resulted in the loss of initial transverse ship stability and added to the growing list of problems. The commanders of the engineering department made an effort to return the ship to an even keel by pumping black oil back to starboard from the port tanks. They also pumped fuel oil from forward tanks aft, in order to decrease the bow down angle.

After attaining an 18- to 20-degree list, the ship rapidly rolled to the left and capsized with its keel up. Only 110 meters of the stern initially remained above the surface. Capsizing occurred at 4:15 a.m., October 29, exactly two hours and forty-four minutes after the explosion. The ship continued to settle into the harbor mud, and by dawn on October 29, eighteen hours after capsizing, she completely disappeared underwater.

The majority of the 608 men lost were killed when the ship capsized. Approximately 1,000 were fished from the water; some managed to swim to the Gospitalnaya Seawall. Many seamen were caught in air pockets in cells in the capsized ship. Only nine of them were rescued. Seven came out through the slot cut into the bottom.

A state commission led by Admiral V. A. Malyshev arrived in Sevastopol ten hours after the ship sank. Two commissions of experts were created in order to analyze technical aspects of the disaster: one studied the explosion; the second, the damage control (seaworthiness) of the ship. The commission concluded that:

- It was an external explosion.
- It was a bottom explosion, that is, the explosive was not attached to the hull but had been located on the seabed.
- Based on the scale of destruction that was caused to the hull's structure, the explosion had a force equal to 1–2 tons of dynamite.

The state commission initially concluded that the explosion had been caused by German influence mines (R type), left on the seabed after the war. While acknowledging that the Sevastopol Harbor security from October 28 to 29 left much to be desired, however, the state commission added that diversionary swimmer action by saboteurs could not be ruled out.

Other officers based in Sevastopol and witnesses of the disaster unofficially claimed that the explosion had been caused by an internal detonation of a charge placed by Italians between the old hull and the newly reinforced bow (which was said to have happened before the battleship was turned over to the Soviet Union). Others stated that there had been a double explosion: an external one caused by swimmers, followed by a detonation of a charge placed internally by the Italians. Some people said that an explosive substance had been sprinkled as a thin layer between the two skins of the ship. Yet another supposition arose that a whole cluster of leftover German bottom mines had exploded after being disturbed by the battleship's port anchor on mooring that evening.

Muru had made a lifetime study of the battleship's sinking and salvage. Soviet navy censors blocked early attempts to print his material, and, he claims, since 1991, so did old-school bureaucrats in St. Petersburg maritime publishing houses, who still oppose promulgating details of naval disasters from the Soviet period.

Muru's written account places responsibility fully on the shoulders of Admiral Gorshkov. Captain Muru concluded that the cause of the explosion was the disturbance of a cluster of German bottom mines by the battleship's port anchor. During the mooring process, after the ship returned from operations at sea at 6:30 p.m. on the day prior to the explosion, the commanding officer used his port anchor to kedge the ship to moor at buoy number three. Three fathoms of chain were put out on the bottom, which consisted of soft mud on a hard clay surface 17 meters deep.

Muru claimed that the mine contact fuses in the cluster in the mud were activated at 6 p.m. when the ship moored but, due to their age, were slow to work and delayed exploding until 1:30 a.m. Evidence that the explosion was not a torpedo or a mine positioned against the actual hull was the amount of mud blown up and into the hull, which covered more than five compartments to a depth of several centimeters.

The large crater found by divers in the sea bottom under the bow also indicated that the explosion had originated far beneath the hull. According to Muru, there was disagreement and confusion among the rescue divers, some of whom reported the single crater twice by misreporting their own locations.

Muru explained that the ship's curious roll and capsizing to port, after being stove in and holed on the starboard bow, was caused by the lack of watertight integrity forward and the progressive flooding that began first to starboard and then shifted with the aggravating free surface effect caused by the open compartments on the port side. Muru attested that poor damage-control preparedness was aggravated by the location of Damage Control Central (the coordination center for fighting the fire and the flooding) deep below the waterline and was thus lost early due to rapid flooding. Sound-powered communications and primary lighting also failed quickly, rendering damage-control efforts extremely difficult. The stability of the battleship had already been severely impaired by the 1937 modernization conducted by the Italians, which raised the waterline a full half-meter and put the equivalent of one additional compartment beneath the waterline.

Rapid flooding of the turret handling rooms and the magazines of the forward two main turrets also accelerated the fast sinking of the bow. Finally, Muru blamed the loss of the ship on the failure of the command to beach her immediately in shallower water. It was not possible to move the ship, however, because the anchor could not be released or disengaged. It thus held the bow fast, despite the possibility of using the main engines, which were in twenty-minute readiness to get underway. Muru also cited the absence of the commanding officer and the chief engineer—both were ashore—as key to the failure of the ship's company to act decisively and to effectively counter the flooding.

Muru is certain that the mines were German bottom mines; he provided data showing that thirteen additional bottom mines were discovered during a detailed search of the Sevastopol Harbor the year after the blast. Three more were found the following year, 1957—several within 50 meters of the spot where the battleship was sunk. This data was suppressed because it indicated incomplete mine sweeping of the harbor area by the former Black Sea Fleet—under the command of Admiral Gorshkov.[2]

The enigmatic circumstances surrounding the sinking of the battleship were only temporarily hidden in the years that the government suppressed information about the event. On May 14, 1988, the first sensational fragments surfaced in the press and pointed to the possibility that the battleship might have been sunk as a result of onboard sabotage in a KGB effort to discredit navy commander in chief Admiral of the Fleet Kuznetsov, or by explosives planted onboard by Italians in unused machinery spaces prior to the turnover of the ship in 1949.

Mysterious Italian Frogmen

Still more bizarre was the theory that the battleship had been attacked by Italian diversionary swimmers employing human-torpedo chariots. This was based on some questionable evidence. Consistent with the Russian penchant for seeing conspiracy in most major catastrophic events throughout history, many Russians still believe that an underwater swimmer commando attack had been staged by Italians, led by the renowned World War II frogman commando and unrepentant fascist "Black Prince" Valerio Borghese.

An Italian navy underwater demolition team at inspection in the 1950s. The Soviet navy unofficially blamed Prince Borghese's Sea Devils for the sinking of *Novorossiysk*.

The debate—German mines versus Italian commandos—continued until 2005, when new evidence was revealed by an alleged witness to the disaster.[3] The source, a former Soviet naval officer named Bar-Biryukov, presents a strong case in a recent book linking the sinking to an act of sabotage conducted by a clandestine group of former WWII Italian commandoes spearheaded by Borghese. In his book *X Hour for the Battleship Novorossiysk*, the author claims that he was approached by the last surviving member of the Italian swimmer team that sank the battleship. In a vague reference (he gives no name), he met one claimed participant in Florida in 1996, who showed him a photo of the whole team.

Borghese had publicly boasted in his own book of his familiarity with the inner harbors of Sevastopol, familiarity that he had gained while assisting the Germans during the 1942 siege of that city. He had worked with teams of the Ten MAS, a renowned commando team that sank the British battleships *Valiant* and *Queen Elizabeth* in Alexandria in 1941. Borghese's colorful background and openly stated intent to gain revenge for the reparations—transferring their beloved battleship *Giulio Caesare* to the hated Bolsheviks—rendered him the perfect suspect.

The new evidence explains that the October 1955 assault was a major commando success. Author Bar-Biryukov also cites evidence that not only was Borghese in the area at the time of the explosion, but he and his team had been dropped in by an (unnamed) Italian merchant ship.[4] The ship was carrying a midget submarine called a Piccolo and entered the Black Sea on October 21. After passing Foros just south of the entrance to Sevastopol Harbor, the ship dropped off the team members to do their dirty business. The source's alleged contact in 1996 with the (unnamed) sole remaining, confessed member of the team that carried out the attack, seems to point to the sinking as the work of the Italian intelligence service, rather than the Italian navy.[5] Borghese, who was later active politically in trying to restore a fascist government and monarchy to Italy, was deported and died in Spain in 1974. But, apparently, the secret of the battleship's sinking did not die with him.

Russian Research and Conclusions

In 1989, a search carried out in long-secret Soviet archives by Russian naval scientific historians unearthed a number of curious facts surrounding the loss of the battleship. While none of these disclosures led to conclusive determination as to the exact cause of the disaster, they do show the extent to which the Russian mind, newly unfettered after years of suppression, will grope for theories of conspiracy. The search concluded that:

- Two explosions occurred, one under the hull and the other adjacent to the hull, instead of the single explosion that was claimed in the initial investigation.
- The blast holed the starboard bow, yet the ship rolled over to port when capsizing.
- Mine-warfare experts testified that German bottom mines of the type known to have been seeded in the Sevastopol area and cleared after the war could not have made the two large craters in the harbor bottom beneath the ship; both craters were reported by divers as too large in diameter and too deep.
- No mine fragments were found under, in, or around the sunken battleship.
- Explosive experts claim that the blast had no characteristics of a torpedo strike.
- The watch officer at the time of the explosion suspected that the blast had originated in unused machinery spaces on board, and he surmised that explosives had been placed there prior to transferring the ship from Italy in 1949.
- One surviving officer from below decks claimed to have heard a low rumble and a scraping sound prior to the blast, possibly indicating the presence of a vehicle manned by underwater swimmers.
- An antitorpedo net guarding the northern bay had been removed for repairs a month prior to the disaster.
- The harbor defense boom and the antisubmarine net protecting the *Novorossiysk* anchorage were opened the evening before the blast to preclude delaying ship arrivals and departures; this was on orders of then Black Sea Fleet commander Sergei Gorshkov.
- The hydro-acoustic antisubmarine warfare (ASW) listening station "Saturn 12," which guarded the approaches to the Sevastopol Harbor, was down for repairs from 8 A.M. to 7 P.M. on the day prior to the explosion and was thus unable to detect a submerged intruder entering the harbor.
- The harbor security patrol, a large ASW hunter BO-427, was moored and not patrolling the harbor as assigned, at the time of the explosion.

These newly revealed facts challenged the secret findings of the original navy investigation, which had been hurriedly completed in November

1955 by the team headed by Military Council members Admirals V. A. Malyshev and N. M. Kulakov. As outlined earlier, their investigation claimed that the fatal explosion had been caused by a cluster of old German bottom mines that were disturbed by the motion of the battleship's anchors. Although the investigation results were suppressed for many years, they were finally published in 1999 by a researcher, the late Boris Karzhavin. Additional long-concealed testimony boosted the theory of an Italian swimmer attack. An Italian sailor, debriefed in Algiers in the sixties, claimed to have been a participant in the operation against the *Novorossiysk*.[6]

Peter Huchthausen published a short account of the battleship sinking in the January-February 1996 edition of the U.S. Naval Institute's *Naval History*. In response to the publication, the author received a letter from the son of a member of Borghese's World War II swimmer team, stating that his father would never have participated in such an action after the war ended.

Despite the claims and counterclaims of how the battleship was sunk, there was sufficient official NATO cover-up of Borghese's actions during the postwar years to give rise to suspicion. The authors Jack Greene and Alessandro Massingnani detailed the obfuscation by Italian government and NATO officials who controlled key archives, in their book *The Black Prince and the Sea Devils: The Story of Valerio Borghese and the Elite Units of the Decima MAS*, which was published in 2004.

Even more curious was evidence provided by an officer in the KGB Maritime Coastal Patrol Forces, Vyacheslav Sergeev, whose small craft had been assigned to patrol the Crimean coast. According to his testimony, on the morning following the explosion in the *Novorossiysk*, his patrol boat was sent to scour Sevastopol Harbor for anything suspicious. He and his crew found a magnetic mine attached to a nearby mooring buoy at which the cruiser *Kerch* was scheduled to be moored. The mine was set to explode ten days later, on November 7, which was the Soviet national holiday celebrating the October Revolution. The cruiser *Kerch*, like *Novorossiysk*, had originally been an Italian warship; her former name was *Emanuele Filberto Duca d'Aosta*.[7]

Soviets Beef Up Underwater Defense

Regardless of the cause of the battleship's loss, the paranoia generated a dramatic surge in Soviet underwater defenses, including the development of special sensors and special forces for harbor protection. In 1953, the Leningrad-based acoustic research institute Morphyzpribor began to devise a fixed underwater sensor system for harbor defense to replace earlier ineffective equipment. Code-named Volkhov, this program was given top priority after the catastrophe. Having completed its state trials in 1956, the Volkhov entered series production, one system being deployed in the Northern Fleet and a second in the Pacific Fleet. Meanwhile, the prototype for a follow-on system—Liman—was laid near Feodosiya to protect the Black Sea Fleet.[8] The most spectacular development, however, was the expansion of underwater special forces.

The Soviet navy had used swimmers during World War II and created its first unit in Leningrad in 1941 to spy on the Germans besieging the city. After the war, the need for combat frogmen had remained, and the Sixth Naval Underwater Reconnaissance Unit was formed in the Black Sea in 1954. One factor behind the decision to focus on underwater defense was the discovery of frogmen's footprints on a beach near an elite Soviet state party sanatorium—a discovery that caused quite an alarm and supported the theory that NATO was conducting special forces operations on the Soviet coast. In October 1956, another naval underwater reconnaissance unit, nicknamed Parusniki (sailing vessels), was established in the Baltic.

An incidence of underwater espionage had just occurred that year in Portsmouth, England, with the Royal Navy swimmer Commander Crabb (detailed in chapter 6). The suspicion that the Black Prince's sea devils were behind the destruction of *Novorossiysk* seemed to further justify the decision to boost underwater reconnaissance units for both defensive and offensive purposes. Accordingly, the special underwater units for reconnaissance and sabotage, or naval *spetsnatz* swimmers, were established in 1957 by order of defense minister Marshal Zhukov.

Another reason for the need to train combat swimmers came in June 1967, when the Six-Day War in Israel caught Moscow and the navy by surprise. Various Soviet naval units were visiting Port Said, and the Soviet high command suddenly realized that its ships were sitting defenseless against a possible underwater attack by Israeli special forces. As a result, the Soviet navy formed the 1st Anti-Diversionary Forces Unit, based in the Black Sea, which was assigned to protect ships and installations against combat swimmers.[9]

In 1969, similar units were created in the Northern, Pacific, and Baltic fleets. Armed with daggers, underwater submachine guns (with twenty-five-inch-long needles, which were able to kill from a distance of 15 to 60 feet), and four-barreled underwater pistols, the swimmers could also act as commandos and mine-clearance experts.

The following year, the GRU (Main Intelligence Directorate) formed its own reconnaissance and sabotage group to collect intelligence and conduct offensive operations against foreign bases. Known as Dolphin, this elite unit imposed the harshest physical and psychological ordeals on its members. Their selection process lasted thirty-three weeks and included running 60 miles in full-battle kit and an additional 6 miles without this encumbrance. Like the U.S. Navy SEALs, the British SBS, and the French Hubert units, GRU recruits would be parachuted day and night, jump from helicopters, exit from submarine torpedo tubes, and become explosives experts. Dolphin combat swimmers specialized in specific geographic areas and underwent a special training in Sevastopol to fight sea animals and underwater saboteurs. They were also taught how to take nuclear charges weighing 27 or 70 kilograms into an enemy port. These they could transport in a Piranha midget submarine. Each Piranha could carry six saboteurs with full equipment. They also used Triton and Triton-2 submersibles and Siren human torpedoes. These vessels each carried two saboteurs and equipment and were released from a submarine torpedo tube. Once a month, Dolphin teams would be tested again in all disciplines.

From 1967 to 1991, Soviet combat swimmers acquired battle experience around the world, notably in Egypt, Angola, Ethiopia, Mozambique, Cuba, Vietnam, North Korea, and Nicaragua. These countries allowed Soviet combat swimmer activities. In fact, two Soviet swimmers were killed in Cam Ranh Bay during the Vietnam War while attempting to mine U.S. cargo ships.

Captain Yuri Pliachenko, the former commander of the Black Sea 1st Combat Swimmer Unit, deployed to Nicaragua in 1984. He stated that their tasks went much further than diving with knives and guns: "In Nicaragua, we did not have to go under water. The Nicaraguans wanted us to tell them which waterways could be open to commercial navigation. The mining incidents there had attracted world-wide attention and our former ally was actually the victim of a mining blockade. . . . We explained to the Nicaraguans how to prevent mine laying operations and how to transform tugs into mine-sweepers. After we left, there were no more explosions."

In 1989, Soviet navy *spetsnatz* swimmers received media attention when they were used to protect the cruise ship *Maksim Gorkiy* in Maltese waters during the Mikhail Gorbachev and George Bush summit.[10]

Following the *Novorossiysk* sinking, the Soviet navy frantically sought yet more original solutions to fulfill its underwater protection and combat needs. Admiral Gorshkov was impressed by the U.S. Navy Underwater Demolition Teams during World War II, and he decided to use sea mammals. The first proposal dated back from World War I. The founder of a dynasty of circus artists, Vladimir L. Durov, had offered to train seals to fight Germans for the Imperial Russian Navy in the Baltic. But Gorshkov was more interested in recent Vietnam War tactics employed by the United States—in particular, the effectiveness of U.S.-trained dolphins in defending the Cam-Ranh base against saboteurs. Dolphins had reportedly killed fifty Vietcong swimmers and two Soviet *spetsnatz*. After detecting a diver, the dolphin would send a signal to its trainer, who then told the animal to attack. The dolphin next pushed the diver to the surface and made a paralytic—and generally lethal—gas injection with a needle attached to its nose. This demonstration convinced Gorshkov to initiate a parallel program.

Sea-mammal training began at the Sebastopol dolphinarium in 1967. Seventy dolphins and several seals took part in this project, which was managed by the GRU Research Center and by V. Kalganov, a legendary Soviet naval intelligence figure. At first, the animals were taught to warn of approaching swimmers and look for objects on the sea floor. Then, the dolphins were taught to attack swimmers and carry explosives under a surface ship or a submarine. In 1991, the Black Sea Fleet conducted a major exercise where mammals simulated sabotaging twenty-three ships, putting 60 percent of the Black Sea Fleet out of action. A parallel training center was created on Vityaz Island in the Pacific. There, sea lions were trained to attack saboteurs, using their teeth. They were powerful enough to puncture aqualungs and devour swimmers. In 1989, the Pacific Fleet war games took place with the participation of trained seals. Twenty-three combat swimmers were allegedly prevented by the mammals from reaching their target site.

After the collapse of the Soviet Union, Iran purchased most of the Soviet combat-trained mammals, and Ukraine took over the Black Sea training center. An international convention in theory now protects sea mammals from being used for military purposes.[11]

6

KHRUSHCHEV AND CRABB
1953–1960

United States and British naval intelligence were surprised by the apparently smooth transition that followed Stalin's death in 1953. Georgei Malenkov took his place at the head of both the party and the government on March 6, one day after Stalin died. There was no sign of a struggle for power. In only a year's time, Lavrenti Beria had replaced Vyasilev Molotov as the number-two man, Marshal Nikolai Bulganin had moved nearer the top, Marshal Georgi Zhukov had risen again under Bulganin, and Marshal Kliment Voroshilov had become the government's figurehead. But despite the fact that the direction of the state apparatus remained in the same hands, the authoritarian Stalinist era had given way to an "enlightened collective leadership." After Beria's elimination in June 1954—perhaps some were afraid that he might trade German reunification against an agreement to end the Cold War—Malenkov appeared to be more than an equal among equals, at least until February 1955. And then Nikita Khrushchev prevailed: he delivered his celebrated February 25, 1956, speech to the Twentieth Party Congress, denounced the Stalinist crimes he had supported, and neutralized his jealous peers the following year.[1]

During the six years that followed Stalin's death, the Soviet Union initiated a so-called peaceful competition with the West. Its disarmament program reflected profound changes in the military strategy and the realization that war was no longer an option to ensure victory over capitalism. Naval port visit exchanges helped defuse previous fears, while traditional colonial interests and remarkable technological developments generated new ones. Naval intelligence was at the heart of this Cold War diplomacy development.

Soviet anti-Western rhetoric was replaced by a softer and friendlier tone. More than ever, the Kremlin was eager to let the world know that it hoped to find a way to settle international differences. During their historical visit to Belgrade in 1955, Bulganin and Khrushchev had admitted that there could be more than one road to Socialism. The old argument that another war would bring the end of capitalism—but not the end of all mankind—was still alive. Marshal Zhukov still made no distinction between the tactical and the strategic use of nuclear weapons. But Khrushchev took the dramatic step of abandoning Lenin's views on "the inevitability of war between Communist and capitalist states as the latter enter the phase of decadence." In Khrushchev's words, "The development of the great camp of socialism" [the Sino-Soviet bloc] and "the emergence of the great zone of peace" [the Asian and other neutrals] had "brought into being forces which were capable of keeping in check the warlike tendencies of declining capitalism." With the assumption that each decolonized territory, as it emerged as a new nation, "would automatically join the forces of peace," Khrushchev demonstrated to his own people that capitalism was no longer a threat.[2]

Material Progress for the Masses

The new Soviet leader seemed eager to push forward his plans for large-scale material progress and cuts in military spending. He had already announced a reduction of more than six hundred thousand troops in the Red Army and seemed interested in improving the living conditions of the Soviet populace. Khrushchev had again dismissed navy commander in chief Admiral Nikolai Kuznetsov as a result of the sinking of the battleship *Novorossiysk*. The unlucky admiral had been dismissed once earlier by Stalin for having allowed sensitive acoustic torpedo technology captured from the Germans to be shared with the Royal Navy. As Khrushchev later explained in his memoirs, "Solving the naval-armament problem was difficult. The admirals voted in favour of a surface fleet. In rejecting the program . . . all of us experienced pain. [Kuznetsov was] a charming and fascinating man . . . but when life itself pushed us into a confrontation the interests of the cause had to be placed higher than the feelings of friendship."[3]

The long-term political-military goals of the navy espoused earlier by Stalin and Kuznetsov were again replaced by a strict subordination to a centralized general staff dominated by the Soviet army. To top it off, Khrushchev abolished the navy ministry. During a conversation at a Kremlin party in May 1956, Khrushchev entrusted Her Majesty's ambassador with a "secret": he was convinced that the navy would have only a very small part to play in any future war.[4] His comments were followed by a decreased naval representation on the Communist Party Central Committee and the purely defensive role assigned to the navy in defense minister Marshal Zhukov's speech at the Twentieth Communist Party Congress.

Assessing Moscow's Psyche

Western nations remained skeptical of Moscow's real intentions. As the U.S. Office of Naval Intelligence (ONI) explained in March 1954, "The problem for the community of the West [was] to decide whether the change [was] a genuine reversal, merely a diversion or a subterfuge."[5] One year later, the Sino-U.S. confrontation made it increasingly difficult for ONI not to predict a new war in which the United States and the Southeast Asia Treaty Organization (SEATO) powers would help Nationalist China's Chiang Kai-shek to resist a Chinese invasion of Formosa supported by the Soviet navy.[6] President Dwight Eisenhower had said that the communists would have to pass the Seventh Fleet to invade Formosa. Furthermore, the rumored Chinese lease of Hainan Island to Moscow seemed to indicate that the Sino-Soviet bloc were planning that very move: "It is conceivable that Soviet submarines based on Hainan with a powerful Soviet air cover could so damage the U.S. fleet [that] it rendered a Chinese attempt to strike Formosa an acceptable risk. The great danger lies in a situation in which neither side appears to be bluffing."[7]

In March 1956, the British Admiralty commented that the "only ultimate hope for a peaceful world" would be "if Communism in the Soviet Union should settle down and become a conservative rather than an expanding force." There was very little prospect for the regime to collapse, "except by a war which would probably be just as disastrous to the free world."[8]

Khrushchev's announcement that 375 warships were to be put into reserve was welcomed in the West as indicating a way to "streamline and improve its overall efficiency [of the Red Navy] by removing some of the many obsolete ships and craft," rather than being evidence of good-will.[9] The Admiralty acknowledged that cruiser building had stopped but the construction of submarines continued apace, with an estimated eighty boats built in 1956 and a planned future fleet of about five hundred submarines by 1960. This size submarine force would be devastating for the British, who had been nearly starved by a submarine blockade twice during the previous fifty years.[10]

Khrushchev's speech about naval reductions, however, made an impression on naval circles. Was it possible that Soviet society would be able to offer the world a better alternative than capitalist society could? The September 1956 issue of the British Admiralty's *Quarterly Intelligence Report*, which was distributed to all naval commands, reflected a certain admiration for the achievements of Soviet education: "In so far as the Communist regime can take any credit for this, it can be said that Soviet education and propaganda have at least encouraged a respect for the cultural heritage of the past, especially for classical literature, and have not pandered to the desire for the corrupted forms of entertainment and literature so common now in the West."[11]

Noting that delegations visiting the Soviet Union from the West frequently gained the impression of a rather puritanical society, the Admiralty's *Quarterly Intelligence Report* warned its readers not to be misled by appearances and lashed at the state of morals in the Soviet Union, while unexpectedly praising the Communist Party and an intrinsic Russian decency:

Erotic advertisements and entertainment does not exist. . . . Women's clothes and general appearances are not calculated to charm. All this gives the transient visitor a misleading impression. . . . Throughout most of Soviet society promiscuity is the general rule. . . . The parks at week-ends swarm with teenage girls who offer themselves for a small price. . . . These unsavoury details are mentioned in order to show that the Soviet Union can boast no superiority in such matters over the "vicious" capitalist world. . . . Virtue, at any rate in the sense of a social conscience, appears to be confined to those whose profession is to be virtuous—to members of the Communist Party. The attitude of "I couldn't care less" is nowhere more common than in the Soviet Union. It is the constant and unrewarding task of the Party to try to overcome this well-nigh universal social and moral apathy. . . . If on the whole a rather unfavourable picture emerges from this report it is only fair to add that in many of the ordinary things of life the Russian people still show a fundamental decency which makes up for many of their obvious moral shortcomings.[12]

Thus, naval intelligence was pursuing the true state of communist control within the Soviet Union, especially as it applied to the attitude of the individual citizen.

The Sverdlov cruisers, considered useless by Khrushchev, performed a series of friendly visits to the West. British commander Lionel Crabb disappeared during his second hull inspection of a Sverdlov in Portsmouth, the Royal Navy's main naval base.

The Frogman Incident

"Join the Navy and see the world" was not a Soviet recruiting slogan and, unlike the U.S. Navy, which portrayed its flag around the globe from the French Riviera to the Western Pacific, the Red Navy had always remained in its own home waters. During the period 1953–1957, however, the Soviet fleet embarked on a remarkable series of naval visits to show the goodwill of the new Kremlin rulers. In June 1953, the Soviet Union dispatched the cruiser *Sverdlov* to Queen Elizabeth II's coronation fleet review. In December 1954, two sister cruisers made a historic visit to the Swedish capital of Stockholm. Then on October 12, 1955, a squadron consisting of two cruisers, *Sverdlov* and *Suvorov*, with four Skorry class destroyers—*Smotryaschy*, *Smetlivy*, *Sovershenny*, and *Sposobny*—was in Portsmouth, while a British squadron paid a reciprocal visit to Leningrad.

The Soviet squadron was under the command of Admiral Arseni Golovko, who had headed the Soviet naval mission in London during

World War II. British naval intelligence praised the remarkably friendly atmosphere of the protocol meetings, as written in the *Admiralty Intelligence Quaterly*, "although it was quite clear that the ordinary Russian officer or rating had to keep a weather eye on his Commissar or political deputy as they are now called." The British were convinced that many of the Soviets, from their commander in chief downward, were moved by the reception they were given, although the British understood that the welcome shown to the Royal Navy squadron in Leningrad seemed even more grandiose.[13] A liaison officer praised the Soviet crew: "The behaviour of the ratings was quite exemplary: they were cheerful without being noisy, extremely polite and never failed to thank the waiters and waitresses, orchestra, etc. . . . Everybody was very friendly toward the Russians, although their reception, judging by reports in the press, did not compare with that which our sailors received in Leningrad."[14] British observers were rather surprised to note that "both officers and men appear reasonably happy." The British concluded that "This [was] only because they [knew] no better."[15]

Besides being under observation by their commissars, the Soviet party was also watched by "outsiders," as a British liaison officer explained, "On Saturday night I saw two men deep in conversation in one of the telephone booths just by the exit from the pub [where the Soviet crew was dining]. . . . I managed to get close enough to them, without them being aware of my presence, to be able to tell that they were speaking Russian."[16]

Moscow was keeping an eye on its sailors. The Soviets, however, seemed to be unaware that a British swimmer and World War II hero, Commander (Special Branch) Lionel KP Crabb, RNVR, GM, OBE, had successfully carried out an underwater inspection of the hull of the cruiser *Sverdlov*. He was accompanied by his diving partner Sydney Knowles.[17] The operation had been coordinated by MI6.

The Sverdlov class of highly maneuverable cruisers was seen as a serious threat to NATO (North Atlantic Treaty Organization) forces and convoys in the North Atlantic. Since 1950, the Admiralty had conducted several design studies for a series of fast 31-knot gun destroyers and light cruisers that were intended to outmaneuver the Sverdlovs on the high seas and neutralize them with rapid fire.[18] Knowles explained that during their nighttime dive at the cruiser's bow, he and Crabb found a large circular opening in the bottom of the hull. Knowles waited at the edge, while Crabb went up inside the hole. There he examined a large propeller, which seemed as if it could be lowered and directed to give thrust to the bow.[19] The two divers then left unseen.

In the summer of 1956, the Soviet navy repeated its friendly visits to European countries: Holland, Denmark, Sweden, Norway, Yugoslavia, and Albania. British intelligence again underlined the remarkably good behavior—"which might be described as unnaturally good"—of the Soviet sailors and "their apparently high morale." The British proposed an explanation: "obviously a deliberately calculated effort, with specially indoctrinated crews, to create a really good impression."[20] But the highlight of the year 1956 was the April visit of the two Soviet premiers, Nikita Khrushchev and Nikolai Bulganin, to Portsmouth aboard the cruiser *Ordzhonikidze*.

Captain A. P. W. Northey, the British naval attaché in Moscow, would accompany the Soviet leaders on their sea journey to Portsmouth. As the first Western official to sail onboard a front-line Soviet-built warship since World War II, Northey was about to get an insight into the Red Navy that seemed at the time improbable for a NATO officer. Soviet captain first rank Solovyev was to be Northey's shadow during his journey, which started in Baltiysk, the Baltic Fleet's main naval base. Before the Royal Navy attaché boarded the cruiser, he had to call on the Soviet base commander. After the usual toasts with the commander, Northey was whisked off to the cruiser through back streets, as he later wrote, "to avoid the enormous crowds which were reputedly massing to say farewell to Bulganin and Khrushchev." Northey was then hustled onboard and down into his cabin and was locked in until the cruiser had put to sea. Thus, he had no opportunity to observe the other ships in port. He later wrote,

> During the voyage, I was treated as an honoured guest—which had its disadvantages. Neither was it worthwhile to dispel the illusion too far by being too interested in snooping. I was told that I could be taken round any part of the ship I liked, but after all engine rooms were just like any other engine rooms! I had the constant companionship of my mentor, though fortunately we were in different cabins[,] and was occasionally allowed the privacy of my own cabin where I could "rest." He also had an inexhaustible supply of brandy which filled in the time when he could think of no other way to interest me. As if by chance, there was an orderly in constant attendance outside my cabin. I therefore did not stray very far.[21]

Captain Northey may have not been in a very good position to gather technical intelligence on a class of cruiser that Khrushchev had dismissed as useless. Instead, Captain Northey enjoyed a rare access to the Soviet leadership and an intimate personal glimpse:

I had a long conversation with Bulganin and Khrushchev, particularly the latter, had supper at their table one evening and attended Khrushchev's birthday presentation. The supper party was most enjoyable. Bulganin was a simple and charming host in a very avuncular way and kept my plate and my glass well filled. Khrushchev was by then disinterested in me and told a lot of dullish stories of the "when I was in X" series. However it seemed that he had a high regard for many of the "Western" people he had met in UNRA [United Nations Relief and Rehabilitation Administration]. Tupolev was the wag of the party and kept everyone amused with long and slightly risqué stories, the points of which were usually beyond my Russian and had to be translated by Sergei [Nikita Sergeyevich] Khrushchev. In any case, Tupolev was laughing so much himself that they were almost incomprehensible. My shadow accompanied me and was almost pathetically grateful that it was through me that he had the honour to dine at the great men's table. I was surprised that the admiral was not present, perhaps he was too busy.[22]

Since Khrushchev had decided to introduce self-service in Soviet military canteens and restaurants, he was most interested in Captain Northey's explanation of the Royal Navy's modern messing system. Northey even thought that his information would influence the future self-service and canteen system across the Soviet Union: "I think that we can watch for the changeover." On military matters, Khrushchev seemed

Captain A. P. W. Northey (far left), the British naval attaché, in Moscow enjoying his journey to Portsmouth with the two Soviet premiers Nikita Khrushchev (far right) and Nikolai Bulganin aboard the cruiser *Ordzhonikidze*.

quite straightforward: he assured Captain Northey "that the Russians had no intention of building aircraft carriers, as their ships[,] having a purely defensive role, would never be outside shore-based air cover." Khrushchev further considered carriers "expensive and vulnerable both from the air and underwater." He made "a big point" of the importance to the Soviet Union of the submarine fleet: "These, he said, properly armed with guided missiles would be what they most required for their defence and would be able to attack the U.S. purely defensively." He explained "that this was the line which they were going to pursue." Captain Northey concluded that "Mr. Khrushchev appeared completely fascinated by the possibilities of the guided missile in any role."[23]

Outside of his privileged moments with the Soviet leaders, Captain Northey had his meals in the wardroom with his shadow and other senior officers, from whom he hoped to have gleaned more than what he had given them about messing in the Royal Navy. On the whole, he was favorably impressed: "My general impression throughout the trip was that everything they did they did very well, but they never had to do anything very difficult." The ship was exceptionally clean throughout, without being in any way "tidy."

In his report Northey described the command relations, saying that the ships were overcentralized, with the admiral making most of the decisions. He found that the ship's commanding officer was "a pleasant but surprisingly young officer." The ship's company appeared "clean, smart, keen, very cheerful, very confident and again very young." His "most striking impression" was the "confidence, both in themselves and in their cause," displayed by officers and men. Northey saw this as a communist trait and a natural result of listening to years of propaganda.[24] The discipline appeared to be very strict, although he was impressed with the ability of officers and men to relax this strict discipline on occasion and become "almost *tovarishy*' [comrade-like] together." As for the commissar, he "was a stinker and no other words fit him."[25] Northey concluded that by making this trip, "I have managed to lift a corner of this particular iron curtain and may have got some insights as to how the Russian Navy works and its possible shortcomings."[26]

Captain Northey noted that signals were passed by flashing light and maneuvering commands signaled by flag hoist. Indeed, the Soviet squadron maintained a strict radio and radar silence during its three-day journey to Portsmouth. No NATO radio communications and radar-signal intercept stations could collect anything emanating from the Soviet ships. Usually, intercepted radio and electronic signals enabled the shore stations to position the ships by triangulation.[27]

The state visit by Nikolai Bulganin and Nikita Khrushchev was a complete success. The Soviet leaders were received by the queen and by Prime Minister Anthony Eden and were given a thorough tour of England during their ten-day visit. The Admiralty first sea lord invited the Soviets to a reception at the Greenwich Naval College. Khrushchev openly discussed his naval ideas: "I gave a speech in which I expressed myself categorically in favour of nuclear missile forces, emphasizing the advantages of missiles and missile bearing aircraft as compared with a surface fleet. . . . I used the expression floating coffins."[28]

Less Said, Sooner Mended

On April 19 at 7:30 a.m., three Soviet sailors sighted a diver swimming between the visiting ships at their mooring at the South River Jetty. The diver, dressed in a black light diving suit with floats on his feet, was seen on the surface of the water for the space of one or two minutes and then dived again under the destroyer *Smotryaschy*. The commander of the Soviet squadron, Rear Admiral V. F. Kotov, informed the chief of staff of the Portsmouth naval base. The latter categorically denied the possibility of a diver appearing alongside the Soviet ships and declared that at the time, no diving operations of any kind were being carried out. On April 29, the Admiralty announced that Commander Crabb had vanished while taking part in trials of secret underwater apparatus in Stokes Bay, Portsmouth. The British press stated that the British naval authorities were actually carrying out secret diving tests in the vicinity of the anchored Soviet ships at Portsmouth.

On May 4, 1956, the Soviet Embassy in London requested an explanation from the British Foreign Office. Four days later, the Foreign Office acknowledged that "The diver was presumably Commander Crabb" and that "his approach to the destroyers was completely unauthorized." It concluded that "her majesty's government [desired] to express their regret at the incident."[29]

On May 9, the prime minister was asked very direct questions at the House of Commons, to which he replied, "It would not be in the public interest to disclose the circumstances in which Commander Crabb is presumed to have met his death."[30] He then suggested that an unauthorized action had taken place by the secret services when he added, "While it is the practice for Ministers to accept responsibility, I think it necessary, in the special circumstances of this case, to make it clear that what was done was done without the authority or the knowledge of Her Majesty's Ministers. Appropriate disciplinary steps [were] being taken."[31]

On May 10, the Foreign Office asked its ambassador in Moscow whether it would be desirable to seek a personal interview with Bulganin or to take further action. The ambassador did not favor this course, noting that since their return, both Bulganin and Khrushchev had been extremely friendly to him: "As late as last night Bulganin, in a short speech at the Czechoslovak Embassy, made a quite friendly allusion to his United Kingdom visit. This, coupled with the silence of the Soviet Press, make me think that they are unlikely to exploit the frogman incident and that the least said soonest mended should be our motto."[32]

The Soviet government, however, reacted through the press. On May 13, Admiral Kotov made the statement in *Pravda* that "Certain circles in Britain took some steps against the Soviet ships paying a good-will visit to Britain, which are not in accordance with the elementary rules of hospitality."

The British ambassador noted that "The treatment was impartial and the comment, in the circumstances and given normal Soviet standards, mild."[33] Neither Bulganin nor the still commander in chief Admiral Kuznetsov alluded to the incident during their conversation with the British ambassador two days later. London was only too happy that Moscow seemed willing to play down the incident.[34]

Scraping the Barnacles from Crabb Lore

On July 4, 1956, a certain Mr. R. Lambert took the initiative to address the following letter to the Soviet Embassy in London:

> Dear Sir,
> It is reported in today's *Evening News* that Lt. Cdr. Crabb is a prisoner in Russia. If this is True, may I suggest that you return him to England immediately and thereby help to strengthen the goodwill created by the recent Russian visit.
>
> If your next Goodwill delegation were to come in an ordinary liner you would be able to invite all the frogmen and secret agents in Britain to inspect her thoroughly, thereby removing secrecy and strengthening Goodwill even further.
>
> They might even scrape the barnacles off, if any, and celebrate with vodka—or better still, orange squash.
>
> May the truth win,
> Yours sincerely,
> R. M. Lambert
> Copy to: Prime Minister, the Home Secretary, Archbishop of Canterbury, U.S. Embassy[35]

For more than half a century following the incident, stories circulated, suggesting that the World War II swimmer hero had been abducted to the Soviet Union. The authorities claimed that the headless corpse found in 1957 near Chichester, southeast of Portsmouth, was the remains of "Buster" Crabb. The pathologist said that the body could have been in the water for six to fourteen months. But not everyone believed it. In 1960, a book allegedly based on a Soviet file detailed how Crabb had been brainwashed into working for the Red Navy as a diving instructor under the name of Lev L. Korablov.

That same year, Commander J. S. Kerans, then a member of Parliament for Hartlepool, opened the matter publicly by saying, "I am convinced that Commander Lionel Crabb is alive and in Russian hands—the Government must reopen this case." In 1964, another MP, Marcus Lipton, also appealed in vain to the prime minister, Harold Wilson. In 1967, a report surfaced that Crabb had been seen in a sanatorium in Czechoslovakia.

More recently, in 1990, the Israeli journalist Yigal Serena interviewed a Russian immigrant named Joseph Zverkin, who claimed to have worked for Soviet naval intelligence and to have lived undercover in England during the 1950s. Joseph Zverkin gave the following account:

> Crabb was discovered when he was swimming on the water next to the ship by a watchman, who was at a height of 20 metres. An order was given to inspect the water and two people on the deck were equipped with sniper guns—small calibre. One of them was an ordinary seaman, and the other an officer, the equivalent of a lieutenant, who was in charge of an artillery unit on the ship, and an exceptional shot. Crabb dived next to the boat and came up and swam—perhaps because of air poisoning. The lieutenant shot him in the head and killed him. He sank. All the stories about him being caught by us or that he was a Russian spy are not true.[36]

This account appears perhaps the most plausible, but there is not enough evidence to substantiate it. The same applies to the November 16, 2007, account published by the *Times* of London that quotes another Russian by the name of Koltsov who claims to have been a frogman protecting the Soviet squadron and to have slashed Crabb's throat with a knife.[37] These stories do not necessarily contradict each other or an earlier theory based on testimonies given by the British traitor Harry Houghton and by the KGB defector Anatoly Golitsyn. Both men claimed that the frogman operation had been betrayed by a KGB mole within MI6 and that Crabb was ambushed. Harry Houghton's handler allegedly told him that Crabb had died of oxygen poisoning after having been taken onboard the cruiser.[38] Houghton may have been given that information to provide an acceptable story to the British should he be captured, as it finally happened.

Sydney Knowles, Crabb's former diving partner, told the January 19, 2007, BBC program *Inside Out* that he had been pressured by British authorities to formally identify the Chichester body as being Crabb: "Crabb had a specific scar on his leg—which he'd got diving near barbed wire. That body didn't have one." Knowles also said that he had reported to MI5 Crabb's disenchantment following his discharge from the Royal Navy: that, recently divorced, Crabb was living in a van, penniless, depressive, alcoholic, and, according to Knowles, joking about defecting to "bloody Russia."[39]

The fact remains that MI6 entrusted Crabb with this second mission to inspect the hull of a Sverdlov class cruiser, and the famed diver seemed happy about it. Ironically, the cruisers were now viewed as expensive floating coffins by the Kremlin, and Khrushchev had ordered a study to convert the remaining cruisers under construction into passenger ships, fishing factories, or floating hotels.[40] But British intelligence was not yet ready to discard them altogether, and Crabb paid the price.

The Suez Alarm

The anticipated "fairly long period of peace" might not have lasted after all. Less than one month after the Soviet premiers' joyful visit to Britain, a new crisis unfolded, precipitating a chain of events that brought the Kremlin to threaten London and Paris with nuclear destruction.

On May 16, the Egyptian president Gamal Abdul Nasser officially recognized the People's Republic of China. Washington withdrew its financial aid for the Aswan Dam project, and Nasser responded ten days later by nationalizing the Anglo-French Suez Canal. For London and Paris, either inaction or military intervention would generate a loss of prestige in the region. France was eager to reduce Nasser's assistance to the Algerian insurgents, while both England and France needed the canal as a route to oil and their colonies. Drawing a direct comparison with the events of the 1930s, London and Paris signed a secret military pact with Israel and embarked on a risky operation. Washington could hardly accept an Anglo-French intervention, which might drive the

Arab world toward the Sino-Soviet bloc, but in the event of a Soviet intervention against the British and the French, Washington would have to stop the Kremlin.

Since August, Captain Poncet, the French defense attaché in Egypt, reported on the disillusionment of Nasser's officers and called for the naval aviation to destroy Radio Cairo—indicated as being very important—the airfields, the navy, and a few more objectives that he would select to limit civilian casualties.[41] Furthermore, the Ethiopian ambassador had assured Poncet —in confidence—that the USSR and Syria were no longer fully committed.[42] Holding to a "worst case scenario," the Second Bureau in Paris came up with terrifying figures on the numbers of warships that the Soviet Union could deploy, and France missed a possibility to unseat Nasser.[43]

Naval intelligence played a key role in helping the British and French invaders monitor weapon shipments to Egypt and follow submarine movements that might precede Soviet intervention. The United States skillfully used its predictions of Soviet submarine movements to scare the British and the French. NATO had first reported three submarines exiting the Baltic on August 31.[44] During the next two weeks, the U.S. Naval attachés in Paris and London volunteered more alarming news to the British

A Soviet-made Whiskey class submarine being transferred to Egypt. Washington manipulated Communist bloc submarine intelligence reporting in another attempt to force London and Paris to reconsider its Suez operation.

and French admiralties: four submarines accompanied by a support ship were spotted in the Atlantic and a possible submarine contact obtained by a U.S. Neptune maritime patrol aircraft south of Sardinia was right in the middle of the Mediterranean Sea.[45] Although the Anglo-French suspected U.S. manipulation, the possible presence of Soviet submarines in the Mediterranean could no longer be ignored and caused consternation in London and Paris. During the following days, Anglo-French forces detected possible non-NATO submarines near their forces.[46] Meanwhile, in Djibouti, French naval intelligence was chasing more submarine rumors. A tip from local fishermen reporting a submarine at Cheikh-Said in Yemen was first dismissed.[47] Then a French officer spotted a real submarine in the Red Sea.[48] The threat was materializing. In the end, two Egyptian submarines did leave Gdynia flying a Polish ensign. They crossed the Danish belts in late October and turned back off Gibraltar to return to Poland when the hostilities broke out on November 1. At this time, two very real U.S. submarines—the *Hardhead* and the *Cutlass*—were playing against the allied fleet assembled off Egypt creating confusion and showing Washington's displeasure.[49]

Soviet Intervention

During the month of October, there was growing concern among the Allies that Egypt had improved its defenses with an influx of weapons and advisers. Early in the month, unverified intelligence indicated that four hundred foreign technicians had arrived in Alexandria. Italian Vampire jet-engine fighters were being delivered via Greece and Syria.[50] In theory, these Vampires could outmatch the French and British carrier-based, propeller-driven fighter-bombers. These alarming reports boosting Nasser's capabilities were proven wrong when Israel successfully broke Egyptian defenses on October 29 and when Anglo-French troops retook the northern portion of the canal on November 6. The Egyptian military collapsed and its new weapons didn't help. While U.S. opposition to France and Britain mounted, Paris and London were clearly worried by Khrushchev's threats to intervene. On November 2, Greek authorities announced that they were expecting Soviet demands to use its airspace. Important military shipments, including heavy weaponry far exceeding Syrian needs, were being unloaded in the Syrian port of Lattakia.[51]

Tension reached a climax on November 5 and 6, when NATO reported aircrafts flying above Anatolia, six submarines located near Crete, and a Soviet squadron about to cross the Bosphorus. At the same time, the Egyptians were actively repairing their airfields, a sign which could announce a Soviet airlift. A British Canberra was shot down at a very high altitude above Syria under circumstances indicating that the adversary could not have been an Arab pilot. In Moscow, Soviet authorities staged a street protest outside the British Embassy.

The demonstration was repeated the next day. The British attaché recalled:

> There was a definite feeling around Moscow that war was imminent, a war for which they did not understand the reasons. . . . By the next day, although nothing of the Cease Fire order in Egypt had appeared in the Soviet press, the buzz had got around. . . . The spontaneous demonstrators lining up outside the Embassy . . . were happy and cheerful. They chanted slogans of friendship between England and USSR and threw paper flowers over the wall.[52]

Meanwhile France contemplated pursuing the war without Britain, but Washington refused its support in case of a Soviet intervention and Paris had to back up. By feeding the Anglo-French dubious submarines reports, moving around two submarines, and spreading alarming signals through NATO on November 5 and 6, the United States had successfully used intelligence in its psychological campaign to stop the Suez operation. Looking back at the whole affair, Vice Admiral Barjot, commander of the French naval forces, doubted the reality of the NATO-U.S. submarine reporting.[53]

Sputnik and Submarines

In May 1957, Admiral Arleigh Burke, the U.S. chief of naval operations, said in his address to the Senate Armed Forces Committee, "In the nuclear missile age, even more than in the past, the side which commands the seas will not be defeated."[54] Three months earlier, on February 15, 1957, the Council of Ministers of the USSR had approved the launching of an earth satellite "for checking the possibility of its observation in orbit and for receiving signals transmitted by the satellite." The Sputnik rocket was launched on October 4. On the Sputnik's first orbit, the Telegraph Agency of the Soviet Union (TASS) transmitted, "As a result of great, intense work of scientific institutes and design bureaus the first artificial Earth satellite has been built."

This news came as a shock to the West and to naval intelligence. The Sputnik launch demonstrated that the USSR had mastered the technology needed to fire an intercontinental missile. Admiral Burke's statement had been proved wrong. Moscow was now in the position to defeat the United States with an intercontinental missile strike. Fitting such missiles onboard submarines would further complicate the new strategic picture. Shorter-range ballistic missiles were already being put to sea. During May 1959, the navy identified the first Soviet Zulu class submarine with ominous characteristics: "Close examination of the pictures taken indicate that an object covered with a canvas tarpaulin may have been a missile launcher."[55]

Later that year, ONI reported "that at least three" Zulu class submarines had been converted to "probably ballistic missile launching boats," with three more anticipated. The Zulu conversions were followed by the 3,200-ton conventionally powered Golf class built for the purpose of carrying missiles. Each was fitted with three vertical SS-N-4 *Sark* ballistic missiles. After beginning production in 1957, the Golf class submarines were first observed at sea in 1960, the same year that the new U.S. Polaris submarine *George Washington* became operational.[56]

Trauma and Intellectual Snobbery

Taking the full measure of the event, the British Naval Intelligence Department tried to understand how the Soviet Union had been able to achieve this milestone:

> The Soviet Union has recently achieved a successful launch of an artificial earth satellite, and has, it is known, fired a large number of ballistic missiles to various ranges, including, in all probability, two to a range of some 3,500 miles. . . . This must raise the question "How is it that the USSR, which in 1945 was regarded, probably rightly, as scientifically backward, has in so short a time been able to outstrip the West in the most modern scientific developments?" Part of the answer lies in the nature of the Soviet State, an extreme socialism in which all activities are directly controlled by a completely autocratic government. . . . It has developed over the last 40 years a system of scientific education which is certainly at least equal in quality to that of any other country and is turning out trained personnel in greater numbers even than the United States.[57]

The Admiralty also stressed the decisive role played by the Soviet Academy of Sciences "to put engineering practice on a sound physical and mathematical basis."[58] The NID concluded that the secret for this successful "marriage between science and technology" probably lay in the absence of intellectual snobbery in Soviet science, an intellectual snobbery that affected Western research.[59]

The fears that consumed Western navies in the late 1940s were becoming true a decade later. The Soviet Union had embarked on a massive submarine-building program. In a staff study released on August 13, 1956, the U.S. Navy wrote that the Soviets had been building long-range submarines "at a rate far in excess of previous estimates." In January 1956, the Soviets were believed to have a total of 421, and by January 1958, this figure was expected to rise to 646, "mostly long range types," The two new classes that caused this concern were the medium and long-range 1,350-ton Whiskey and 2,500-ton Zulu. A more ominous development, however, was that the Soviets would soon have nuclear-powered submarines.[60]

In 1958, the U.S. Navy had predicted that a Soviet submarine fleet would consist of between 445 and 450 boats, with an annual construction capacity of 160 units. In the navy's estimates, the Soviets had launched almost 300 first-line, snorkel-equipped submarines, with 260 of these being long-range, offensive types. While admitting that the Soviets had historically considered the submarine a defensive weapon, the U.S. Navy concluded that they now realized the offensive potential of the submarine.[61]

Even more worrisome, Khrushchev had stated that he wanted guided missile-carrying submarines equipped with atomic warheads. Many sightings had been made in the Northern Fleet areas, the Baltic, and the Pacific of submarines "with large hangar tanks on deck, ramp-like structures for launching, and even airplane-like objects on deck." These indicated that in 1957, the Soviets probably had a supersonic turbojet missile with a range of 500 miles. By 1962, that range could be extended to 1,000 miles. Although there was no evidence of a ballistic missile that could be launched from submarines, such as the Polaris type, the U.S. Navy asserted that this could be "an intelligence deficiency rather than a lack of Soviet effort in the field."[62] Sputnik came as a stark confirmation that the Soviet Union was farther ahead in some fields than Western intelligence had predicted.

By 1959, there had been a shift in U.S. Navy thinking about the Soviet submarine threat. Although the Soviets were not expected to deliberately provoke a general war, ONI pointed to the increasing likelihood of limited wars in which the Soviet submarine force would play an important role.[63]

In that context, reliable East German sources reported to British intelligence that Moscow was building a major naval and submarine base in the Adriatic port of Saseno Island, Albania. The reports said that the Soviets were also enlarging the ports of Durazzo and Valona for larger draft ships, while Saranda, Albania, was being turned into a modern naval base. Between five and seven thousand Russians, East Germans, and Czechs were said to be working in twenty-four-hour shifts to build the new Soviet naval base and develop all Albanian seaports. Two squadrons of Soviet jets were already stationed near Valona, where the Soviets had built an airstrip two years earlier. Half a dozen submarines were said to have been or were about to be transferred to the Albanian naval and submarine base on Saseno. The Admiralty concluded by stating that the Soviet Union had virtually detached the island of Saseno, the ports of Saranda and Valona, and the cape of Linguetta from the Albanian Republic and placed them under the direct supervision of the Soviet Ministry of Defense.[64]

Enlisting Franco

The creation of a permanent Soviet submarine base in the Mediterranean and the expansion of its submarine fleet forced Western nations to counter the new threat and seek new partners. In late 1955, London had decided to aim for a rapprochement with Spain. Meanwhile, Washington had embarked in a security assistance program to modernize the antisubmarine capabilities of the Spanish fleet.

To mark this new policy, the newly appointed commander in chief of the British Mediterranean Fleet made a formal visit to Barcelona in October 1957. During the visit, General Francisco Franco, unaccompanied by advisers or interpreters, received the commander in chief. Franco suggested that it was time that the major powers agreed on the standardization of the equipment and urged all of the Western powers to join in preparing plans since there would be no time for this when an emergency arose. The commander in chief answered that purely from the naval point of view, the more ships and ports in the Mediterranean that were available to the West, the more efficient the West would be in deterring an aggressor.

The generalissimo expressed his concern over the transatlantic sea lanes to Western Europe being cut at the very outset of war. He felt that relatively few Soviet submarines, strategically placed, together with well-organized sabotage merchant vessels proceeding to the Western hemisphere, could isolate Europe in the first few weeks of the war. The commander in chief of the Mediterranean Fleet noted that recent scientific developments—for instance, the satellite and the intercontinental ballistic missile—had made him realize that the German scientists whom the USSR had acquired after the last war were now fully integrated into the Soviet system. He further stressed that the nuclear submarine was likely to be the most deadly weapon in the future and might in itself constitute a perfect deterrent to an aggressor.

The generalissimo then praised Soviet crews onboard merchant ships and fishing vessels. He had observed that unlike the fishing vessels of other nations, where the crews slop around in any old clothes, Soviet seamen were as disciplined as those onboard a warship.

Enlisting Generalissimo Franco ensured Spanish cooperation in the key Mediterranean passage of Gibraltar, where Soviet submarines regularly passed; in the mid-Atlantic around the Canary Islands; and in the Mediterranean. As Rear Admiral C. E. Weakley commented in 1959, "During the last war, every U-boat came to the surface and transmitted very nicely almost once a day and we had a location for them. In some respects the ocean surveillance system substitutes for what promises to be a very difficult intelligence area the next time."[65]

U.S. naval air and missile strike capability was viewed by the navy as the most effective way of neutralizing Soviet submarines before they could depart their bases. Yet fighting snorkel-equipped, diesel-electric submarines in the open ocean was still difficult. One of the most pressing problems was initially detecting a submarine. With the development of the Sound Surveillance System (SOSUS), it was possible to locate Soviet submarines, including those carrying missiles, by means of coordinated passive arrays mounted on the sea floor.

Invented in 1950, SOSUS first became operational in 1956 in the Atlantic and in 1958 in the Pacific. As part of a U.S. continental defense concept, SOSUS assisted a growing number of "Hunter-Killers," antisubmarine aircraft carrier groups on both U.S. coasts, by providing initial locating information on submarines. These contacts could then be attacked by antisubmarine aircraft fitted with Julie and Jezebel sonobuoy systems, which were floating sonars. By 1956, twelve SOSUS stations were under development in the Atlantic, with seven in the Pacific.[66] During February 1959, eight U.S. Guppy class submarines, playing the role of Soviet missile-carrying submarines, were used to simulate missile launches against U.S. West Coast defenses. SOSUS detected them all, proving its value in the warning role.[67]

SOSUS was effective but it was also expensive, and the U.S. Navy was looking for other intelligence sources such as electronic intercepts and coordinated surveillance patrols at key choke points in sensitive geographical areas. A closer cooperation was required between the United States and the United Kingdom.

Mountbatten of Burma in America

One year after the launch of Sputnik, First Sea Lord Louis Mountbatten toured Canada and the United States. Besides securing U.S. cooperation for the British nuclear submarine program, one of the main purposes of Mountbatten's visit was to discuss joint anti-submarine detection and the establishment of two lines of surveillance, in Iceland and in Gibraltar. The initial proposal for a fixed Greenland to Norway system had been discussed at policy conferences in 1955 and 1957. Britain was not convinced by SOSUS and its passive fixed arrays. British passive systems had not performed sufficiently well off the Shetlands, and the advent of more silent submarines would prove them useless.[68]

Mountbatten, while visiting the Washington, D.C., Naval Laboratory, learned that at optimum depth, U.S. submarines' sonar could detect a submarine at ranges up to 100 miles. Both for antiair and antisubmarine warfare, the U.S. Navy was developing and improving its Naval Tactical Data System computer for sorting out and sharing the submarine-detection data collected by the radar and the sonar, and it could present a clear picture of the tactical situation to the operators. It is the same kind of computer that the U.S. traitor Joseph Barr would later design for Admiral Gorshkov.

Meeting with SACLANT (Supreme Allied Command Atlantic) in Norfolk, Mountbatten discussed D-Day of a future war.[69] The first priority would be the destruction of Soviet Northern Fleet naval bases with 170 naval attack aircraft. The second priority would be the defence of North America against guided missile submarines in a perimeter extended to 500 nautical miles from the East coast. If the guided missile threat did not develop, the forces allocated for this defence would be transferred to the third priority of shipping protection, the first convoys being not scheduled to arrive in Europe before D-Day +30. At the Pentagon, the idea was aired that batteries of POLARIS ballistic missiles might well be mounted in merchant ships, which could move overnight and evade Soviet naval intelligence.[70] Despite the Suez episode, blood was thicker than water and the Anglo-American U.S.-U.K. axis remained pivotal in the Western strategy to defeat Khrushchev's rockets and submarines.

7

ANATOMY OF TREASON
1958–1964

With elaborate schemes of intelligence proliferating on both sides in the 1950s, certain individuals turned against their own countries for reasons that varied from ideological disappointment to personal motivations, such as professional stagnation, money, or love—and in some instances, all of the above. Several of the most notable cases of treachery and disloyalty have only recently been divulged in full.

A Commander Abandons His Ship

In 1959, the defection to the West of a Soviet naval officer, Captain Second Rank Nikolai Artamonov, sparked a mystery that remained unsolved until the end of the Cold War. Only after 1991 did the tangled tale appear to unravel, as new evidence emerged when former Soviet intelligence officers publicized personal accounts of their operations.

Artamonov, the destroyer commander in the Soviet Baltic Fleet, was known in the West as Nick Shadrin. His defection precipitated one of the longest and most bizarre intelligence dramas in the naval Cold War.

Lev Vtorygin, a naval academy classmate of the defector, related the story. Vtorygin completed the Higher Naval School in Baku a year after Artamonov had finished his studies in Leningrad. They served at sea together in the wardroom aboard a minesweeper in the Baltic Fleet. Junior Lieutenant Nikolai Artamonov was a mine warfare officer, and his cabinmate Lev Vtorygin was the gunnery officer. The two men grew close as shipmates in the confined spaces of the small combatant. During that period, they shared their political views, personal ideals, and goals and spent many evenings ashore together in their homeport of Baltiysk near Kaliningrad.

After two years as shipmates, Artamonov and Vyorygin parted ways. Artamonov went on to destroyer duty, while Vtorygin attended the Military Diplomat Academy in Moscow, where he developed excellent foreign language skills. Vtorygin then entered military diplomatic services as a GRU (Main Intelligence Directorate) military intelligence officer.[1]

Ten years later, in 1958, Vtorygin was assigned to the Soviet Embassy in Argentina. It was there that he first learned from another intelligence officer in the embassy that a Soviet navy captain second rank had disappeared from his Baltic-based ship in 1959 and ended up in the United States as a defector. At the time, Lev had said, "I knew the moment I heard the news that it was my friend and former shipmate, Nikolai Artamonov. He had been promoted quickly and had gained a reputation as a rising star in the fleet. But something had happened to him mentally, I'd heard. It was said he developed a leak in his morals."

Captain Artamonov had abandoned command of his Project 30b (Skorry class) destroyer while in Polish waters. The ship was being prepared for transfer to the Indonesian navy, as were a number of Soviet units including a cruiser, destroyers and submarines, as part of a large aid program. At the same time, Artamonov, a now thirty-five-year-old commanding officer, had his ship's motor whaleboat modified for the dash from the Polish Baltic port of Gdynia to Sweden.

At first glance, the occurrence appeared to be a simple case of a disgruntled naval officer fleeing a system that he felt was stifling his promising career. Later, after disappearing, he declared that he had sacrificed his Soviet naval life in exchange for freedom and love. Eventually, the Swedes turned him over to U.S. intelligence, and he made his way to live in the United States under the alias Nick Shadrin.

Popular accounts of Artamonov's biography appeared in the Western press after 1975. They ascribed to him a number of senior Soviet naval mentors, such as Baltic Fleet commander Admiral Fyodr Golovin, the wartime naval hero Admiral Arseni Golovko, and supposedly a father-in-law—navy commander in chief Admiral Sergei Gorshkov himself. In fact, Artamonov was married to the daughter of the naval political commissar of the Baltic Fleet, who at that time held the rank of colonel. Artamonov's son, Nikolai Nikolaevich, was born in Leningrad in 1953.

A detailed written account of the whole affair by Artamonov's longtime friend Admiral Dimitry Krasnov, a schoolmate and neighbor from Leningrad, was provided to Peter Huchhausen by Lev Vtorygin in 1995.[2] It describes Artamonov's year-long preparations for his flight by sea to Sweden. On taking his destroyer into the shipyard for a scheduled overhaul, commanding officer Artamonov secretly ordered shipyard workers to install an auxiliary fuel tank—a fifty-five-gallon drum—into his ship's motor launch. He personally attached a full-size ship's magnetic compass, allegedly so that he could use the launch as a fishing platform. Artamonov planned his escape to the West more than a year in advance and, according to Krasnov, skillfully evaded the prying eyes of the ubiquitous naval security branch of the KGB. Krasnov stated that after the defection, the KGB investigation into every aspect of Artamonov's escape was falsified. This was done to conceal the KGB's failure to recognize in advance the signs of Artamonov's preparations to defect.

After Nikolai Artamonov and his Polish lover, Eva, fled to Sweden aboard his launch, through the teeth of a Baltic gale, they sought political asylum in the United States. They were welcomed with open arms by the Office of Naval Intelligence (ONI), which soon thereafter granted Artamonov status as a special consultant and a source of much insider information about the Soviet navy. That Artamonov had been the commanding officer of a second-rate destroyer, shunted out of the mainstream operational Soviet navy, and assigned to train Indonesians to operate obsolete Soviet equipment somehow eluded U.S. intelligence officers, who treated the defector's words as absolute doctrine.

Meanwhile, as Artamonov was establishing his bona fides as a truly conscientious political defector in the United States, his former shipmate and friend Lev Vtorygin left Buenos Aires and was assigned to the Soviet Embassy in Washington, D.C., as an assistant naval attaché. Knowing of Artamonov's presence in Washington, Vtorygin was initially plagued with uncertainties about how he might react should he be sought out by his close friend and former shipmate while living in Washington. Vtorygin was so frightened by the prospect that Artamonov might contact him and attempt to lure him into defection that Vtorygin concealed his former close relationship with Artamonov from Soviet security officials, for fear that it would result in cancellation of the coveted Washington assignment.

Vtorygin began his attaché duties in the Soviet Embassy in 1960, working in the Belmont Avenue military attaché annex until 1965. The feared meeting with Artamonov occurred, according to Vtorygin, completely by chance on the crowded streets of Washington, D.C., during the pre-Christmas shopping rush of 1960. Lev described the encounter:

> What I feared all along finally occurred. I was entering a large department store on F street with another assistant attaché from our embassy staff. It was a crisp cold day and the streets were filled with American Christmas shoppers. I swung the outer glass doors open and stepped into the space between doors where a large fan blew warm air at the entrance with a loud roar. I saw a large man laden with

packages[,] with a woman at his side, step through the inner doors into the same space to depart the store. He was large and burly and wore a heavy overcoat. The woman was dressed elegantly and wore a fur hat resembling the *shapkas* worn in Eastern Europe. The tall heavy-set man wore an American cap. The couple spoke Polish, and the very tall man looked familiar. I asked my friend if we knew him, and he said that he was probably an employee from the Polish embassy we had met during social events. At that point I shouted, "This is Artamonov!" and ran outside the department store after him. I did not know what to do, really. But I could see no trace of him. He must have recognized me first and ran. My old shipmate was now swollen and overweight. When I returned to the embassy I reported to the naval attaché what had happened. The attaché was a very clever man. He decided not to inform Moscow of the incident. He felt that I would certainly be recalled and under suspicion of having had a personal contact with Artamonov. I believe that Artamonov reported our encounter to his new masters fearing that I had come to shoot him. And people later made up that story.[3]

Several written accounts of the Artamonov story recorded that the defector feared that Lev Vtorygin, his old shipmate and friend—an excellent hunter and sharpshooter—had been sent to eliminate him on American soil. When Peter Huchthausen showed Lev the statements, he laughed and said that was impossible. "Certainly, the KGB had their trained people who did such things, called wet operations. But they would never have used an assassin whom Artamonov could recognize to do the job. Plus," said Vtorygin, "I could never have accepted such an assignment against an old friend, despite how I loathed his traitorous actions."

While Vtorygin was looking over fences and skulking near rail and ship-loading ports along the U.S. East Coast during the Cuban missile crisis, his friend Artamonov was busy giving U.S. naval intelligence his personal analysis of Soviet intentions of Operation Anadyr. This was the code name for the missile insertion into Cuba, about which Artamonov, as a former skipper on an obsolete destroyer, knew absolutely nothing. Said Vtorygin, "Perhaps he was feeding the American intelligence services the contents of our navy books on basic maneuvering and navigation or our division officer's political guidebook. Certainly, he was no communications or radar expert, and his knowledge of ASW and sonar was limited to that of the obsolete class ship he commanded before he defected."

Nevertheless, Artamonov became a live trophy for the CIA and ONI, and he was paraded at U.S. military installations in forums for officers as a sample of disgruntled Soviet leadership. As a student at the Defense Intelligence School, the future U.S. Navy attaché for Moscow Peter Huchthausen had heard lectures by Artamonov and had lunched with him twice, along with other students who were eager to talk with an actual Soviet naval officer.

After sixteen years in the United States, the defector Artamonov, still known as Nick Shadrin, was apparently recruited by the FBI to carry out a mission in Vienna. Whether he was being turned as a double or triple agent is not really known. But while on the mission in Austria, he apparently was approached by Soviet intelligence in what appeared to be an attempt to recruit him again. He was scheduled to meet a Russian on the outskirts of Vienna for an autumn hunt, one of his favorite pastimes. Then he disappeared into thin air and was initially thought to have re-defected to his native USSR.

One popular U.S. theory about his fate was that Artamonov was approached in the United States by the KGB, which passed him a letter from his Russian wife begging him to make amends for his crimes. The emotional letter appealed to Artamonov to do the patriotic deed of returning to the Soviet ideological camp for the sake of their son, Nikolai Nikolaevich. In 1972, young Nikolai had been denied entrance to the prestigious Nakhimov Naval Preparatory School in Leningrad because of his father's shameful record. In this scenario, the defector Artamonov, overcome by remorse and guilt, acquiesced to KGB pressure and began work as a double agent. He traveled to Vienna, Austria, in December 1975 on a ski trip with Eva and mysteriously disappeared. This version of the story has it that he was murdered in 1975 by heavy-handed Soviet intelligence services that tried to kidnap him near the Austrian-Czech border.

Lev Vtorigyn had heard a similar story. After receiving a letter written by his son, Artamonov had agreed to cooperate with the KGB with the apparent desire to re-defect to the Soviet Union. When he was asked to provide a photocopy of the ONI's phone directory, his illegible reproduction arose suspicion that he was being used as a triple agent by the United States. He was easily convinced, though, to come to Vienna and opposed no resistance when a KGB team abducted him. Oleg Kalugin, the chief of the first directorate, had taken the unusual step to participate in the operation. Overruling the KGB female doctor, he administered a fatal dose of sedatives to the defector. Upon their arrival by car at the Austro-Czechoslovakian border, the car got stuck in the mud and the "goons" symbolically balanced Artamonov's body across the "Iron Fence."

The report of Artamonov's killing seemed to be corroborated by Oleg Kalugin, a born-again anticommunist who suddenly saw the light during

Gorbachev's glasnost period. After Kalugin defected, he claimed that he had been sent on a mission to kidnap and drug Artamonov and spirit him over the Czech border. He said that his men accidentally killed Shadrin/Artamonov while attempting to abduct him and return him to the Soviet Union.[4]

Lev Vtorygin had another explanation:

> Kalugin was afraid that Artamonov could give away some details of Kalugin's contacts with the FBI and the CIA; his initiative to go on the Czech border and kill Artamonov was to cut short this possible compromising. He [Kalugin] had been working for the U.S. since the time when he was studying in Columbia University. He was an agent of influence; like Yakovlev, his mentality was more American than Soviet. People like him tried to help American ideals penetrate into the Soviet Union. Gorbachev, from the very beginning, was a puppet in the hands of those people.[5]

Yet, Shadrin brought divide to the U.S. intelligence community. In his book *Soviet Naval Strategy*, Commander Robert Herrick, a naval intelligence officer and a one-time Artamonov watcher, expressed the diverging views of the defector. Herrick describes the Soviet navy's central doctrine as being defensive in nature and predicated on a multilayered ring of protective zones extending around the periphery of the USSR, with each zone being defended by surface and air-launched cruise missiles of ever-increasing range. The book contradicted the official U.S. Navy doctrine—that Gorshkov's new blue-water navy posed a serious challenge to Western domination of the world's oceans.

Another book, *Widows*, by U.S. Army intelligence officer William Corson, claimed that Artamonov was sighted at retired former commander in chief Fleet Admiral Gorshkov's bier, following the admiral's death in May 1988. The U.S. and Royal Navy attachés visited the bier as well and thought they recognized Artamonov, in the uniform of a Soviet navy captain first rank, standing with Gorshkov's mourners. This has never been confirmed. Artamonov's deserted Russian wife often makes statements in the Russian press that she has had clandestine visits with her husband, who was reportedly still living near St. Petersburg. Other stories have him secretly buried near Lofortovo Prison in eastern Moscow, with a Latvian cover name on his tomb. The true story of his demise will no doubt never be known.

Another extraordinary case of Soviet espionage, long kept under wraps by both East and West, carried the seeds that sprouted the Soviet misadventure in Cuba.

Murat Explosive

General Charles De Gaulle could scarcely have been pleased by the 1967 book written by Leon Uris, the controversial author of *Exodus*. The new novel, *Topaz*, which Uris claimed was based on facts taken from undisclosed diplomatic sources, was even more explosive than *Exodus*. *Topaz* describes a Soviet spy ring that infiltrates so deeply into the heart of the French government that even the president is being briefed by a Russian agent. One prominent U.S. publisher dropped *Topaz* for fear of libel, but another decided to go ahead with autumn 1967 publication.[7]

It is now documented that the French government had been deeply penetrated by Soviet intelligence. One traitor, Georges Paques, an aide of President De Gaulle's navy minister from 1940 to 1945, was recruited by the NKVD. Paques later served as a high official in various French government ministries before going to work for NATO as a public affairs officer. Paques was arrested by the Direction de la Sureté du Territoire in 1963 after passing documents to a Soviet diplomat. Paques was tried, convicted of espionage, and sentenced to twenty years in jail. Paques, a devout Christian, had proclaimed to the tribunal during his trial, "I beg you to believe me. I have never been a Soviet agent. Everything I did was to assure France's survival."[8] Six years later, he was pardoned by General De Gaulle.

Soviet military intelligence infiltrated a number of other French circles, most notably the French air force. In the aftermath of their World War II fraternity of arms, several senior air force officers held genuine sympathy for the USSR. One in particular, a World War II fighter ace, cast his lot with the Soviets during the crucial late 1950s/early 1960s period of French military turmoil. The GRU handlers of this French general code-named him Murat, after a famous Napoleonic cavalry general.[9] Details of Murat's activities were first revealed in a 2003 book by Mikhail Boltunov. The book is primarily a biography of Viktor A. Lyubimov, a talented Soviet naval officer and spy who, during his assignment in Paris from 1961 to 1965, controlled three other agents.

Born in 1926, Lyubimov joined the Soviet navy and graduated from the naval academy in 1948. Soon afterward, he was recruited by the GRU and initially worked in the naval headquarters and on the main general staff intelligence directorate. His first operational posting was to the United States in 1953, as a door keeper at the Soviet naval mission. When ordered to monitor ships' movements, he impressed his superiors by obtaining data from classified programs at the U.S. naval ordnance laboratory. Lyubimov was then admitted to the military diplomatic academy, where he studied French contributions to NATO in antisubmarine warfare.

He arrived in Paris in October 1961, assigned undercover to the Soviet trade mission. His double life included a position as a representative of the Soviet Merchant Marine Ministry. There, he was in charge of liaison with the French state-owned shipping company Compagnie Generale Transatlantique, which operated, among other routes, the Leningrad–Le Havre line. Murat had no diplomatic status or immunity.

Pleasing Louise

The French air force officer code-named Murat was initially recruited in 1958 in a Brussels cinema by the World War II hero of the Soviet Union Valentin Lebedev. The details provided in Boltunov's book may have been altered to prevent the positive identification of Murat. The book claims that Murat was a fighter ace but said that he did not belong to the famed Free French Normandie-Niemen squadron operating on the Soviet front during 1942–1945. His birth date given in the book, his NATO postings, his promotion to general in 1962, and his death in 1968 do not correspond with any French air force officer's true biography. But if one or two details are changed, then several officers can match his description. In a case related or unrelated to Murat, one of the highest ranking air force generals had been suspected of being a Soviet agent.[10] Murat was apparently motivated by a vitriolic hatred of the United States and in later years by the love of a woman, Louise, who was his Soviet agent courier.[11] He rationalized that he was not passing any information that was specifically about France; rather, it dealt primarily with NATO and U.S. capabilities and plans. According to Boltunov's book, only after 1962 did Murat's intense love for the forty-seven-year-old Louise surpass his initial motivations to assist the Soviet Union.

Khrushchev Shocked

Murat traveled widely and had apparently been smuggled into the Soviet Union twice for more complete debriefings, using forged papers and complicated routes. The material he compromised was considered sensational by the Soviets. Primarily, he passed them the details of U.S. and NATO plans to deploy to Europe strategic and medium- and intermediate-range missiles targeted against the Soviet Union and its crack army divisions that were poised against NATO in East Germany. This data, when added to the unfolding news of the U.S. submerged-launched Polaris missile, gradually convinced Soviet leaders that they were seriously behind the West in strategic missile capability. In fact, they believed that a missile gap existed, but it was the reverse of the view then promulgated by the United States and their allies. Boltunov detailed this entire concept of perceived Soviet inferiority and said that it was presumed to have had a definite impact on Premier Nikita Khrushchev's gamble in Cuba. Boltunov explained,

> From all the documents marked Top Secret [provided by Murat], two were extremely important. The plan issued by the Supreme commander of the NATO air forces in Europe number 110/59, dated 10 January 1960, on nuclear strikes. The second document dealt with the emergency plan of defense for tactical operational units on the Central European Military Theater. The first document also included ten appendices. Its corpus described the main features of a future conflict, defining the tasks and planned forces cooperation. The appendices contained information about the nuclear forces to be used and their alleged location on the theater as well as targeting lists set in a prioritized order. Now everybody knows about the war plans, Drop Shot, and Foul Play, among others that were designed in the Pentagon and NATO. But in the 1960s nobody could imagine that the U.S. was planning for the total destruction of the USSR. And Murat was the one who provided us with this war plan and with the other war plans created by NATO. When Premier Khrushchev saw the plans he was in a state of shock. It was as if the Soviet leader was reading the death sentence for his fatherland. Nowadays many liberal political scientists consider that Khrushchev had overreacted [in Cuba] to stolen U.S. plans. They accused him saying: that's him who took us to the brinks of catastrophe in 1962; that's his fault. But it seems quite different to me.[12]

Murat was so well trusted that he persuaded the Soviet high command to launch a full-scale Warsaw Pact exercise that he would monitor from his NATO headquarters. The purpose would be to enable the USSR to better understand the Atlantic alliance's intelligence capabilities. On November 28, 1960, the commanding general of the GRU reported to the Soviet chief of general staff Marshal Andrei Grechko on Murat's suggestion, saying,

> At the last meeting with our agent who transmits precious documents, he was tasked to inform us on any future planned attack by NATO on the USSR. He told us about his ideas.

1. The last NATO war games organized in Autumn of this year [1960] revealed some . . . lack of organization with some tasks not being fulfilled.
2. By conducting frequent flying exercises on the borders of the Soviet Union and its neighboring countries, NATO forces are trying to accustom the Soviet armed forces to their presence. NATO generals think that by acting that way they will weaken the vigilance of the Soviet armed forces.
3. The agent's idea is to test the preparedness of NATO air forces by organizing war games in East Germany. We have to inform him about the beginning and character of the war games so that he could observe the reactions and make evaluations.

Following Murat's suggestion, the Soviet Union organized a strategic exercise in the first half of 1961.

In October 1962, Murat was posted in the NATO headquarters in Paris, Porte Dauphine. He was ideally placed to assist Moscow while the Cuban crisis was unfolding. Louise provided Murat with twenty-five rolls of film for his Minox camera, which he used to photograph lists of 1,093 NATO nuclear targets in the USSR and Warsaw Pact countries. The information was evaluated by GRU as "very important."

From October 10 to 30, the chain from Lyubimov to Louise to Murat worked around the clock. On October 24, Murat warned Lyubimov that U.S. forces in Europe were on full alert. According to Murat, the tension within NATO was extreme. Handler Lyubimov was convinced that the information provided by Murat helped the Soviet leadership to make key decisions during the crisis.

Murat rarely delivered written material to Louise. Given the urgency of the situation in October 1962, however, he broke his rule and supplied her with papers detailing the organization and the locations of NATO nuclear forces at that time. He also passed on detailed plans for nuclear strikes, code-named Kangaroo, Bombast, Peacock, and Dingo.

More Recruits from Paris

Murat was not the only successful source exploited by the clever Lyubimov. He was also responsible for handling three more successful agents code-named Gouron, Arthur, and Bernard.

Gouron served with the U.S. forces stationed in France. He allegedly provided details on the U.S. command structure and the location of nuclear weapons stored in Europe. In February 1964, defense minister Marshal Rodion Malinovsky and chief of GRU Marshal Andrei Biriyuzov wrote a top secret memorandum to Soviet premier Khrushchev, praising and summarizing the spy Gouron's work: "For the period 20 January 1963 to 30 January 1964, we have received about 200 documents related to delivering, keeping and moving NATO and US nuclear warheads on the European theatre. As a result of this information, six stockpiles were found in West Germany, the Netherlands and Greece; twenty-seven nuclear bombs were located in Turkey and sixteen in Greece. We also found nuclear stockpiles at twelve air facilities: one in England, five in West Germany, one in the Netherlands, two in Greece, three in Turkey."[13]

Lyubimov's second source was no less important. Lyubimov described Arthur as "a great scientist." Arthur passed "precious information" on to the Paris GRU station, about NATO missile technology and equipment such as Nike, Sidewinder, Falcon, Safeguard, Blowpipe, and Hornet missiles and the latest naval antiair warfare defenses.

The third source, Bernard, was also productive. He was the deputy director of a French company that was 70 percent owned by U.S. investors. Lyubimov and Bernard had concluded that there were two ways of obtaining data on U.S. technology. The first was espionage with the help of an employee like Bernard inside a French company that was doing business with the United States. The second way was perfectly "legal": the USSR would buy certain technologies with the help of a front-man like Bernard acting in his official capacity. There was one critical area where Bernard's contacts made a major contribution. The Caribbean crisis had revealed weaknesses in Soviet nuclear forces. The Soviets wanted detailed information on U.S. solid fuel for ballistic missiles. During 1962 and 1963, Lyubimov procured twenty-four essential documents that contained data on chemical formulas used for the Minuteman ICBM's solid fuel. He obtained the documents by using Bernard as the contact inside the company, who copied or stole them.

At that time, Soviet scientists were deciding which direction to take in developing future rockets: should they use highly dangerous liquid fuel or more stable solid fuel? In early May 1965, U.S. secretary of defense Robert McNamara reassured the public that the USSR had no long-range solid-fuel missiles to match the U.S. solid-fueled Minuteman and Polaris missiles. Using Bernard's information, however, rocket expert and inventor Sergei Korolev's design bureau had already developed the RT-2, a 900-kilometer, solid-fueled ballistic missile. At the 1965 World War II victory parade in Red Square, Western military attachés were stunned to catch their first glimpse of a Soviet missile that apparently burned solid propellants. Lyubimov and Bernard had done more than their share in assisting the rocket genius Sergei Korolev.[14]

Shelving Murat

In 1964, the main Moscow intelligence center at the KGB instructed the GRU residency in Paris to release Lyubimov from the task of recruiting more agents. This was done both to protect him from French counterintelligence and to give him more time to work with his most productive agents.

At this point, GRU Moscow appointed a new chief of the Paris station, who looked carefully at Murat's successes. Murat, still motivated by his deep love of Louise, had reached his peak in spying productivity. In mid-June 1964, the new Paris GRU chief reported to Marshal Biriyusov, the chief of the GRU, that Murat's latest report detailed the location and the firing positions of the new U.S. Pershing nuclear missiles and a list of a hundred of their priority targets in Central Europe. As Boltunov recalled in the book, "Murat was very active during July–August 1964: during this period we received thirty-one top secret documents consisting of 2,300 pages. [More] important than the volume was that his material was extremely high value." Murat's July delivery to the GRU contained information on a wealth of subjects that ranged from strategic planning, antitank tactics, and infrared technology to details of many Western military automated systems.[15] This would be Murat's swan song, however. The French air force reassigned Murat to a post where he had little access to classified material, and Louise deserted the general. It was of little use for the Soviets to take any additional risks with their successful source. With Louise gone, Murat became a broken man.

In Boltunov's book, the handler Lyubimov reviewed the significance of Murat's work:

> In the 1960s when the situation in the world was complex and tense it was hard to overestimate the importance of the work done by Murat. It could only be compared with the information provided by Sorge [during World War II]. It was even more important if we consider the fact that Murat provided documentary and not analytical information. When fulfilling my requirements, Murat took enormous risks and sacrificed his family's well being and his personal situation in the bourgeois society, acting as if the USSR was his second fatherland.

Later, it was planned to make Murat a hero of the Soviet Union, but this honor was never bestowed on him.[16] In 1965, Lyubimov was ordered to return to the USSR, his Paris operations terminated. Murat is said to have died three years later, a date that was most likely falsified.

Profumo

In the early 1960s, Britain suffered a succession of spying scandals with a naval connection. Revelations by three CIA sources led to the arrest of an Admiralty clerk and a petty officer who were found to have been lured into spying for the KGB. Anatoly Golitsyn, a high-ranking KGB officer who had worked inside the KGB's First Chief Directorate, defected to the CIA in Helsinki in 1961.

Golitsyn, whose name appeared in Leon Uris's *Topaz*, told his handlers of two British naval spies attached to NATO who were revealing U.K. submarine capabilities. The first naval spy had been recruited in Moscow by the KGB's Second Directorate (which was responsible for foreign intelligence collection). The second spy was allegedly a more senior figure run by the KGB residency in London, who had provided at least three top secret Admiralty documents detailing plans to expand the NATO Polaris base Holy Loch on the Clyde River.

Just before defecting, Golitsyn had been writing a report on NATO naval strategy. When he was shown the three Western documents from his MI5 (Britain's national security intelligence agency) contacts, he recognized the reports at once. MI5 then looked at the distribution list and came up with a number of suspects. But the decisive evidence came from another senior KGB officer, Yuri Nossenko, who defected to the CIA in June 1962. Golitsyn claimed that the KGB residency had used homosexual blackmail to recruit an important British agent. Nossenko's information led MI5 to conclude that the two naval spies were one and the same man. He was John Vassall, who had been stationed in Moscow before joining the Admiralty. Rolls of films containing 176 classified Admiralty documents were found at his home.

In reality, Vassall had consented to work for the KGB primarily to indulge his high taste in luxurious living. Vassall was sentenced to eighteen years in prison for espionage.

Another celebrated case was that of Gordon Longsdale, an illegal Soviet agent posing as a Canadian, who had recruited a ring of spies within the British Admiralty. Author Peter Wright explained how Sniper, a Polish CIA source, had told him of two spies in Britain that MI5 had code-named Lambda 1 and Lambda 2. In March 1960, Sniper gave MI5 a tip that led to the arrest of former Royal Navy petty officer Harry Houghton. This man had been posted in Warsaw in 1952 and had later worked in the underwater weapons establishment at Portland. After that, he was in naval intelligence.

The most spectacular Soviet spy scandal involved British secretary of state for war John Profumo, a Conservative Party politician and a potential future prime minister. Profumo's compromising led to the collapse of the Conservative government of Harold Macmillan and to the victory of the Labour Party at the subsequent general elections.

John Profumo had an affair with twenty-year-old Christine Keeler, a British subject who apparently was also concurrently involved with the Soviet naval attaché in London, Captain Yevgeny Ivanov. The British and foreign press were quick to trumpet that British high society was debauching in a new "pornocracy." Strangely enough, the sensational story had been broken to the public by a little-known journalist writing in a political newsletter titled *Westminster Confidential*. This journalist was the former U.S. naval reserve lieutenant Andrew Roth, who had fled the United States in August 1948 after having been accused of passing secret information to the Soviets.[17]

Training a New Handler

The Soviet attaché in question was Yevgeny Ivanov. During World War II, Ivanov had joined the navy and had been accepted at the Baku naval academy, where he met Lev Vtorygin. Later, Ivanov served in the Pacific and the Black Sea. In 1949, the young Ivanov studied at the military diplomatic academy.[18] At the academy, Ivanov also socialized with fellow officer Oleg Penkovsky who later spied for the West and was caught, tried, and, in 1962, shot. Yevgeny graduated in 1953 and was posted as the assistant naval attaché in Oslo.[19]

During his four years in Norway, Ivanov ran three agents. The two most important were senior naval officers working in the Norwegian naval headquarters at Horten, Norway. Both officers had similar profiles and motivations and both volunteered their services at about the same time. As Ivanov recalled in his memoir *The Naked Spy*, "It was love at first sight in both cases, as I have already explained. My interest in delicate information suitable to my position as deputy naval attaché was apparent to both officers; their desire for money was equally transparent. Both . . . gave me to understand soon enough that they had information from NATO sources and were ready to share it with me—for payment. To all appearances they had no conscience about this, probably because the information they supplied concerned not Norwegian but U.S. and NATO forces."[20]

As senior staff officers, they had access to top secret material. Lax security procedures enabled them to fill in forms stating that certain documents had been destroyed; then they passed these documents over to their handler. One Norwegian officer traveled regularly to NATO conferences in Europe or the United States and afterward provided Ivanov with the latest NATO briefings. To meet his contacts without attracting attention, Ivanov had chosen a place in the Norwegian countryside where he went every weekend to ski and lunch. Every third Sunday he left the restaurant through the back door for five minutes to meet his first contact or took a short walk on the road to meet his second contact, in the process always losing his Norwegian surveillance team, who grew bored by his yearly routine. "It was rather all un-dramatic," Ivanov said. "I would put a roll of money and instructions safely hidden in the package with the remains of what I had eaten in the café into a garbage can. Simultaneously, I picked up the package left for me. . . . To minimize risks, my backup watched me from the window of a hotel across the road. A lamp on the windowsill meant that he had put the package in the prearranged place."[21]

In addition to handling his agents, Ivanov watched NATO military bases and visited the bars near the bases. On one occasion, Ivanov claimed to have had a stroke of luck. He noticed a brown envelope on the table that a U.S. colonel had just left for a couple of seconds to order another drink. Ivanov snatched it and fled. It turned out to be a top secret document describing the deployment of NATO forces and NATO's objectives in Scandinavia.

A Valuable Friend

Dr. Stephen Ward was a famed British osteopath who had made his reputation in the 1950s by treating the back problems of celebrities, such as Prince Philip; Winston Churchill, the former prime minister; the millionaire J. Paul Getty; and Lord William Astor. Dr. Ward's house was always open for friends. He enjoyed the company of attractive women, including then teenagers Christine Keeler and Mandy Rice Davies, who often stayed at his home. Ward was also a talented artist who, after having painted portraits of the royal family and of politicians for the *London Illustrated News*, decided to sketch members of the Soviet Politburo. Dr. Ward, however, had difficulties obtaining a visa to visit Moscow.

One of Ward's notorious patients and a potential publisher was Sir Colin Coote, the one-time editor of the *Daily Telegraph* and a former Secret Intelligence Service (SIS, also known as MI6) officer who had served in Italy in the 1920s. In the early 1960s Coote had hosted Captain Ivanov, the newly appointed Soviet assistant naval attaché in London, during an attaché corps tour of his newspaper. Coote introduced the Russian to Dr. Ward, allegedly to ease Ward's visa difficulty. Following a chummy lunch with Ward, Coote and Ivanov became friends. Ivanov explained in his book that the friendship with Dr. Ward opened up his access to the royal family and to the highest circles of British society. "Stephen was a true find for me," wrote Ivanov. "He knew a great deal and was a good companion." Dr. Ward introduced Ivanov to his relations and took him to visit Winston Churchill. The latter, then in retirement, lectured Ivanov on the Anglo-Saxons as being like a modern version of the Greeks and the Russians, the modern barbarians.

Soviet attaché Ivanov impressed his new friends with his intellect and clever conversation. Ivanov explained in his memoir:

> At the top of my list were secretary of war Jack Profumo and his wife Valerie Hobson. . . . Next came the Astors. . . . Third on my list were Princess Margaret and Anthony Armstrong Jones. I was introduced to them by Dr. Ward, who also equipped me with certain details about their life. . . . This connection held out the possibility of getting information through provocation and blackmail. The fourth was billionaire J. Paul Getty, a man whose business connections could lead me to arms producers and dealers. Fifth place was occupied by two men connected with the press. . . . And the last was Captain Souls [HM Coast Guards] whom I recruited.[22]

To show his appreciation, Ivanov invited Dr. Ward to the Soviet Embassy to attend the cosmonaut Yury Gagarin's reception during his visit to England.

Swimming Pool

On July 8, 1961, William Astor hosted a social weekend at his estate in Cliveden, England. Included among the many celebrities invited were Lord Louis Mountbatten, the first sea lord; Secretary of State for War John Profumo and his wife; and the prime minister of Pakistan. On that same day, Dr. Ward was with Christine Keeler and some friends in his adjacent cottage. When Lord Astor and Profumo walked past the swimming pool by chance, they met the naked nineteen-year-old Keeler trying, but not too quickly, to cover herself with a towel. Profumo was enthralled with her beauty and gave the young lady a tour of the house. That same evening, however, Keeler left for London to fetch Ivanov, whom Ward had invited to Cliveden for Sunday. Ward, Keeler, and Ivanov then mixed with the lord's other guests and joined a swimming contest organized by Astor. Profumo and Keeler soon became a team in the water play. Later that day, Ward suggested to Ivanov and Keeler that the two should drive back to his apartment in London, where he would rejoin them after attending to Lord Astor's back problems.

Dr. Ward did not return to London that evening. Nevertheless, he showed up the next day and met with Christine about Ivanov. She said that Ivanov had gotten drunk, fallen asleep, and then departed before she woke up. But the fact that Ward had left Ivanov and Keeler alone in his London apartment was apparently part of an MI5 plot to use Keeler unwittingly to compromise the Soviet attaché, a plot that went awry. Ivanov, being married to the daughter of an important Party official, could hardly have indulged in an extramarital affair. MI5 had put him to the test.

The following Tuesday, Profumo called the young woman and drove her around London in his official vehicle. Later in the week he invited her to his house and, according to Keeler's subsequent testimony, made love to her for the first time.[23] The two continued to meet regularly at the doctor's apartment, where Ivanov was a frequent visitor.

Knowing what had transpired, MI5 approached Profumo to seek his-cooperation in a plot to frame Ivanov. The minister refused MI5's offer and apparently distanced himself from Keeler, under the pretext of traveling abroad.

Moscow's Back Channel

On August 7, 1961, Nikita Khrushchev made a speech on television concerning the situation in Berlin. He called for a rapid buildup of Soviet forces. At this point, Dr. Ward took the astonishing step of volunteering his services to the British Foreign Office as an intermediary in the East-West relationship. When at first he was unsuccessful, he again approached the foreign ministry through Lord Astor. Later, he invited one of his patients, Conservative MP Sir Godfrey Nicholson, to dine with attaché Ivanov. During the lunch he offered a back channel exchange between the foreign office and Khrushchev, through Nicholson, himself, and Ivanov. As the biographers Philip Knightley and Caroline Kennedy wrote in their book *An Affair of State*, "Ward was discussing international political issues with an important GRU officer . . . acting as an unofficial pipeline to convey information between the Russians and the British."[24]

In his memoir, Ivanov explained how Moscow appealed to him to try to establish some sort of back channel with the British government, during both the Berlin and the Cuban crisis:

> I was a daredevil then, just like Khrushchev. The most important thing was that I was impudent and self-confident, I thought. This would teach the Americans not to encircle us with their bases! Deploying missiles right in their underbelly meant that no early warning system would save them from a missile attack because the missile flight time would be more than three or four minutes. In five days of one crucial week [Ward] organized my meetings with Lord Astor, Lord Arran and Sir Godfrey Nicholson. Prime Minister Harold McMillan could make a major step towards peace if he suggested a Soviet American summit with the aim of settling the Cuban crisis. I assured them that Khrushchev would be prepared to accept that proposition. . . . I was politely informed that the Prime Minister would be informed of my proposition.[25]

Andrew Roth Breaks the News

For more than a year, the news of secretary Profumo's liaison remained in the closet. In December 1962, however, the London GRU chief summoned Ivanov to his office and said, "A major scandal is brewing here, involving Christine Keeler, Profumo, and yourself. In short, start packing."

Ivanov was already in Moscow when Andrew Roth—the former U.S. Navy lieutenant who had fled to England after being indicted as a Soviet spy in 1948—broke the news of the Profumo-Keeler-Ivanov affair, precipitating the downfall of the Conservatives. Was Moscow turning the case to its advantage in the end?

When Profumo died in March 2006, Roth wrote Profumo's obituary in the *Guardian*. In it, Roth detailed the following chronology of the affair and his own justification for reporting it:

> Although MI5 had known about Profumo's dalliance with Keeler for many months, the first politician to learn about it was John Lewis, the former Labour MP for Bolton, who mistakenly thought Dr. Stephen Ward had seduced his wife. In January 1963, he informed George Wigg,

Harold Wilson's intelligence chief, that Ward's protégé Christine Keeler, previously the girlfriend of property speculator Peter Rachman, had had a four-month affair with Profumo, after a brief fling with the Soviet naval attaché Yevgeny [sic] Ivanov. In February 1963, Macmillan's private secretary, John Wyndham, later Lord Egremont, was told about the Profumo-Keeler relationship. About then, Bob Kerby passed on to me a copy of an affectionate note Profumo had sent Keeler that she was unsuccessfully trying to sell in Fleet Street, where its content was widely known. Editors, two of whose journalists had been jailed at the time of the scandal concerning the Soviet spy John Vassall, were reluctant to cross the Macmillan government again. I started checking the background of the Profumo-Keeler letter. A press gallery colleague working for Tass, the Soviet news agency, confirmed that Ivanov had formed connections in Tory circles. Then suddenly, I had to use the material in my newsletter "Westminster Confidential" because my other potential lead—the plan of then chancellor of the exchequer, Reginald Maudling, to float the pound—had been aborted. So, in March 1963, I printed the Profumo letter in Westminster Confidential. All hell broke loose and my mail stopped coming. After Bob Kerby showed my newsletter to the Conservative chief whip, Sir Martin Redmayne, his face becoming even redder than usual and he rushed it to Harold Macmillan, demanding I be deprived of my lobby privileges. Sitting in the press gallery I heard a rattled Macmillan attack me. Then I heard Profumo, flanked by Macmillan and the leader of the Commons Ian Macleod, threaten the likes of me with a libel action after reading out a whitewash concocted overnight by Redmayne, the solicitor general, the attorney general, and information minister William Deedes—a meeting from which the stiff necked home secretary Henry Brooke was excluded. Profumo insisted: "There was no impropriety whatsoever in my acquaintanceship with Miss Keeler." It was after this that the outspoken Tory MP Nigel Birch voiced all cynics' suspicions by saying: "What are whores about?"[26]

The devil had been turned loose. Keeler was said to have been instructed to ask Profumo about the rearmament of West Germany. Stephen Ward was accused of "living wholly or in part on the earnings of prostitution." He committed suicide on August 3, 1963, apparently distraught by the betrayals of his many friends.

Queen Elizabeth II finally intervened to allow Profumo to resign and avoid the further humiliation of being dismissed.

Sleeping . . . or Not Sleeping with Christine Keeler

In Keeler's written introduction to Captain Ivanov's memoirs, published in 1993, she described the dramatic chain of events that followed her alleged night with Ivanov: "If I hadn't gone to bed with Ivanov that night it is very likely that none of it would ever have happened. My love affair with John Profumo would never have developed overtones of treason. Harold Macmillan's government might not have been rocked to its socks. Stephan Ward would never have killed himself." She then added, "Yet to this day, I still believe no-one except Yevgeny [*sic*] and I know the real truth. He tried to convince his Soviet masters that he had seduced me as part of some brilliant, spy plan."

In this book written thirty years after the facts, Ivanov also admitted to having slept with Christine Keeler, giving a professional rationale and explaining how sex was being used by his masters:

> Yet the very fact that Christine had been both my and Jack Profumo's lover was a potential trump card in any possible future game involving the ill-fated Secretary of State for War. Sex has always been, and will most probably remain, a powerful weapon of political blackmail and espionage. The GRU did not use it as frequently as did the KGB. In this sphere the Lubyanka [headquarters of the KGB] was the undisputed champion. It suffices to recall the case of William Vassal, a British military diplomat and homosexual who was recruited by Moscow Centre.[27]

After having exploited Dr. Ward's tapes, researchers Knightley and Kennedy doubted this version. They noted that the young Keeler, who had entrusted the doctor with details of her love affairs, had never told him about her alleged night with Captain Ivanov, an occasion that the doctor had apparently created by leaving the pair in his apartment. Ivanov's former classmate and colleague Lev Vtorigyn gives a different interpretation to the whole affair:

> It was known among the members of the Soviet Embassy [in London] that Dr. Ward and his friend Keeler were "agents provocateurs." Embassy personnel knew that they should avoid them at all costs. In the case of Ivanov, it was different. He met Keeler once or twice at a reception in the high society. He was looking for such contacts. But I can't say that he had such close relations with that girl. Ivanov was well too smart to fall in the trap and be seduced by the young Keeler. There was

no intention and no part on the Soviet side to compromise Profumo. It did no good whatsoever to Ivanov's career. Later in retirement, Ivanov worked for the APN publisher and was involved in a propaganda effort to write a book about his affair in London for the foreign public. This book, *The Naked Spy*, did not reflect the truth.[28]

In the end, it appears that what was originally a British plot to blackmail a well-connected Russian naval attaché and Nomenklatura figure through a liaison with a young lady may have become a Soviet scheme to undermine the conservatives by exposing the coincidental liaison of the defense minister with the same young lady. Coincidence it was. However, is it coincidental that the journalist who broke the news to the press was a former U.S. naval officer previously accused of serving Moscow? Is it coincidental that he wrote his article once Ivanov was out of the country? Whatever the answers, the part of the Soviet naval attaché had not been lost on newly elected Labor prime minister Harold Wilson. When Wilson visited Moscow in 1964, he allegedly asked for an interview with his benefactor, Captain 1st Rank Yevgeny Ivanov.

8

CUBA
1962

One day in early April 1962, Soviet defense minister Rodion Malinovsky and Soviet general secretary, premier, and Communist Party chairman Nikita Khrushchev walked together on the beach at the posh Black Sea resort called Omega. The minister pointed seaward and said, "There, just across the Black Sea in Turkey NATO has installed Jupiter rockets with ranges capable of hitting cities throughout the Soviet Union without advanced warning." The party general secretary walked in silence for a bit, then suddenly stopped and said, "Well, we have a potential missile platform now in America's backyard. Yes, Cuba. We could deploy rockets with similar ranges as theirs in Turkey and Italy. Install them secretly and it then would be a fait accompli."[1]

The Cuban missile crisis may have been the most spectacular display of brinkmanship of the Cold War era, although some historians argue that the Berlin crisis of 1961 was even more serious. In 2002, Peter Huchthausen's book *October Fury* revealed details from the Soviet side that showed for how close the incident had come to igniting a calamitous confrontation between U.S. Navy antisubmarine destroyers and Soviet submarines in the Atlantic.[2] A single shot fired by any one of several warships could have led to an immediate nuclear exchange.

Arming Revolution

Fidel Castro's takeover in Cuba on New Year's Day 1959 was neither sponsored by the Kremlin nor directed against the United States. Yet the new regime's economic policy and the nationalization of U.S.-owned companies had antagonized President Dwight Eisenhower's administration and eventually led to the rupture of diplomatic relations in March 1961.

Meanwhile, U.S. intelligence received continuing indications that Moscow had proffered military assistance to Cuba. In July 1960, Nikita

Khrushchev stated, "The USSR was raising its voice and extending a helpful hand to the people of Cuba." The Soviet premier then added that in case of necessity, "Soviet artillerymen could support the Cuban people with rocket fire." The next day Che Guevara, already a firebrand, hinted at Moscow's commitment when he declared that Cuba was "defended" by the Soviet Union, "the greatest military power in history."

Within two months, the National Security Agency (NSA) had learned that the USSR was exporting weapons to Cuba. The NSA's eavesdropping stations intercepted a series of communications linking the trade official Leonid Yastrebov with Cuba. For more than six years, this man was known to have been associated with exporting Soviet military cargoes to Indonesia, Egypt, and Syria—and now to Cuba.

Two Soviet ships, the *Ilya Mechnikov* and the *Solnechnogorsk*, had already delivered military equipment to the Caribbean island in September. Intelligence discovered that the cargo ship *Nikolaj Burdenko*, loaded with 3,234 tons of unknown cargo in the Black Sea port of Nikolayev, would arrive in Cuba in mid-October. The NSA also intercepted communications indicating that the merchantman *Atkarsk* had loaded 2,454 tons of "Yastrebov's cargo" at Nikolayev and was scheduled to arrive in Cuba in late October. Then the intercept stations learned that Yastrebov, the arms trader himself, had applied for a visa for Havana.[3] In January 1961, intercepts revealed possible military cooperation between the communist bloc and Havana's armed forces. A Spanish voice was overheard on Czechoslovak air-to-ground VHF communications divulging that a Spanish-speaking pilot was probably undergoing flight training of an unidentified type at Trencin airfield in the Soviet Union.[4]

Until diplomatic relations were ruptured, the last U.S. naval attaché to serve in Havana, Commander Joseph H. Floyd, had actively watched Soviet bloc shipping and cargo, using his own network of informers: disillusioned Cubans, other military attachés, and prostitutes. Late in the preceding year, Floyd had sighted what appeared to be crated missiles in Havana harbor, but he was unable to confirm this. A Cuban radioman also provided the busy attaché with the full spectrum of radio frequencies used by Castro's military, which was an extremely useful tool to help the NSA monitor the island.[5] After March 1961, the United States had to rely more on the British and the French attachés for on-site reporting. With the closure of the U.S. Embassy in Havana, monitoring communications was the main source of information on Soviet-Cuban dealings.

The ill-fated April 1961 Bay of Pigs operation failed to provoke a popular uprising to oust Fidel Castro. The Office of Naval Intelligence

(ONI) had served as the middleman between the navy and the CIA in arranging naval support to the operation. The CIA's operational security on the Bay of Pigs landing was so tight that only five officers in ONI were aware of the support arrangements. The plan called for the navy to protect the landing force at sea. Yet the navy was concerned that if the landing were to go wrong, there would be no way to recover the anti-Castro fighters. And this is exactly what happened.[6] The CIA inspector general's subsequent inquiry concluded that the agency had committed "serious mistakes" by not inviting the individual military services, in particular the navy, to give "a cold and objective appraisal" and, most egregious, by not immediately telling the president "that success had become dubious."[7]

The arms dealer Yastrebov continued to be the NSA's key target. His name appeared frequently in the radio traffic associated with shipments to Cuba, and once his name was mentioned, the nature of these shipments became clear. By monitoring Cuban air force operator chatter, the NSA learned that Cuban military personnel had been instructed to learn Russian.[8] In Cuba, the military gossip overheard by Americans listening in helped identify new equipment and training requirements: "How are things there? Good. Let me tell you that things are hot here. We now have radar here. . . . How is that? Where is it? Here beside the station; under some mango tree."[9]

In total, during the first quarter of 1962, forty-three Soviet dry-cargo ships made voyages to Cuba and delivered at least 228,000 metric tons of cargo, excluding military cargo. These ships also transported approximately 370,000 metric tons of Cuban sugar and 2,238 tons of nickel concentrate back to the USSR. Although the number of voyages during the quarter remained on a par with voyages for the first quarter of 1961, the tonnage delivered in 1962 represented an approximate increase of 50 percent over the known tonnage for that same period in 1961. Of special significance were eight motor torpedo boats and five Kronstadt class patrol vessels coming from Murmansk. The *Tbilisi* and the *Kolkhoznik* each carried four motor torpedo boats (MTBs) on deck and towed a patrol vessel. The *Baltijsk*, the *Bratsk*, and the *L'gov* each towed a patrol vessel, in addition to carrying other cargo.[10]

Cuban military plaintext communications continued to betray a wealth of information on the country's new toys. In early April 1962, radar-tracking activities were mentioned, along with references to equipment designators of known Soviet radar sets.[11] In May, tactical training was confirmed to be underway. Linguistic analysis of intercepts from the radar operators established the participation of non-Cuban control personnel, apparently "Slavic or Middle European."[12]

Mongoose

After the Bay of Pigs humiliation, President Kennedy realized that his low-profile policy had had no effect on Cuba's march toward Communism. The refugee flow out of Cuba now included lower-middle classes who expressed some vestiges of popular opposition. This fact prompted Kennedy to undertake a special effort "in order to help Cuba overthrow the Communist regime."[13] On January 18, 1962, the president gave his brother Robert Kennedy, the attorney general, a straightforward directive: "It's got to be done and will be done."

The next day Robert Kennedy convened a special session that included the Joint Chiefs of Staff, CIA director Richard Helms, and other CIA and Office of the Secretary of Defense (OSD) officials.[14] The attorney general made it clear that "a solution to the Cuban problem [carried] the top priority." The U.S. military was ordered to engage in provocative action against Cuba, named Operation Mongoose. Its ultimate goal was disrupting the sugar harvest and inciting a Cuban response that would lead to overt hostilities and the overthrow of the Castro government. Meanwhile, the hoped-for social unrest would keep the Castro government from meddling in the internal affairs of other Latin American countries.[15] On March 14, 1962, guidelines were issued for covert action. The navy was not involved, but the CIA had planned maritime operations of its own. Nineteen attempts were made to infiltrate sabotage teams, out of which eleven had to be aborted.[16]

By July, however, weather conditions and newly delivered Soviet patrol boats had suddenly made maritime infiltrations more difficult, and the infiltration plan failed to achieve its objectives.[17]

Maskirova

After the Soviets successfully launched their Sputnik satellite, U.S. intelligence services had assumed the existence of a missile gap between the two nations—in other words that the United States was lagging far behind in the total number of operational intercontinental ballistic missiles. With the evolution of newly acquired satellite imagery in 1961, however, the supposed missile gap suddenly became incorrect. U.S. Air Force intelligence analysts now estimated the number of Soviet intercontinental ballistic missiles to be five hundred. The CIA had calculated their number to be only about fifty. Yet the first satellite imagery revealed that the Soviets had just four ICBMs. In February 1961, Defense Secretary Robert S. McNamara told the U.S. public that there was no missile gap. His evaluation was confirmed two months later by the GRU defector Oleg Penkovsky.

Washington now knew the real strength of the USSR and the Kremlin knew that its bluff was over. Because of its fear of the missile gap, the United States had planned for the total destruction of the communist bloc by delivering about five thousand nuclear weapons within several hours. As former Soviet GRU officer Captain Lev Vtorygin later explained, "The possibilities were not [equal] for the United States and the USSR, and the information communicated [to Moscow] by the French general—Murat—about NATO war planning was precise and therefore of utmost importance to the Soviet Union."[18] Murat's betrayal of the NATO Air Force war plan number 110/59, dated January 10, 1960, had made such a strong impression on Khrushchev that it may have been a decisive reason that the Soviet leader deployed missiles to Cuba, in order to reduce the very real missile gap now faced by the USSR. Murat was for Moscow what Penkovsky had been for the West: an eye opener into the adversary's real capabilities.

In July 1962, Moscow's defense minister, Marshal Malinovsky, issued verbal orders activating the Soviet covert shipment of missiles, atomic warheads, and tons of military hardware to Cuba in an operation code-named Anadyr. Khrushchev's plan to install the medium- and intermediate-range missiles covertly in Cuba called for extreme secrecy. *Maskirova* is the Russian word for ruse. Measures to keep the deployment under wraps ranged from the practical to those of a comic opera. For example, all troops that were being deployed as part of a mechanized infantry regimen wore only civilian clothes. Military staff members flying to Cuba were sent first to Conakry, Guinea, and thence to Cuba to avoid attracting undue attention. Military hardware was carefully crated, and even the concrete launchpads for the missiles were prefabricated in the USSR and sent by ship to Cuba. The choice of the operational name Anadyr was also intended to cover the ultimate focus on the Caribbean. Anadyr had been the code name of an old post–World War II Soviet contingency plan to invade Alaska. On the one hand, arctic clothing was ordered on the same supply requests as tropical khaki uniforms, yet according to some submariners, every navy wife in the Northern Fleet was sure that she would soon be moving to the wonderful Caribbean climate.[19] The plan transformed Cuba into a strategic Soviet missile platform in America's backyard, with key U.S. cities easily within reach of medium- and intermediate-range missiles.

In August, eighty-six Soviet merchant ships that were earmarked for transporting troops, weapons, and matériel began to shuttle to Cuba from ports around the Soviet Union. The means of deception were elaborate but somewhat clumsy. Soviet aircraft, including MiG-21 interceptors and naval Il-28 medium-range bombers, were dismantled and transported in crates aboard the cargo ships. The Soviets were unable to fully disguise their cargos, though, and Western intelligence analysts were soon tracking and counting the Soviets' hidden treasures.

Increased Listening

Just as Marshal Malinovsky was giving his instructions to initiate Operation Anadyr, U.S. secretary of defense McNamara had ordered the secretary of the navy to increase its participation in the SIGINT (signal intelligence) collection effort against Soviet-Cuban activities.[20] Under the control of the chief of naval operations, the Naval Security Group Command managed several specialized ships to collect communication and SIGINT for the National Security Agency. This increased effort against the Caribbean island required one more ship and an additional twenty officers and men to replace the USS *Oxford* (AGTR-1) which had contributed much of the earlier Cuban air force gossip intercepts on the island. The navy and the NSA agreed to have a new intelligence ship partially manned by army personnel on station and to replace the *Oxford* with an interim platform. On July 19, the NSA director sent the following message to the chief of naval operations, underlining the highest priority given to the mission:

1. NSA has been directed by Sec Def to establish a SIGINT collection capability in the vicinity of Havana, Cuba as the highest intelligence priority. They initially diverted the USS *Oxford* for this purpose. Director NSA has initiated action to obtain the services of a Military Sealift Command (logistics support) ship for this purpose to be operational not later than 1 December 1962.

2. Since full capability of *Oxford* cannot be utilized effectively in Havana area, request CNO assign ship to relieve *Oxford* as an interim measure until a ship subordinated to MSTS [Military Sea Transport Service, the military subordinated, civilian manned ships] is available.[21]

Unusual Number of Passenger Ships
en Route to Cuba

On July 24, at the very outset of Operation Anadyr, the NSA reported
the first movements of the passenger ships destined to Cuba. It noted
that there were "at least four, and possibly five Soviet passenger ships en
route Cuba with possibly 3,335 passengers aboard." Three of these ships
had departed from Baltic ports, while two came from the Black Sea.
The *Khabarovsk* and the *Mikhail Uritkij* were said to be making voyages to Vladivostok; however, the NSA suspected "deceptive measures"
to conceal voyages to Cuba. The *Admiral Nakhimov* and the *Latviya* had
both declared for Conakry, Guinea, when departing the Black Sea. The
Latviya, however, was tracking directly toward Cuba. Thus, the NSA
considered it likely that the *Admiral Nakhimov* would follow suit. High-
precedence messages sent to both the *Khabarovsk* and the *Mikhail Urtskij*
betrayed "other than routine activities."[22]

The Soviets used these large passenger ships, such as the *Admiral
Nakhimov* (which later sank in a storm in the Black Sea, with a heavy
loss of life) to transport the personnel of three mechanized rifle brigades
to support the Cubans. Although COMINT (communications intelligence) allowed the United States to monitor the Soviet ships' movements
to Cuba, the real purpose of their activity still remained obscure to NSA
analysts. On August 7, the NSA volunteered a series of explanations, all
of which failed to recognize the Soviet military buildup in Cuba:

> Some of the apparently extraordinary aspects of this merchant activity
> could reasonably be explained: (a) the light loading of the vessels might
> be a reflection of a Soviet inability to furnish industrial/agricultural
> materials to the Cubans; (b) the military equipment already in Cuba
> was in excess of what was needed or could not be paid for; therefore,
> the ships were needed to export the material from Cuba (an unprec-
> edented occurrence); and (c) little . . . traffic would be expected if the
> ships were not carrying anything but stated cargo tonnage.[23]

The possibility that these ships could carry military equipment
seemed less likely, although it was not totally rejected:

> There is some evidence to substantiate an assumption that at least
> some of these ships may be carrying military cargo. It has been past
> experience that the Soviets generally attempt to conceal military equip-
> ment by such means as declaring 'varied' or 'general' cargo when

required by passing through the Bosphorus. At least nine of these ships made such declarations, which permitted the assessing of a certain amount of the cargo to be possibly military.[24]

Meanwhile, USS *Oxford*'s monitoring revealed an unprecedented amount of new Soviet equipment in use by the Cuban military. On three occasions in August 1962, the ship intercepted for the first time "a signal displaying characteristics of the Soviet AA fire control radar WHIFF," confirming the deployment of this equipment to Cuba.[25] Despite the Soviets' efforts to speak Spanish, COMINT sources continued to reveal Russian and non-Cuban voice activity on Cuban Revolutionary Air Force tactical frequencies. The *Oxford* interceptors reported, "Concentrated efforts have been made by Bloc pilots and controllers to converse entirely in Spanish but, on occasion, they have reverted to their native tongue to convey a difficult command or request to other Bloc pilots or controllers. In only one recorded instance has a Bloc pilot requested a Cuban controller to speak more slowly because he was unable to understand."[26]

The CIA's Quiet Clairvoyance

On August 10, 1962, at a meeting in Secretary of State Dean Rusk's conference room, the CIA director John McCone speculated that the Soviet material being shipped to Cuba could be "electronic equipment for use against U.S. space base Cape Canaveral, Florida and/or military equipment including medium range ballistic missiles."[27] Ten days later, the CIA reiterated its warning on the increased activity in Cuban ports. By then, Thyraud de Vosjoli—the French SDECE (Service de Documentation Extérieure et de Contre-Espionnage) resident in Washington and Havana—may have played his pivotal role, as depicted in Leon Uris's book *Topaz*:[28]

> It appears that between 4000 and 6000 Soviet/Bloc personnel have arrived in Cuba since 1 July. Many are known to be technicians, some are suspected to be military personnel; there is no evidence of organized Soviet military units, as such. . . . The unloading of most ships takes place under maximum security, with the Cuban population excluded from the port areas. Large equipment is noticeable; large crates have been observed which could contain airplane fuselages or missile components.

The CIA was not yet absolutely positive of the nature of the shipments and still considered the wishful possibility of "an increased technical assistance to Cuban industry and agriculture." The military implications were the "possible establishment of surface to air (SAM) missiles sites" and the "possible establishment of Soviet Comint-Elint facilities targeted against Cape Canaveral and other important U.S. installations."[29]

On August 21, CIA director McCone gave Secretary of State Rusk definite information on surface-to-air missiles and again speculated on whether medium-range missiles were present in Cuba. The following two days, he repeated the same information to President Kennedy and questioned the need for the extensive SAM installations unless they were to help conceal medium-range missiles. During that meeting, McNamara expressed strong feelings that "Every possible aggressive action in the field of intelligence, sabotage and guerrilla warfare should be undertaken." McCone reported that so far, the efforts had been disappointing, "due to a very tight internal security situation."[30]

On August 23, 1962, President Kennedy asked for an analysis to be prepared of the probable political and psychological impact of basing missiles in Cuba that could reach the United States—either surface-to-air missiles or surface-to-surface missiles. To decide on his plan of action, he also ordered a study of the advantages and the disadvantages of issuing a statement that the United States would not tolerate the establishment of military forces that could launch a nuclear attack from Cuba. And if making a statement was not advisable, he wanted to know the possibilities and ramifications of various military alternatives.[31]

It was becoming more tempting to risk obtaining low-level photography by using air force RF-101 or navy F8U aircraft to fly over Cuba.[32] On September 29, high altitude U2 photography confirmed extensive Soviet deliveries, including eight Komar class guided-missile boats. According to CIA estimates, the radar-guided missiles—each carrying a 2,000-pound high-explosive warhead—were effective against surface targets in ranges of between 15 and 17 miles. U2 photography also revealed that eight surface-to-air missile sites were being set up, while thirty-seven MiG fighters had been located, out of an estimated total in Cuba of sixty. Sixteen Soviet dry-cargo ships were at that point en route to Cuba, of which at least ten were thought to be carrying military equipment.[33]

The Right to Defend Cuba

By September 1, the Soviets had shipped a full reserve of ammunition, spare parts, and missiles into the port of Mariel. Then the merchant ships transported large missile boats of the Komar class for coastal defense.

These were impossible to conceal and were carried on the weather decks. The first shipment of the sixty-seven-foot-long, medium-range missiles was loaded aboard the merchant ship *Poltava* in the closed port of Sevastopol. Once the missiles were carefully placed in the hold, the five-ton concrete launchpads were stowed below. Thousands of troops from three missile regiments were crammed below decks and ordered to remain there for the entire voyage. They were permitted on the weather decks only after dark. The overall missile division commander, Major General Igor Statsenko, was deeply concerned that his troops and cargo would be discovered prior to their arrival in Mariel.[34]

In early September, two Soviet cargo ships, *Indigirki* and *Aleksandrovsk*, sailed from Severomorsk with the first shipments of nuclear warheads for the missiles that were already in Cuba. The warheads were stowed in the ship's main deck superstructure. *Indigirki* carried eighty cruise missile warheads for the Komar class missile patrol boats, six nuclear warheads for the naval Il-28 medium-range bombers, and a dozen atomic warheads for the 25-mile Luna coastal defense missiles. *Aleksandrovsk* carried twenty-five atomic warheads for the R-14 intermediate-range missiles, which remained in the ship's hold in the port of La Isabela, awaiting the missiles. The R-14 missile had a range of more than 2,000 miles carrying one nuclear warhead. These two ships made their eighteen-day transit and were sighted and photographed by U.S. and British surveillance but caused no alarm.[35]

The orders to the masters of *Indigirki* and *Aleksandrovsk* read, "Regarding the self defense of your ships . . . during the transport of special cargo you have been equipped with two 37mm, automatic antiaircraft guns with 1,200 rounds for each. Open fire only in event of an attempt to seize or to sink your ships, and report the same attempt simultaneously to Moscow."[36]

U.S. intelligence analysts had made and distributed Soviet crate-recognition guides to the fleet to assist ships and aircraft in identifying the military hardware, based on the size and the shape of the crates. Some Soviet cargo ships were designed with large cargo hatches into which they loaded mechanized vehicles, tanks, and the concrete-slab launch platforms for the missiles. In order to minimize the chance that the United States might discover the construction of these platforms in Cuba, they were prefabricated, shipped in merchant hulls, and off-loaded at night in Cuban ports. The entire plan called for rapid construction and erection of the missile sites, since it was inevitable that they would eventually be identified by U.S. reconnaissance aircraft. All military personnel, including the commanding general of all Soviet forces in Cuba, Issa Pliyev, and his staff, traveled in civilian clothes, using circuitous routes to hoodwink the Americans. The general and his staff flew to Conakry, Guinea, to have their passports stamped before resuming their flight to Havana.

It was not too difficult for Western intelligence analysts to recognize the difference between dark-complexioned Cubans and their pale, mostly blond Soviet mentors. Cubans nicknamed the Soviets "night crawlers" for their proclivity to move and build only under cover of darkness. Often, when the Soviet ships were challenged by Western navies and asked about their destination and what cargo they carried, the Soviet skippers' stock answer was "agriculture equipment and medical aid to Havana." In all, the operation had been hastily organized and the cover thinly applied. In fact, the Soviets' deception was at times laughable to Western analysts, but not for long. The threat of confrontation left little time for humor.

On September 4, the United States protested the construction of air defense missile sites and the delivery of the cruise missile boats. Being unable to hide the magnitude of its military shipments any longer, Moscow issued a communiqué claiming "the right to extend military assistance to Cuba while danger of invasion persists."[37]

By mid-September, the nuclear warheads for the Soviet R-12 medium-range Sopka missiles were in place in Cuba. From then on, the USSR had at its disposal a squadron of six nuclear-armed Il-28 bombers and three battalions of Luna short-range missiles (six launchers, twelve missiles, twelve special warheads, and twenty-four conventional missiles). On September 8, Defense Minister Malinovsky had authorized the commander of the group of Soviet forces in Cuba to use these new weapons at their discretion against an expected American landing:

> When the destruction of the enemy is delaying further actions and there is no possibility of receiving instructions from the USSR Ministry of Defense, you are permitted to make your own decision and to use the nuclear means of the Luna, Il-28 or FKR-1 as instruments of local warfare for the destruction of the enemy on land and along the coast in order to achieve the complete destruction of the invaders on the Cuban territory and to defend the Republic of Cuba.[38]

To further complicate the situation, an armed Cuban exile group called Alfa-66 apparently unilaterally attacked a British and a Cuban cargo ship off the northern coast of Cuba.[39] Soviet ships were now susceptible to harassment by irregular elements. On September 27, the following message was transmitted to all ships in the Soviet Maritime Ministry:

> In case it is impossible to defend against armed attack of your ship by foreign personnel boarding your ship by force, the head of the embarked military unit will destroy all documents aboard that contain military or state secrets. Upon threat of seizure of your ship by foreign ships the master and head of the embarked military unit will resist the attack and boarding, and if it becomes necessary, to sink the ship using all means to save the crew in accordance with directives of the Maritime Fleet.[40]

Another twenty-six Soviet merchant ships were now en route to Cuba, and a CIA source within the Cuban navy reported that more torpedo boats and two types of antisubmarine ships were expected to arrive in Cuba in late 1962 and early 1963.[41] The CIA was also reassessing its estimates on a suspected missile site at Banes, Cuba. It concluded that the seaward-orientated site, located about 300 feet above sea level, could probably launch cruise missiles against target ships at fairly close ranges: SS-N-1 or SS-N-2 missiles with effective ranges of 20 to 30 and 10 to 15 nautical miles, respectively, were likely to be used. But the larger anticarrier SS-C-1, with an estimated range of 150 to 300 nautical miles, could not be excluded.[42]

On September 16, 1962, McCone insisted on carefully studying the prospect of the USSR secretly importing several Soviet MRBMs (medium-range ballistic missiles) and placing them in Cuba. The MRBMs could not be detected if Cuban defenses denied U.S. intelligence permission to fly over the island.[43] The CIA director did not wish to be overly alarming, but he now calculated that a total of 4,200 Soviet military technicians had been brought to Cuba. Since late July, Soviet passenger ships had made nine unpublicized trips, with two more vessels en route.[44] The CIA director also estimated that the current Soviet deliveries had substantially improved Cuba's air and coastal defense and would tend to ensure the continuation of the Castro regime, "with the likely consequence that it would become more aggressive in fomenting revolutionary activity in Latin America." At this stage, however, McCone saw the establishment of a Soviet submarine base in Mariel, Cuba, as even more probable than the anticipated installation of land-based medium- and intermediate-range ballistic missiles.[45]

On September 20, the CIA was back to studying the ballistic missile option. Castro's personal pilot, Claudio Molinas, had said on September 9,

"We have forty-mile-range guided missiles, both surface to surface and surface to air, and we have a radar system which covers, sector by sector, all of Cuban air space and [beyond] as far as Florida. There are also many mobile ramps for intermediate range rockets."[46]

More reports were also coming from agents who had apparently seen trailers similar to SS-N-4 missile tractors.[47] Action was urgently needed. On October 4, Robert Kennedy expressed the president's "concern about progress on the Mongoose program and [felt] that more priority should be given to trying to mount sabotage operations." In view of the "meager results," he urged "massive activity." They considered several options, such as sending navy aircraft to fly low above the Caribbean island, or mining Cuban waters. It was pointed out that "Non-U.S. attributable mines, which appear to be homemade, were available and could be laid by small craft operated by Cubans." The attorney general concluded the meeting on a consensus: "All efforts should be made to develop new and imaginative approaches to the possibilities of getting rid of the Castro regime."[48]

Operation Kama

The Soviet navy's role in the Cuban Anadyr plan was called Operation Kama and was headed by the submariner Kontr Admiral Leonid Rybalko. Its objective was to move a squadron of warships with eleven submarines to Cuba, including seven Project 629 (Golf class) diesel-powered submarines, equipped with new intermediate-range, atomic-tipped missiles. The surface combatants included two gun cruisers of the Chapayev class, two missile destroyers, and two squadrons of mine-warfare craft. These submarines and their supporting surface warships and logistics ships were to make the deep-water port of Mariel their permanent home base, 25 miles west of Havana. Then the orders were changed, as Captain Ryurik Ketov recalled:

> In mid-1962, the 20th Squadron (*Eskadra*) of diesel-electric submarines was created for permanent basing in Cuba. The squadron, to be based at Mariel, Cuba, consisted of the tender *Dmitri Galkin*, a division of missile-carrying submarines, and one brigade of torpedo-carrying submarines. As was typical, as the world political climate changed, so too did the views in the highest echelons of the government and navy regarding the shape and mission of this submarine force. Thus, the force was reduced from the original scheme described above to a brigade of torpedo-carrying boats supported by the *Dmitri Galkin*, and finally—on

the eve of departure—to a mere four Foxtrot submarines. The mission set before us changed accordingly—from permanent basing in Mariel, to temporary basing on tender *Galkin* and—a week before departure—to simply a transfer to Cuba. It is notable that the initial arrangements were for the boats to undergo the passage openly on a designated route, but the final operational orders called for the movement to be covert in nature. . . . Most importantly, the boats were to be equipped with "special-weapons" i.e., nuclear torpedoes. All of these orders were a source of anxiety not so much for the staff officers of the brigade, as for the commanders of the submarines. This was strongly felt [extremely worrisome to the crew should it have become necessary] by the personnel aboard the boats when it became necessary.[49]

The brigade of four long-range, diesel attack submarines deployed on October 1 for the unusual mission.

On October 16 and 18, the CIA had reached the conclusion that Soviet MRBMs were in the early stages of deployment in the Sierra del Rosario Mountains about 50 nautical miles southwest of Havana. The dimensions of the trailers indicated that either the SS-3 or the SS-4 ballistic missile systems were involved. Intermediate-range, 2,200 nautical miles (SS-5) ballistic missiles were also expected to be installed near Havana. But Operation Anadyr had been successful. The ballistic missiles had reached Cuba without being detected. The only certainty was that agents on the ground in Cuba had reported, "Offensive missile systems [were] being introduced into Cuba primarily through the Port of Mariel." But the CIA acknowledged that it had "[no] evidence from shipping coverage or other sources to indicate definitely when the missile units arrived in Cuba."[50]

While the Soviet naval forces steamed south, U.S. Navy destroyers and aircraft carriers in Atlantic fleet antisubmarine warfare Hunter-Killer groups prepared to participate in the sudden blockade announced by President Kennedy on October 22. Four tense encounters resulted, with each of Rybalko's submarines being challenged by the U.S. antisubmarine Hunter-Killer groups. Three of the four submarines were forced to surface after long chases, to charge their batteries. During one encounter, between submarine B-130—which was suffering mechanical failure of all three diesel engines—and the destroyer USS *Blandy*, a misconstrued maneuver by the destroyer's commander caused his Soviet adversary to panic. The submarines were armed with "special" (that is, nuclear) weapons. But the captain could not make the decision to fire this nuclear-tipped torpedo without the consent of the officer who had been especially entrusted by Moscow to supervise the weapon. Only the unshakable patience and maturity of the four Soviet skippers and the U.S. submarine hunters kept the two nuclear superpowers from blundering into an atomic holocaust.

Yet the unsung heroes of the crisis might have been the handful of quiet young men hailing from various regions of the vast Soviet Union who performed their duties aboard each submarine. The most deserving of hero status were the English-speaking radio monitors who listened carefully to the chatter of U.S. sailors and pilots. This was their only source of true information about what was transpiring. Moscow told them nothing. These men, in their persistent scanning, gleaned that a state of war had not yet been reached, and that although a tense standoff existed, there was hope for avoiding what seemed to be inevitable war. Using antennas thrust toward the ocean's surface from their cramped, steel-encased compartments, they often experienced heat exhaustion in their oxygen-starved environment. These men scanned the airwaves and plucked invaluable nuggets of information from America's military communications and commercial radio and television frequencies to determine the true state of the conflict. Had it not been for the information these men gathered from the airwaves, their commanders might have blundered into a needless confrontation with U.S. ships.

The crews on the four Soviet submarines suffered in the deplorable heat and uncertain conditions, with little direction from Moscow. Yet their efforts proved fruitless for them in the end. They returned to a cold welcome and a reprimand in their home port and were cast as scapegoats for the failed mission to move covertly to Cuba.

Three Soviet ships that the United States suspected of being ballistic missile carriers continued to retreat toward the USSR, as did thirteen other dry-cargo ships that had initially been carrying military cargoes to Cuba. Six dry-cargo ships, however, were still heading toward Cuba.[51] In fact, the U.S. forces in the blockade never boarded a single Soviet merchant ship. They stopped and inspected only one ship, the ex-U.S. *Liberty* ship SS *Marucla*, which had been charted by the Soviets but was flying the flag of Lebanon.

On October 24, 1962, Cuban intelligence gave Fidel Castro its estimates on the chances of a U.S. invasion. Capitan Pedro Luis, from the Estado Mayor General 2nd Bureau, did not anticipate an imminent U.S. landing and correctly assessed that the first echelon would be made by two to three army or marine divisions, which would be transported by 120 to 130 ships and would be detected in time by Cuban—and Soviet—intelligence.[52]

Banking on his past success in strafing the Bay of Pigs invaders with antiaircraft batteries, Fidel Castro again planned to distribute his guns in three distinct groupings to use them against low-flying aircraft and amphibious forces.[53] On the most tense day of the crisis—October 26—Castro wrote to Khrushchev that he considered aggression to be imminent within the next twenty-four to seventy-two hours, although an invasion was in his view less likely than a massive air attack "against certain targets." Castro went on to suggest to the Soviet leader a nuclear strike against the United States, should it attempt to invade Cuba. He concluded by expressing his infinite gratitude and recognition to the Soviet people.[54]

Upon receiving this information on October 27, the Soviet general staff authorized its forces in Cuba "to use all available air defense resources" but "prohibited the use of nuclear weapons, missiles, cruise missiles and aircraft without orders from Moscow."[55] On that same day, the CIA noted that Cuban military forces continued to be mobilized at a high rate but remained under orders not to take any hostile actions unless attacked.[56]

Who Won?

The secret annex to the October 28, 1962, agreement between Premier Nikita Khrushchev and President John F. Kennedy contained a U.S. pledge not to invade Cuba and Soviet assurances to keep secret an agreed-on removal of the U.S. medium-range missiles already in place in Turkey and Sicily. This, and the fact that no U.S. invasion took place, is used by the Russians as a rationale that the whole affair was a Soviet victory. Completely absent in the Russian reasoning is the fact that the U.S. medium-range missiles were already slated for withdrawal from Europe. The MRBMs became redundant because of the growing fleet of Polaris submarines with nuclear-missile-launching capabilities that had taken to the seas.

The Soviets held several conferences in subsequent years in Moscow and Havana to commemorate their success, inviting senior participants of both sides. Retired Soviet general Issa Pliyev, the 1962 commander of the Soviet forces in Cuba, and general of the army Anatoli Gribkov, the originator of plan Anadyr for Khrushchev, reiterated that at the time, the rules of engagement allowed General Pliyev to launch nuclear weapons without Moscow's permission had the United States invaded Cuba. In 1989, Defense Minister Dmitry Yazov—who, in 1962, commanded one of the mechanized rifle regiments deployed to Cuba—told Captain Peter Huchthausen in an interview that given the same situation, each senior Soviet officer in Cuba agreed that he would have launched atomic weapons had the United States attacked any Soviet forces.[57] Yazov was jailed following the abortive 1991 coup against Soviet president and Communist Party chief Mikhail Gorbachev.

In retrospect, Captain Vtorygin recalled his impressions from that time when he was serving as an assistant naval attaché in Washington:

> The Cuban crisis was the worst of all the Cold War crises. Being atta-chés, we knew nothing about the deployment of rockets. We learned about it when Minister A. I. Mikoyan flew to Cuba and on his way back came to Washington where he spoke to us about the situation, the troops in Cuba, and the [equipment]; only at this moment, we understood the scale of our involvement in this event. Mikoyan flew to Cuba to convince Castro to accept the Soviet proposals to withdraw the Soviet rockets from Cuba. Castro was against it. It was very unpleasant. At the begin-ning he refused our proposal. He depended on that weaponry. It took several days for Mikoyan to convince him. When his mission was over Mikoyan made a stop in Washington. I don't know why, but he released the information to a limited number of people in the Soviet Embassy for us to understand what was going on. I attended that conference.[58]

The secret deal between Moscow and Washington had not yet solved all of the issues related to Operation Anadyr. On November 5, 1962, CIA director McCone briefed the National Security Council on the information communicated by Moscow. The Soviets claimed to have sent forty-two missiles to Cuba, whereas the United States had estimated that there were forty-eight, thirty-three of which had actually been seen. The Soviets had agreed to provide all of the shipping information so that the United States would be able to count the missiles leaving Cuba, and McCone seemed satisfied.[59]

But the question remained about the other Soviet weapons brought to the Caribbean theater by Operation Anadyr. President Kennedy wanted a guarantee that no Soviet submarine base would be built in Cuba. He also wanted the nuclear-capable Il-28 bombers taken off the island.[60] At about the same time, on November 5, Malinovsky had telegraphed General Pliyev to consult with him on the final destination of these weap-ons. He suggested transferring to the Cubans within eight to ten months the coastal and air defense missiles, the Komar missile boats, and the various other ships and aircraft. Malinovsky then added, "With regard

Missile transporters withdrawn from Cuba in November 1962 were made clearly visible onboard *Ivan Polzhunov* to comply with the U.S.-Soviet agreement.

to warheads for the 'Luna' coastal defense missile and Il-28 bombers, so far their withdrawal has not been discussed. They should be left in Cuba under your command."[61] At that point, the Soviets were planning to keep their tactical missiles and nuclear warheads on the island. The United States would be less likely to find out because Secretary of State Dean Rusk had successfully opposed increased reconnaissance flights for political reasons, leaving the surface navy to conduct most of the verification process by monitoring merchant shipping and monitoring communications.[62]

By the end of November, after reconnaissance flights had resumed, carrier- and shore-based naval aircraft reported the dismantling of Soviet missile batteries ashore and verified the presence of missile tubes on Soviet ships departing Cuban ports.[63] Unaware of the remaining tactical nuclear missiles issue, the United States' main concern remained the Il-28 bombers. Washington was not yet ready to commit to not invading Cuba.

During a meeting with the president on November 22, the Joint Chiefs of Staff argued that the quarantine should be lifted in exchange for the removal of the Il-28s. Should the quarantine fail in obtaining their removal, the chiefs called for "taking them out by air attack." They also expressed concern about the large stocks of modern army equipment that could be used against Guantanamo "or any other invasion attempt," which clearly indicated that this option was still open, despite the promise made to Moscow.[64]

Moscow's main concern was of a different nature. On November 22, the Soviet ambassador in Havana sent an urgent message to Mikoyan, the Kremlin's special negotiator who was now in the United States:

> The directive from the Cuban Foreign Ministry to their representative in New York, Lechuga, contained the following phrase: "we have tactical weapons, which we should keep." . . . It would be advisable for our Cuban friends to correct urgently the directive given to Lechuga in that part, and to tell him clearly that there are no nuclear weapons in Cuban hands. It is important to give such directive to Lechuga immediately so that he will not be able to make some statement in a careless conversation which could be eavesdropped upon.[65]

U.S. intelligence apparently remained unaware of these additional tactical nuclear weapons. The Soviet Union finally consented to remove all the Il-28 bombers, and Kennedy made good his promise not to invade Cuba. And this may have been Operation Anadyr's longest-lasting success, from both a Soviet and a Cuban perspective.

9

TRANSITION TO WAR: VIETNAM
1961–1975

On December 14, 1960, the National Security Agency Director, Vice-Admiral Lawrence H. Frost, instituted a SIGINT (signal intelligence) readiness condition Bravo (the second highest) for all U.S. electronic eavesdropping stations throughout the Far East. Electronic detection of a Soviet airlift to support the communist factions in Laos had sparked the alert issued from Washington. This subsequently led the Kennedy administration to file a diplomatic protest to Moscow. There was genuine concern in Washington that either the Soviets or the Chinese Communists or both would intervene in the conflict in Laos and bring former President Dwight Eisenhower's domino theory to fruition. If Laos seemed to be the strategic key to Southeast Asia, communist infiltration in South Vietnam, under the guidance of North Vietnamese leader Ho Chi Min, soon became another concern for President Kennedy.[1] It wasn't long until the North Vietnamese strategy entered the maritime arena.

The initial maritime infiltration began in 1963, when North Vietnamese navy rear admiral Tran Van Giang organized a maritime infiltration unit code-named "Group 759." Its 603rd Special battalion, located at Haiphong, was assigned to move military personnel and supplies down the east coast of South Vietnam. A second organization, the Communications Section of the Lao Dong Party's Research Office, transported Communist Party agents along South Vietnam's 1,200-mile coastline while hiding amid the large fleet of local fishing boats that normally worked in those waters. U.S. signal intelligence found it almost impossible to track sea-born infiltration craft. A combination of communication monitoring and other human intelligence sources, primarily from captured enemy crews and documents, were not nearly sufficient to keep pace with the influx of agents to the south.

In December 1961, the U.S. Navy established a patrol line to seaward of the 17th parallel demilitarized zone. This cordon was covered continuously by five ocean-going minesweepers and several army and navy reconnaissance aircraft. Despite the U.S. commander in chief Pacific Fleet's initial skepticism over the seriousness of the North Vietnamese maritime infiltration, Secretary of Defense Robert McNamara ordered the patrol line extended south and westward in the South China Sea along Cambodia to Phu Quoc Island. By March 1962, the navy again concluded that their counterinfiltration efforts had not been productive: "From results attained to date it must be concluded that the patrols have not been effective in capturing infiltrators if significant infiltration is indeed taking place, although the patrol's presence may have discouraged attempts."[2]

Covert Support for South Vietnam: The Real Story of the Gulf of Tonkin Incident

Recently declassified information gives a new view of exactly what transpired in the Gulf of Tonkin in the summer of 1964. National Security Agency historian Robert J. Hanyok's study titled *Spartans in Darkness: NSA, American SIGINT, and the Indochina War, 1945–75* was partially declassified in December 2007. It provides for the first time a complete SIGINT version of the events in the Gulf of Tonkin between August 1 and 4, 1964. His research is based on the discovery of 122 SIGINT documents together with watch center notes and oral history interviews. His study includes message traffic among the various SIGINT and military command centers involved in the Gulf of Tonkin incidents. This new information shows that a SIGINT blunder led President Johnson to commit the United States into its bloodiest conflict since World War II.[3]

Similar to its apparent inability in earlier years to control the infiltration by sea, in 1964 the new regime in Saigon was not faring well in thwarting the overall North Vietnamese–sponsored Vietcong penetration south. The new Saigon government had taken over after President Diem's assassination. Secretary of Defense McNamara and the newly inaugurated President Lyndon B. Johnson were determined to provide covert and effective support to their South Vietnamese ally. They ordered the escalation of military pressure on Hanoi. At the beginning of 1964, the Department of Defense assumed control of all covert missions in the area from the various agencies that had been involved earlier in separate and uncoordinated operations.

A new plan called for twelve months of selective attacks that would be "attributable" to the South Vietnamese to "convince the Democratic

Republic of (North) Vietnam leadership that their continued direction and support of insurgent activities in the RVN and Laos should cease." The plan further called for "unspectacular" harassment attacks to oblige Hanoi to allocate major forces in response. The next "attritional" level targeted important military and civil installations whose loss could cause "temporary immobilization of important resources," which, in turn, might provoke popular unrest in the north. The third "punitive" level was intended to damage or destroy critical economic or military facilities. The final step of the plan would authorize an aerial bombing campaign to force Hanoi to abandon its support of the Vietcong. The same plan initiated new aggressive maritime operations using fast Norwegian-built Nasty class patrol boats (PTFs), provided and directed by the U.S. Navy. These PTFs carried South Vietnamese special forces troops, augmented by Nationalist Chinese and South Korean mercenaries. The Nasty craft were further tasked to shell North Vietnamese radar stations at night and, by landing saboteurs, to spread terror on the shoreline, cut bridges, and otherwise damage the North Vietnamese coastal infrastructure.[4] However brilliant such operations sounded, the North Vietnamese proved able to react successfully. Most damaging to the operations was the lack of reliable intelligence in advance on the targeted objectives.[5]

As a result of the paucity of concrete intelligence, Secretary McNamara ordered U.S. naval forces to conduct special intelligence-gathering patrols in support of the raids on the North Vietnamese coast. "DESOTO" was the cover name given to a navy and marine signal intelligence–collection program, in which naval SIGINT direct-support units were placed on board U.S. destroyers patrolling the Vietnamese coastline to collect intelligence. DESOTO patrols had been initially designed to assert the right of freedom of navigation in international waters. A small communications van was usually lashed to the ship's helicopter deck to house intercept positions for personnel. They analyzed voice and manual Morse communications and recorded radar emissions. The assigned detachment of naval security group personnel was linked in communications with the other monitoring stations afloat and ashore in the area.

In mid-January 1964, the U.S. Navy command in Vietnam tasked the USS *Radford* (DD-446) to monitor the North's ability to resist the navy's projected commando operations. However, the mission was canceled so that it wouldn't interfere with a planned South Vietnamese raid. Then in early July, the U.S. forces command again requested intelligence on Hanoi's defenses in those areas targeted for July raids.[6] In mid-February, General Harkins, the commander of the U.S. Military

Assistance Command, Vietnam, also requested radar photography of the North Vietnamese coasts. The destroyer USS *John R. Craig* (DD-885) was dispatched to the Gulf of Tonkin, where she was directed to remain 15 nautical miles off the Chinese coast and 4 nautical miles off the North Vietnamese shore. Fog restricted visual observation, but precious electronic intelligence was collected. The patrol went unchallenged, although a Chinese Kronstadt class patrol boat and an unidentified plane shadowed the destroyer. The *Craig* returned to Taiwan in mid-March, as Communist China issued its 280th "serious warning" asserting that the *Craig* had violated its territorial waters.[7]

In mid-July, the destroyer USS *Maddox* (DD-731), under the command of Squadron Commander Captain Robert Herrick and ship's commanding officer Commander Dan Ogier, was tasked to locate all coastal radar transmitters along the North Vietnam coastline and to scout maritime infiltration routes. The *Maddox* was ordered to stay 8 nautical miles from the North Vietnamese coastline, but only 4 nautical miles from any of the islands where attacks by South Vietnamese commandos were under way.[8]

It Has Been Decided to Fight the Enemy Tonight

SIGINT intercepted by U.S. forces soon revealed that Hanoi was determined to defend its coastline against the nightly PTF raids and DESOTO intelligence patrols. On June 8, the NSA reported that the level of North Vietnamese tactical radio communications had increased almost fourfold from the previous period in May. The NSA naturally assumed this to be in reaction to the attacks and the intelligence-gathering patrols along the coast. The SIGINT support to these navy operations was code-named "Kitcat." Thus, in reaction to the indications of opposition, the SIGINT support to the navy was intensified. The 660 military cryptological personnel positioned in South Vietnam were reinforced to 790. The Naval Security Group at San Miguel, Philippines (called USN-27), was also tasked to undertake additional coverage of North Vietnamese naval communications. During this period, the local U.S. Air Force Security Service also focused primarily on naval communications between the coastal surveillance posts and the patrol boats. In addition to the special detachment aboard the *Maddox* (called USN-467N), a marine SIGINT detachment (USN-414T) joined the U.S. Army Security Agency intercept site at Phu Bai (USM-626J).[9]

The first significant raid took place on the night of July 30–31, when South Vietnamese commandos attacked the radar station on Hon Me

Island, off the central coast of North Vietnam. After being driven off by North Vietnamese defenders, the patrol boats machine-gunned coastal defenses and retreated, passing close to the USS *Maddox*. Meanwhile, intelligence gained by SIGINT showed a single North Vietnamese Swatow class patrol craft T-146 communicating tracking data on the destroyer *Maddox*. The Chinese-built T-146, which was capable of making 44 knots, was equipped with a surface-search radar called Skin Head. The craft was apparently acting as a radio relay between the coastal command centers and the Soviet-built P4 torpedo boats, which had a top speed of 50 knots and were stationed in the nearby ports.

Shortly after 11 a.m. on August 1, SIGINT reported a North Vietnamese message sent to an unidentified patrol boat, announcing, "It has been decided to fight the enemy tonight," and further asking for the "enemy's position." The North Vietnamese patrol boats were ordered to concentrate near Hon Me Island later in the morning of August 2. The NSA issued a warning of a possible attack, but the USS *Maddox* was not on the warning distribution list.[10] At 11:44 a.m. local time, the marine SIGINT group attached to Phu Bai intercepted a message from the Swatow T-142 to its base reporting on a pending torpedo attack: "Have received the orders, 146 and 142 use high speed to parallel the enemy track following a launch of torpedoes." Intercepts also indicated that two more vessels, T-166 and T-135, were involved in "Tracking and following the enemy."

A half hour later, the other marine SIGINT detachment at San Miguel, Philippines, intercepted a retransmission of the same message. Meanwhile, the destroyer *Maddox* had sighted three and then five North Vietnamese patrol craft. At 1 p.m. local time, San Miguel intercepted a new message ordering the T-146 to "leave squadron 135." That squadron consisted of three P4 torpedo boats, which previously had been wrongly identified as a single boat. They intercepted the further order: "And turn back to the path of the enemy." One hour later, the San Miguel site intercepted a message that seemingly called off the attack: "Order 135 not to attack by day." Six minutes later, the marine detachment deciphered another message identifying the "Enemy" as a large ship bearing 125 degrees, at a range of 19 miles, speed 11 knots. Thereafter, the North Vietnamese were probably given the order to attack with the intercepted but garbled phrase "Then determine."

The conflicting orders from two different command centers at Port Wallut and Haiphong gave the impression to the listening monitors that the North Vietnamese had lost control of the situation. The North

Vietnamese patrol boats were now moving at 45 knots, with a 25-knot advantage over the speed of the *Maddox*. Forewarned by intercepts of the Skin Head radar, the *Maddox* increased her speed from 11 to 25 knots, obliging her foes to run another thirty minutes before reaching an attack position. By 2:30 p.m. local time, the *Maddox* was at battle stations. Ten minutes later, her commanding officer requested air cover from the carrier USS *Ticonderoga*, located 280 miles to the southeast. Four F-8E Crusaders were launched from the ready alert fighter air patrol, while the destroyer USS *Turner Joy* (DD-951) sped toward the scene. At 3:05 P.M. local time, the *Maddox* opened fire. Undeterred, ten minutes later the boats closed in for attack and were hit by the *Maddox*'s fire and by the Crusaders' strafing. Patrol craft T-339 was left dead in the water, with four killed and six wounded, while T-333 and T-339 slowly retreated with significant damage.[11]

On learning of the attack, President Johnson tried to downplay the incident. He issued a warning to Hanoi that "more unprovoked actions would result in grave consequences." Washington had underestimated the response to its covert operations against the North Vietnamese coast, and Secretary McNamara viewed the incident as a "miscalculation" or "the impulsive act of a local commander." SIGINT facilities remained on full alert, with the brunt of the effort handled by the navy unit in the Philippines and the army and the marines at Phu Bai.[12]

Nothing but Black Sea and American Firepower

Despite the increased tension on the North Vietnamese shore, the United States decided to launch another covert operation on the night of August 3–4, which coincided with a reinforced DESOTO intelligence-gathering patrol. Again, a four–Nasty class PT boat expedition shelled a radar site—this time at Vinh Son—while the *Maddox* and the *Turner Joy* were ordered to remain clear of the area. The two ships steamed to the vicinity of the island of Hon Me. Meanwhile, the North Vietnamese were still trying to recover their missing boats from the earlier action of Squadron 135. At the same time, T-142, a Swatow class patrol boat, had taken a position to the north of the two destroyers and started to report on their movements.[13] At this point, the SIGINT station at Phu Bai misinterpreted the North Vietnamese search-and-rescue operation as another attack in the offing. At 6:15 p.m. local time, the Phu Bai SIGINT site issued a warning stating, "Possible DRV naval operations planned against the DESOTO patrol tonight 4 August." Forty minutes later, Phu Bai reported "imminent plans of DRV naval action possibly against DESOTO mission." The CINCPAC (the joint commander in chief Pacific) was then warned of an imminent attack, and the two destroyers were ordered to steam southeast at 20 knots. The North Vietnamese Swatow class T-146 reported to Haiphong that the tugboat *Bach Dang* should avoid the DESOTO mission. But again the SIGINT analysts translated the sentence as "T-146 supply fuel for the 333 to enable her to return to operations." The salvage operation was not understood by the SIGINT sites.

Twenty minutes after the first warning, the *Maddox* reported two surface and three air radar contacts 40 miles to the northeast. At 9:08 p.m. local time, the *Maddox* detected another series of intermittent radar returns that it evaluated as a tight formation of patrol boats only 15 miles away to the southwest and coming closer. Twenty-eight minutes later, the *Maddox* detected a single craft coming toward it at 40 knots, while the *Turner Joy* also reported a moving object. Three minutes later, the object appeared to make a sharp turn 6,200 yards from *Maddox*. It seemed too far away for a torpedo launch, but the *Maddox*'s sonar detected the acoustics of a running torpedo and convinced the *Maddox* Combat Information Center that it had been attacked. Both destroyers were now firing rapidly at their radar returns, which soon disappeared. The *Maddox* and the *Turner Joy* concluded that two of the boats they presumed had attacked them had been sunk. The destroyers' five-inch star shells illuminated the area where the supporting aircraft from the USS *Ticonderoga* had been sent in vain. They could not find any boats.

James Stockdale, then a navy pilot at the scene, who had "the best seat in the house from which to detect boats," saw nothing. "No boat, no boat wakes, no ricochets off boats, no boat impacts, no torpedo wakes— nothing but black sea and American firepower."[14] In fact, the *Maddox*'s gun director was never able to acquire a target. It had been determined that the distinctive high-speed sound of torpedoes was caused by the two destroyers' rudders reflecting the turbulence of their own propellers. Captain Herrick, the squadron commander aboard *Turner Joy*, expressed his doubts in his report about the action: "Entire action leaves many doubts except for apparent attempted ambush at beginning . . . never positively identified a boat as such." The carrier *Ticonderoga* also reported, "No visual sightings of any vessels or wakes other than *Turner Joy* and *Maddox* visible from 2,000–3,000 yards."[15]

In Washington, the SIGINT carrying the news of an imminent attack had prompted Secretary McNamara to call the president at 9:25 a.m. Forty-five minutes later, the destroyers' news that they were actually under attack seemingly came as a confirmation. Within four hours, the president approved a retaliatory strike to be carried out at 6 a.m. local time, August 5. The subsequent doubtful reports led the president to seek confirmation. Then came the attempted translation of a second SIGINT intercept, which appeared to be a North Vietnamese after-action report: "We shot at two enemy airplanes and at least one was damaged. We sacrificed two comrades but all are brave and recognize our duty." This transcript was soon followed by what appeared to be a second portion: "They did continuously see with their own eyes enemy aircraft fall into the sea. Enemy vessel perhaps is damaged. Report this news to the mobilized unit."[16]

After speaking with Admiral U. S. Grant Sharp, the Pacific commander in chief, Secretary McNamara discussed the transcript with the Joint Chiefs of Staff. The latter SIGINT intercept had convinced the president and the joint chiefs that this was hard evidence in the form of a North Vietnamese after-action report. At 7 a.m. on August 5, the commander in chief Pacific ordered the USS *Ticonderoga* to launch massive air strikes against North Vietnamese naval facilities.

The Consequences of a Mistaken Transcript

The distorted translations and the amalgam of two SIGINT intercepts had convinced the Johnson administration to start a war. Unfortunately, the original decrypted Vietnamese-language version of this intercept is missing. President Johnson later admitted in his memoirs that the real meaning of "comrades" was unclear: "Our experts said that this meant either two enemy boats or two men in the attack group."[17] The experts themselves probably sensed the magnitude of their mistake. Hanyok explains:

> The NSA version says they lost two boats. I'm not a Vietnamese linguist but I did look up in the code charts for this system—and I did look in dictionaries. (*BDongchi*), which means comrade, is a noun used specifically only in a personal context. In other words, when you talk about a comrade, you are talking about people. I'm not doing some sort of metaphorical or poetic license thing saying comrade-boat. I'm saying comrade. The word for boat is (*Btao*), which is often abbreviated to "t" when you see it in messages—you'll see "t" followed by a number, which basically means boat so and so. They're not even close to each other, one begins with a five, and the numbers aren't even close to get a Morse garble. So the question has remained in my mind and those of people who have looked at it. One, how did you get from comrades to boats? Two, what happened to all the original Vietnamese [texts]?[18]

Bogeys and Skunks Again

The Gulf of Tonkin incidents were to be replayed on September 18 of the same year when another DESOTO patrol consisting of the USS *Morton* (DD-948) and the USS *Richard S. Edwards* (DD-950) acquired radar contacts shadowing them. This time, SIGINT did not report any hostile intent by the North Vietnamese. Intercepts indicated that the North Vietnamese had warned their ships to be on the alert for "provocations" by the Americans and "to avoid provocations." Shortly after detecting the shadower, the two destroyers increased speed and maneuvered to escape the boats that were suspected of trailing them. Forty-five minutes later, the destroyers counted five presumed craft on their radar scopes that were trailing them, always matching their speed. The *Morton* opened fire, and for an hour the destroyers expended almost three hundred rounds at the contacts and claimed to have sunk as many as five of the elusive vessels. The Joint Chiefs of Staff ordered a search for debris to confirm the attack. On the following day, the air force and the navy flew vainly above the area, but not even an oil slick was spotted. The DESOTO missions were suspended and never resumed.[19]

The Electronic Wizards and Pilot Survival

On July 24, 1965, during an air strike over North Vietnam, a U.S. McDonnell-Douglas F-4 Phantom was shot down by a Soviet-built SAM-2 surface-to-air missile. This event marked the first success of Soviet missiles in Southeast Asia. Until then, Vietnamese air defense had consisted of MiG-17 and MiG-21 fighter interceptors and radar-controlled antiaircraft guns.[20] U.S. Air Force and naval aviation losses so far had been light. But now it was acknowledged that the only way of dealing with the new antiaircraft threat was to develop airborne electronic warfare systems to neutralize the radar that was used to guide the surface-to-air missiles. Indeed, electronic intelligence and electronic countermeasures became key to survival in the Vietnam War because of the increasing variety and sophistication in the enemy's weapon systems.

ONI (Office of Naval Intelligence) and the Naval Scientific and Technical Intelligence Center were intimately involved in designing the ARM, an antiradiation missile engineered to destroy enemy air defense. This missile would home in on the air defense radar and run right down the beam, destroying the emitting radar.[21] At the same time, great efforts were devoted to gathering intelligence on the Soviet SAM-2 system

(NATO code-named "Guideline"), and the missile tracking *Fansong* radar. One radar transmitter was capable of guiding three missiles simultaneously. Mounting six launchers, the whole system was transported by towed trailers and could be set up in about six hours. ELINT (electronic intelligence), however, had revealed several shortcomings in the system. It took the Guideline missile a full six seconds after launching to be picked up by the tracking radar that would guide it onto the target. Exploiting this six-second delay, the pilot could nose-dive in the direction of the battery as soon as the missiles were launched. Then the pilot of the targeted aircraft would suddenly pull up as hard as possible into a steep climb, flying inside the trajectory of the missile to escape. This evasive tactic would not work when the skies were cloudy and the departing missile could not be seen.[22]

Intelligence Support to the Fleet

The commander of U.S. Naval Forces in Vietnam controlled three U.S. Navy task forces and the Vietnamese navy. His staff had no specific sea craft dedicated to intelligence collection, however, and SIGINT alone proved inadequate to provide timely information on maritime infiltrations.[23] The North Vietnamese and the Vietcong simply did not chatter much on the air.

Despite U.S. and South Vietnamese efforts at interdiction, in 1964, the communists used as many as twenty-six large trawlers, displacing more than sixty tons, to smuggle the agents of a unit named Group 125 southward. The U.S. and South Vietnamese navies responded with Operation Market Time, a coastal screening force made up of destroyers, destroyer escorts, mine-warfare craft, and armed junks, to stem the flow of weapons and supplies to the Vietcong.

Nevertheless, despite the increased surveillance forces, in 1965, it was estimated that the communists received nearly 70 percent of their supplies by sea and only 30 percent by land, a portion of the latter being also shipped by sea to Cambodia. After studying their interdiction efforts with available intelligence, the South Vietnamese navy clearly showed that Sihanoukville in Cambodia was the transshipment point through which supplies were being infiltrated to the Vietcong in South Vietnam. This was despite Prince Sihanouk's vigorous denial. The material was normally off-loaded in Sihanoukville and carried by truck to the Vietnamese-Cambodian border.

In the spring of 1968, a U.S. submarine was stationed off Sihanoukville to check merchant ships entering that port. The rules of engagement, however, required the submarine to remain 15 miles off the coast, and it was therefore unable to verify that the ships had actually entered the harbor.

In the Gulf of Tonkin, however, the interdiction was more successful. Operation Market Time managed to stop the flow of infiltrators, and by mid-1968, the North Vietnamese had been forced to shut down their maritime smuggling operations for more than a year.[24] As naval historian Edward Marolda explained:

> Destroyers, mine warfare ships, Coast Guard cutters, gunboats, patrol craft, shore-based patrol planes, and high-powered coastal radars made it almost impossible for the North Vietnamese to slip one of their munitions-laden, 100-ton supply ships past the Market Time patrol. Allied naval forces destroyed or forced back to North Vietnam all but two of the fifty steel-hulled trawlers that tried to run the blockade between 1965 and 1972.[25]

This success was brought about partly by a breakthrough in SIGINT.

In October 1967, an RC-130 flight (Commando Lance) intercepted a low very high frequency (LVHF) voice communications network, which was providing information on the number of soldiers infiltrating into South Vietnam, including their destinations. Nicknamed the "Vinh Window," this success forced the United States to rely almost entirely on SIGINT for intercepting infiltrators. Collateral information, such as from POW interrogations, was used to fill in gaps. To meet the communist communications, the commander in chief of the Pacific fleet augmented the naval intelligence effort with two technical research ships.

The first of these ships, the USS *Oxford* (AGTR-1), arrived at Subic Bay in May 1965. Before taking its station the following month, the USS *Jamestown* (AGTR-2) soon followed. Both ships were capable of intercepting Vietcong communications south of the Mekong Delta, as well as Cambodian communications. Intelligence-collection missions were conducted by other specialized ships. On November 15, 1966, the USS *Banner* (AGER-1) was harassed by a large number of North Vietnamese fishing boats near Chinese waters. The destroyer *Everett F. Larson* (DD 830) managed to extricate the intelligence ship in time.[26] The *Jamestown* and the *Oxford* spent the next several years working in Vietnamese waters.[27] Other ships were also involved, to provide tactical SIGINT or to rescue downed pilots.

On the opposite side, North Vietnamese air controllers could detect air force and navy strikes early enough and could often vector their MiGs into tactical advantage.[28]

One of the Greatest Intelligence Lapses in the Post–World War II Era

The Tet Offensive, which began on January 31, 1968, changed the nature of the war and forced President Johnson to announce a cessation to the bombing of North Vietnam and his desire to open negotiations. Once again, naval intelligence had been caught by surprise. In reality, the NSA had provided a specific warning six days before the attack. This was not sufficient, however, given the profuse volume of these routine-type alerts received almost daily by U.S. forces ashore or in the rivers of South Vietnam. The intelligence ships *Jamestown* and *Oxford* were too far away in the south to provide effective tactical SIGINT support against the Tet Offensive.[29] Various explanations have been offered for the failure to predict and prevent the Tet Offensive:

- Insufficient control of the waterways by overstretched forces
- Lack of air support to confirm agent reports of infiltration
- Predictable patterns in the U.S. Navy's inspection practices
- Poor language training and inability to make use of the South Vietnamese navy as an intelligence organization to "read the people"

As historian Glen Helm explained, "All source intelligence available to Admiral Veth, then Commander of all naval forces in Vietnam or operating off shore, on 27 January . . . indicated that a major enemy offensive was imminent."[30] Widespread communist attacks in the northern portion of the Republic of Vietnam during the night of January 29–30 were a final warning, which had been ignored in most locations. At his home in Saigon, Admiral Veth believed that the enemy forces would not run the risk of attacking the major cities even during the most important Vietnamese holiday. As Helm concluded, "Had the allied intelligence community placed greater emphasis on intelligence collection, and acted properly on the intelligence that they did receive, the Tet Offensive may never have found its way into the history books as one of the greatest intelligence lapses in the post–World War II era."[31]

Vietnamization and Catastrophe

With the election of Richard M. Nixon and the advent of Vietnamization, the NSA reinforced the South Vietnamese SIGINT organization, however distrustful it was of an ally that had obviously been deeply penetrated by the Vietcong. The Naval Security Group maintained fleet support detachments at Da Nang and Cam Ranh Bay, while the marines remained at Da Nang. The Phu Bai SIGINT site was closed in 1972. The Philippines and Thailand stations were forced to assume a larger share of the total SIGINT effort.[32]

In December 1967, the Red Crown controllers were again actively supporting the naval participation in the resumed heavy bombing campaign chosen by President Nixon as a means to unlock the negotiations. On January 28, 1973, the Paris Peace Agreement officially terminated the U.S. phase of the Indochina War.[33] U.S. SIGINT assets were progressively dismantled and evacuated before the fall of Saigon in April 1975.

Escalation to World War?

Always lurking during the Vietnam War was the concern that the Soviet Union or China might be drawn into the fray. Throughout this period in the Western Pacific, Soviet and Chinese submarine contacts varied in number and location. Until the end of 1968, these contacts were evenly distributed from Vietnam north through the Sea of Japan, mostly in the latter area. In 1969, Soviet submarine activity dropped significantly, and what little remained was concentrated around the Japanese home islands. Most of the submarines that were detected were en route to or from Petropavlosk to operational holding areas and to cover the tracks employed by the U.S. carriers coming to and departing bombing stations in the South China Sea.[34]

Captain First Rank Lev Vtorygin, who served as a Soviet intelligence adviser in Vietnam in the 1980s, thought that the situation during the war was not in danger or escalating:

> The fear of a World War existed during Vietnam. But generally we understood that the Soviet borders were not at risk because practically it was a fight between the Americans and Vietnam. We only assisted Vietnam. For me, it resembled the situation of the Civil War in Spain in 1936. Americans tried to defend their zone of interest in the area, but local people under the leadership of the Communist Party were fighting

to defend their land, not to permit the Americans to dominate in this area and interfere with the construction of the new society under the local communist leadership. We did not think that it could be turned into a big war. When Soviet intelligence ships were attacked by South Vietnam, there were natural casualties. Our ships were involved. We could expect that the other side would not be very polite: usual casualties. There was no competition between China and the USSR. We were defending the interests of our country and armed forces. We later received in exchange very good bases in Cam Ran. Our navy was able to use them. China supported its own interests. Our interests were not coinciding with the interests of China.[35]

Benefits accrued to the Soviet Navy by the use of the air and naval bases at Cam Ranh Bay did not last long after the end of the Cold War when the Russian Navy withdrew from its foreign bases. Ironically, suggestions were made that the United States had been offered the lease of Cam Ranh to counterbalance China's rising power and reassure Vietnam, now an unexpected strategic partner. These suggestions were officially denied by the Vietnamese government.

10

A SUBMARINE IS LOST AND FOUND
1968–1974, 1989

The year 1968 was a bad one for submariners. The Soviet navy lost a Golf class, diesel-powered, strategic missile–carrying submarine in March, and the U.S. Navy lost the nuclear attack submarine USS *Scorpion* in May southeast of the Azores, following an explosion. Its cause is still uncertain at this writing.

What follows is Captain Peter Huchthausen's account of the disappearance, the recovery, and the aftermath of the loss of the Soviet K-129 in March 1968. He served as the U.S. naval attaché in Moscow from 1987 to 1990.

In August 1989, the phone rang in my Moscow embassy office. The unmistakable voice of Soviet captain Valentin Serkov boomed on the line in Russian: "Comrade attaché, your presence is requested at Griboyedovo Street—Admiral Makarov's office at eleven o'clock." As a naval attaché in Moscow, I was accustomed to performing unusual tasks in the name of diplomacy and intelligence, but the encounter that day with Soviet naval officers was the most bizarre of all. I arrived at their navy headquarters, the single tidy-looking building in an otherwise run-down section of Moscow. I bolted up the stairs past a bust of Lenin and two uniformed guards who stood saluting with their heads raised in exaggerated tilts—arrogant, I thought, although I knew they were merely meant to look proud. I was whisked into the admiral's office by a nervous aide who reeked of salami and tobacco.

The mood at the headquarters was cordial. Admiral Valentin Makarov, the chief of the main navy staff, greeted me sternly in the center of a room full of nautical trappings. He was a solidly built man with a seaman's direct manner. A half-dozen blue-clad naval officers stood behind the admiral, dutifully looking as stern as he. The meeting commenced surprisingly without the usual intelligence representative from the defense ministry liaison office, nicknamed the "spook warren" by Americans. We sipped Russian tea from glasses in metal holders with ships emblazoned on the sides.

"Our sailors were treated well in Norfolk last month," the admiral began without smiling. "Yours will be treated the same in Sevastopol." An exchange of warship visits between the two countries had resumed that summer—the first in seventeen years. Our ships would be the first visit to the Crimea since 1921, when U.S. ships helped evacuate remnants of the White forces defeated in the Russian civil war. The pending visit was not the main theme of the gathering. The conversation continued in an unusually relaxed ambience until a Soviet captain burst through the door.

Captain Valentin A. Serkov was the Soviet navy's top law-of-the-sea expert, a long-time negotiator, and an original participant in the Incidents at Sea Prevention talks that had begun in 1972, between the U.S. and Soviet navies. I knew Serkov well, as had every American naval attaché in Moscow over the last twenty years. He was one of the few naval officers still serving who had taken part in the initial negotiations and the signing of the agreement with then secretary of the navy John Warner and Soviet admiral Peter Navoytsev. Serkov's arrival at the meeting came as a surprise and immediately signaled a shift in the room to a stiffer and more legalistic atmosphere. I suspected that Serkov intended to table some terrible mishap that might have occurred at sea, even though incidents at sea between the two navies had decreased notably during that period of improving relations. Instead, Serkov carefully outlined the thrust of recent Soviet press coverage of a U.S. operation conducted fifteen years earlier in the Pacific Ocean called Operation Jennifer.

Serkov abruptly came to the point. "The Soviet navy, acting on behalf of the families of eighty-nine of our sailors lost in a submarine in the Pacific in 1968, requests you confirm that the crew members' remains, retrieved when you salvaged the hull from the ocean bottom, have all been re-buried at sea." I was astounded.

Serkov continued, "We request further that you disclose the exact position of their re-interment to satisfy the families for Russian Orthodox religious reasons." Adding to the sense of intrigue, Serkov said darkly, "The request is warranted, given the American navy's involvement in the sinking of that very submarine." The meeting was over. I agreed to report their request, stood up, grabbed my briefcase, and departed past the same sentries, whom I now felt looked angry. I hurried across Moscow to the embassy.

The meeting and the request for assistance followed the appearance of a sensational article in the outspoken magazine *Ogonyok*, which reported the 1974 salvage by the United States of a Soviet Pacific Fleet submarine. The article, which Serkov had outlined at the meeting, stated that the submarine had disappeared in 1968 and was found and raised during an American recovery effort called Operation Jennifer. The article, citing Western sources, reported that the U.S. government had retrieved the remains of a number of the Soviet submarine crew members and had subsequently buried them at sea.

One of the most remarkable and still unexplained incidents of the undersea Cold War was the loss of that Soviet ballistic missile submarine. The sinking occurred on March 8, when a Soviet Project 629M diesel-powered ballistic missile submarine (NATO name Golf II class) disappeared in northern Pacific waters 16,000 feet deep. The cause of sinking of the submarine, which Soviets called K-129, has never been revealed publicly by either side. Although details of the story have been told in numerous speculative books, the U.S. Navy still refuses to comment officially on the cause.

In 1995, while I was conducting research in Russia, I met the former commander of Pacific Fleet Rybachy Submarine Division Twenty-Two, Kontr Admiral Victor A. Dygalo, in whose division the ill-fated K-129 had been assigned. The admiral related the story of the disappearance on a cold Moscow evening in his tiny, dimly lit apartment. Dygalo, a long-time assistant editor of the Soviet navy digest *Morskoi Sbornik*, spoke in careful Russian and provided his personal account of the story neatly printed in his own hand. To this day, the kindly and deeply religious former submariner believes the U.S. Navy was behind the disappearance of his submarine with all ninety-eight crewmen.

Dygalo's account alleges that K-129 was proceeding on a covert transit to a mid-Pacific patrol station, snorkeling intermittently, when it was intercepted and trailed by a 637 class U.S. nuclear attack submarine. While tracking the Soviet submarine, the U.S. sub maneuvered actively and passed beneath the target at critically short distances to study and photograph the hull. According to Dygalo, a collision occurred on March 8, ostensibly when the Soviet sub turned suddenly and exposed its broad side. The maneuver was not promptly noticed by the trailing American boat. The U.S. submarine then unintentionally struck the bottom of the Soviet submarine's central command post with the upper portion of its sail while at a depth of 150 feet.

K-129 went down after struggling for nearly thirty minutes. Dygalo maintains that U.S. Pacific Fleet command representatives denied that a collision took place. They claimed that the crushing sounds of the sinking Soviet submarine were heard and pinpointed by the U.S. undersea passive acoustic listening arrays, called the Sound Surveillance System (SOSUS), which explained America's knowledge of the precise location of the sinking. Until 1997, the identity of the U.S. submarine involved in that incident was closely guarded as a politically sensitive and potentially damaging secret. (The Soviets claimed that it was the USS *Swordfish*.) In October 1997, a witness, who wished not to be named, told me that he had been aboard the U.S. ship and that the loss indeed resulted from a collision or an accidental weapons launch by an unnamed U.S. Navy ship.

Admiral Dygalo was convinced from the beginning that the U.S. attack submarine began to trail K-129 during the latter's special reconnaissance operations off the Pearl Harbor submarine base in Hawaii. The sequence of communications events and the fact that the submarine was found on the sea bed with the bottom of its hull torn open, and with periscope, antennae, and snorkel in stowed positions, proves, according to Dygalo, that it was involved in an underwater collision with another submarine or was the target of an ASW (antisubmarine) weapon. Otherwise, claims Dygalo, its snorkel would have been extended as it had gone down.

U.S. officers have refuted the early Soviet charge that the U.S. nuclear attack submarine USS *Swordfish* was the submarine involved—a charge that was based solely on the latter's reported arrival in the Ship Repair Facility, in Yokosuka, Japan, on March 12, 1968, with a badly damaged sail. On August 31, 1994, there was a meeting of a joint U.S.-Russia commission seeking answers to questions of Cold War and previous missing prisoners of war. Retired U.S. Navy admiral William D. Smith informed Dygalo by letter that the allegation of *Swordfish*'s involvement in the sinking of the Soviet submarine was not correct and that *Swordfish* was at least 200 miles from the site of the submarine sinking. (Incidentally, this is the same distance cited by the U.S. Navy when it described the distance of the nearest Soviet warships from the USS *Scorpion* wreck later the same year.)

The joint commission, headed by General Dmitry Volkogonov and Ambassador Malcomb Toon, informed the Russians that there had been no U.S. submarines within 200 nautical miles of the site where the K-129 was found. In a book published in 2005 in Russia by Mikhail Voznesenskiy, titled *Theft of Submarine K-129*, the author claims that the Golf II was sunk by an ASW weapon launched (possibly by accident) by a U.S. Navy destroyer, submarine, or aircraft. During Peter Huchthausen's interview with Admiral Dygalo in 1994, the admiral stated that the Soviets began to suspect that the K-129 had sunk, after reading U.S. Navy Pacific Fleet communications. Had there been an incident or an accident with a U.S. unit, there certainly would have been reports and incident accounts in the radio traffic.

Ed Offley, in his book *Scorpion Down*, assumes that the convicted spy John Walker had begun to pass U.S. Pacific Fleet crypto keys to the Soviets before 1968, implying that Moscow was reading U.S. naval radio communications by then. The claim by Dygalo that the K-129 hull was holed beneath Compartment Three indicates that either a weapon or a collision below the hull caused its demise. The fact that the U.S. Navy still will not release photos of the K-129 hull may support the theory of a U.S. involvement. It is also curious that the U.S. Navy refuses to release photos of USS *Scorpion*'s entire hull. In a conversation with Wood's Hole deep submergence expert Charlie Hollister, who has visited the USS *Scorpion*, the USS *Thresher*, and Soviet K-219 wrecks in the Atlantic numerous times aboard several deep submergence vehicles, I was told that the photos of *Scorpion*'s hull show twisted metal with Cyrillic writing enmeshed with the hull.

According to Admiral Dygalo, as soon as the K-129 failed to communicate during two designated reporting sessions, it was declared missing. As the division commander, Dygalo was responsible for reporting the loss to then Pacific Fleet commander in chief Admiral Nikolai Amelko, who was absent from headquarters. He was aboard the cruiser *Dmitry Pozharsky*, which was bound for a scheduled port visit to Madras, India. Captain First Rank Lev Vtorygin provided the following account of Dygalo's meeting with his fleet commander in chief:[1]

"Admiral Dygalo flew to meet the fleet commander in chief in Madras. I was aboard the cruiser accompanying Amelko as an English interpreter for the visit. The mood on the flagship as it steamed toward India was gloomy and morose. We already knew K-129 was overdue for a communications session; we all feared the worse.

"The Pacific Fleet commander in chief, Admiral Amelko, sat alone in his flag cabin and spoke with no one after first receiving the message that a ballistic missile submarine was unaccounted for and feared lost in the mid-Pacific. I was scheduled to review the Madras visit protocol program with him.

"It was a day I'll never forget. Amelko appeared sad and already resigned that the lost communications with one of his Pacific Fleet submarines was not only the end of his career but possibly a harbinger of something worse—an encounter with the American navy that might portend a heightening of tensions."

Vtorygin paused, as if feeling the uneasy sorrow of the day. It was mid-March 1968, and the U.S. superiority in nuclear submarine technology was becoming a glaring handicap to those in the know of the Soviet navy's senior echelons. They had already experienced their first major loss of life at sea in 1961, with the accident aboard the K-19, caused by poor safety and the lack of redundant features that are found in newly constructed nuclear submarines. Men were dying in increasing numbers at sea, due to the operational forces' lack of understanding of the nuances of basic nuclear safety. The most damaging to the morale of the burgeoning Soviet submarine force was their glaring inferiority to the Americans in sound quieting. This was common knowledge among the men of the submarine force who had survived close encounters with U.S. nuclear attack submarines, either off the Soviet coastal waters or while deployed near U.S. home waters.

Vtorygin continued, "Admiral Amelko knew and understood the handicaps shrouding his submariners, and he lost sleep knowing that his boats, nuclear and diesel powered, were deploying with major disadvantages.

"We had entered port in Madras and were in Admiral Amelko's cabin awaiting the first protocol call at the Indian naval host's headquarters, when Kontr Admiral Dygalo entered to report the situation regarding K-129, missing since March 8. Dygalo had flown into Madras ahead of the Pacific commander's arrival aboard his flagship. The moments when Dygalo described the situation were burned indelibly in my memory," Vtorygin said.

"As Dygalo briefed Admiral Amelko about the sequence of events surrounding the failed attempt to receive scheduled reports from K-129, since its first reporting date of March 8, Amelko stared at the porthole in his cabin. 'What was her mission just prior to her scheduled report?' the admiral asked. 'Reconnaissance of the departure channel of Pearl Harbor, Comrade Admiral,' Dygalo replied. 'Why are we using a strategic unit for that mission?' Amelko asked."

"I'll never forget the response," said Vtorygin sadly. "Admiral Dygalo shifted uneasily and then responded that it was impossible to use any other submarine because of the high probability of counterdetection by U.S. submarines guarding the area. Dygalo continued, saying that their few nuclear-powered units with the Pacific Fleet were too easily detected because of their noisy engineering plants to conduct close surveillance of the U.S. submarine departure zone off Pearl Harbor.

"Division Commander Dygalo explained further in the deepening aura of gloom that there was suspicion that American units had detected

K-129 and tracked her until March 8. 'We had one signal from one of her deployed slot buoys that she was being pursued by American ASW forces; we presume either air, surface, or submarine. There was no doubt that an incident with those forces accounted for her failure to send a scheduled signal. We fear she has been lost, Comrade Admiral,' said Dygalo, looking totally defeated.

"I can still hear the ticking of the clock in Admiral Amelko's stateroom after Dygalo spoke," Vtorygin said darkly. "It was the first time I had witnessed such a report and an admission that our forces were in such mortal danger at sea. It was unheard of in our propaganda, which showed only the positive side of everything, that our much-vaunted ships and submarines may be outclassed at sea by the Americans. After an awkward pause, Admiral Amelko dismissed me from his cabin while he and Dygalo sat together looking ever so sad," Vtorygin recalled.

Later, following the initial calls on the host Indian navy commander, Admiral Amelko, accompanied by his aide and Captain Vtorygin, boarded the Pacific Fleet commander's special aircraft. They flew to Moscow together. According to Vtorygin, "Little was said during the flight, but we knew the fleet commander was expecting to be relieved of his command by the navy commander in chief, Admiral of the Fleet Gorshkov, known as a ferocious commander given to periods of blind rage in front of subordinates. Gorshkov had been the navy commander in chief since 1955 and was feared by all naval officers, junior and senior alike."

Vtorygin continued, "Admiral Amelko was for some reason not relieved of his fleet command, but, according to Kontr Admiral Dygalo, the mission of K-129 was the last close surveillance mission in the Pacific to be conducted by a Soviet diesel-electric submarine against Pearl Harbor. Since then, surveillance missions would be conducted only by nuclear-powered units."

The search for K-129 continued fruitlessly for several months. The U.S. Pacific Command quietly guarded its knowledge of the exact position where K-129 went down, which allegedly had been pinpointed by SOSUS. If the position of the lost Golf II had been known by other means, possibly by the unit that caused its sinking, the U.S. Navy still guards the information as classified.

While fragments of the records concerning the CIA-sponsored recovery of that submarine have been released to the public, the U.S. Navy, specifically the submarine force, will not release its accounts of the contact and the pursuit of the Golf II by antisubmarine warfare (ASW) forces of the Pacific Fleet. The U.S. Navy, however, ostensibly knew the exact location of the sinking and, after a decent pause, dispatched forces to search for the wreck. An oceanographic research ship named USNS *Mizar* and a specially configured, deep-sea reconnaissance submarine, called the USS *Halibut*, were sent to locate and survey the sunken Soviet sub.

The USS *Halibut* (SSG [N] 587) was originally designed to carry the Regulus I surface-launched, 80-mile-range cruise missile (designed by the German scientist Herbert Wagner; see chapter 1). It was extensively reconfigured in 1965 as a deep-sea reconnaissance-and-recovery submarine and was redesignated SSN-587. The *Halibut* had huge compartments for housing spools of cable to deploy observation equipment miles below the ocean's surface. It was reportedly even capable of retrieving small items from great depths while submerged and thus free from observation.

The *Halibut* found the lost Soviet sub on the bottom four months after it sank and reconnoitered the wreck, then obtained photographs and retrieved some items from the hull. This was an incredible accomplishment in 1968, which justified its first presidential unit citation. After studying the site and the exterior condition of the sub's remains, the CIA planned a salvage operation. A deep-sea mining vessel called *Glomar Explorer*, which was built and operated by the CIA under cover of the Summa Corporation, returned to the site in 1974 and retrieved portions of the submarine from a depth of 3 miles. The remarkable salvage operation was leaked to the Western press during its progress but continued until completion, despite the arrival of several Soviet ships. They watched intently at close range but did not interfere.[2]

The recovery ship was built to resemble an underwater mining platform. It lowered a clawlike device on the end of a 3-mile pipe-string, grasped the largest hull section of the submarine, and began to hoist it to the surface. The physics of the feat can best be described as trying to remove a hat from a man's head on Fifth Avenue by dangling a hook on a string from the top of the Empire State Building. As the submarine neared the surface in the grip of the mechanical jaw, some of the wreckage broke free and plunged back into the sea. The forward portion of the submarine's hull was recovered and placed into a floating barge for inspection. After careful preparation, submarine analysts, forensic experts, and engineers descended into the twisted wreckage and carefully scavenged through the grisly remains. They expected to emerge with precious code books and warheads—the prized intelligence booty. Instead, they were appalled by what they found.

The interior of the submarine, although badly mangled from the immense pressure of the depth, distinctly showed very crude construction.

Steel hull plates were of uneven thickness and welding was rough and irregular. One hull section was found to be reinforced by wooden shoring; between the inner and the outer keels, salvagers found ballast in the form of hundreds of lead weights that had apparently been used to adjust trim on the submarine by hand. The presence of such primitive devices was shocking to U.S. submarine analysts, who assumed that the crude measures were required to offset serious defects in the submarine's obsolete design. The flaws found in the sub's hull contrasted sharply with the technically sophisticated warheads found on the torpedoes and the missiles.

The remains of at least six crewmen were collected from inside the hull and were later buried at sea with appropriate military honors. The salvage personnel videotaped the procedures to prove that the burial was accomplished with dignity; the Soviet national anthem played as the crew's remains were interred in the deep.

The salvage operation recovered other valuable equipment, including publications relating to secret communications and portions of the nuclear warheads from two torpedoes. The interior of the hull and the human remains bore significant traces of radioactivity from the deterioration of the nuclear torpedo warheads.

In 1989, the disclosures in the Russian press of the K-129 sinking provided a new opportunity for the Soviet navy to seek an official U.S. response and thus gain further cooperation in resolving scores of other submarine incidents. Soviet naval representatives at the annual review of the Incidents at Sea Prevention accord—an agreement between the two navies signed in 1972 to prevent the growing number of dangerous incidents between ships and aircraft of both sides—tried in vain to gain information about the K-129. The Soviet navy regularly tables a request to include submarine operations under the agreement; the U.S. Navy consistently refuses to consider that proposal. American officers insisted that the pact would continue to apply only to surface ships and aircraft; they claimed that placing restrictions of any kind on U.S. submarines would violate their traditional autonomy of operations.

In 1975, an early attempt to acquire confirmation of the K-129 salvage from the U.S. government resulted in only limited success when Secretary of State Henry Kissinger passed the names of three of the deceased seamen to Soviet ambassador Anatoli Dobrynin. The remains of only three of the total unknown number of crewmen recovered had been identified. That was the last information offered by the U.S. government regarding the incident before the Soviet Union collapsed in 1991. The following year, during initial sparring for joint cooperation, the CIA and the former KGB began to meet on a limited basis. During one of the early meetings, CIA officials gave the Russians some documentary evidence, including the videotape showing the burial at sea ceremony. In 2007, additional material was released by the United States to the Russian navy, including K-129's bell and a sailor's camera.[3]

Regardless of the motives behind the Russian attempts in 1989 to find out more about the missing submarine from the U.S. naval attaché, public reaction following its appearance had focused on my desk in the Moscow embassy. Before the summons to navy headquarters, I had received a series of telephone calls from Soviet citizens and several from the press asking me to comment.

For days after the article appeared in the Russian press and after the meeting with Admiral Makarov, I received more calls from Soviet citizens who claimed to be families of the crewmen of the sunken submarine. One call was from a disabled Soviet navy warrant officer in Leningrad who had lost his sight while serving on a nuclear submarine. He said that his father had been among the missing crewmen of the lost submarine. The former sailor insisted that we meet because he had important data to show me. Given the drama of my earlier encounter at navy headquarters and the request to divulge the location of the interred sailors' remains, I agreed to see the young man from Leningrad. At his request, we would meet on the square in the park in front of the Hotel Ukraina, across the Moscow River from the embassy. It was not unusual for Americans from the embassy to walk across the bridge to the Ukraina to shop in the well-stocked press kiosk in the lobby or visit with guests in the hotel.

I approached the park with trepidation, as bells of caution rang in my mind. It was 1989, and the Cold War was not yet over; any prearranged meeting between a Soviet citizen and a Western diplomat was still risky.

The park was crowded with summer strollers. After scanning the throng, I noticed a man with dark glasses standing near the statue in the center of the park. A woman held his arm and watched me approach. She had high cheeks of Tatar or Georgian origin. I recoiled when I first saw the man's face. It was a grotesque mask of multicolored tissue, the nose badly deformed despite signs of attempted reconstructive surgery. He wore a scarf to conceal much of his face. After exchanging silent signals of recognition, we shook hands and began to walk together through the park. The sky was bright with the high puffy clouds of late summer, but a low bank of dark clouds in the east threatened a thunder shower.

The blind man, who gave his name merely as Igor, walked slowly but with the gait of a sailor. It was impossible to discern his age, but he was tall and erect. He wore a sailor's blue-and-white-striped jersey under a dark jacket. Like many former Soviet military men, he wore a small metal badge indicating prior service—his with a submarine silhouette on a blue background over the words *dalni pohod,* meaning a patrol abroad.

Igor spoke in slow, careful Russian from the side of his mouth. He had a noticeable slur. "My father was an engineer aboard the missile submarine we lost in the Pacific in 1968. I went to sea following my father's example and served aboard a nuclear submarine as a seaman engineer, working my way eventually to the rank of warrant officer." He turned as if looking around, and I wondered if indeed he was blind. He continued. "I served aboard a Northern Fleet cruise missile submarine we called the K-131." Igor stopped walking and turned to face me. He unwrapped the scarf from his chin and neck, revealing ghastly scars and disfigured flesh.

"I was burned in a fire that killed fourteen of my shipmates in the Barents Sea in June 1984." He replaced the scarf and started to walk again.

As we strolled, Igor related the deplorable state of nuclear power and submarine safety in the Soviet navy. "I'm working with a group of former submariners, many of them survivors of accidents. With the help of the former author and former navy *zampolit* Captain Nikolai Cherkashin, we are compiling a full history of the submarine catastrophes for publishing. So far, however, we have been repressed. I'll give you a summary if you help us tell the whole story."

Igor motioned for us to stop. He produced a plastic folder from a red shoulder bag and showed me the contents—a thick sheaf of handwritten papers. He quickly returned the papers to the folder and gave it to me. "I request, Mr. Attaché, that you please try to obtain the location of my father's remains. We're not interested in whether you Americans salvaged his submarine; we only want to know the location of reburial, and whether there was any evidence found to identify the dead." I promised to try to help and thanked him for the material. Before parting, the woman with Igor, identified only as Yelena, handed me a file in a plastic cover.

"Here's another account of disaster, this one suppressed for more than thirty years." she said solemnly. These were the first in a long series of accident accounts that I was given over a period of more than five years by Igor and other former submariners and their families.

Igor's folder contained a list of submarine incidents that had occurred between 1956 and 1989 and gave the submarine designations, the dates, the names of the commanding officers, and other data in neat handwritten script. The material was in Russian and had been related by different people, apparently former crew members of the submarines described. I later learned that a finished version of the information was published in 1992 in Russian, in a book called *Investigating Submarine Catastrophes.* A table in the back of the book contains an expanded list of submarine accidents, including the number of casualties, the commanding officers, the location and the cause of each accident, and where the dead were buried whenever this was known. The first pages of the document gave the graphic details of an early accident in July 1961 that caused the first deaths of crewmen who had been exposed to radiation at sea; it was recounted by survivors and their families.[4]

11

OCEAN SURVEILLANCE
1962–1980s

During the first years of his tenure as commander in chief of the Soviet navy, Admiral Sergei Gorshkov had been a circumspect executioner of Khrushchev's naval policies. But in the aftermath of the Cuban missile crisis, Gorshkov appeared to gain political support for overseas naval deployment, especially after Leonid Brezhnev took over the Soviet leadership in 1964. By February 1968, Admiral Gorshkov's picture was on the front cover of *Time* magazine with the title "Russia's Navy: A New Challenge at Sea." The article expanded on a quote attributed to Gorshkov: "The flag of the Soviet Navy now proudly flies over the oceans of the world. Sooner or later, the U.S. will have to understand that it no longer has the mastery of the seas."[1]

In the early 1960s, the attitude of the Soviet political leadership toward its naval forces had radically changed. The ground forces were still given the task of destroying the "probable enemies" in Europe and moving to the coast of the Atlantic Ocean in the shortest possible time to prevent the landing of U.S. reinforcements. The naval strategic nuclear forces, however, were becoming an important component of the Soviet strategic nuclear triad. Pivotal to the naval forces were the nuclear-armed submarines with ballistic missiles of the Zulu, Golf, and Hotel classes, later to be joined by the Yankee class. Soviet long-range naval bombers were also being equipped with nuclear weapons. Thanks to these new capabilities, the Soviet navy was assuming a greater strategic role within the Soviet armed forces than the one envisioned by Khrushchev. An increasingly key aspect of the strategic balance was monitoring the U.S. nuclear-capable submarines and aircraft carriers and deterring an eventual attack on the Soviet Union.[2] In turn, this increased Soviet maritime activity justified greater U.S. reconnaissance and surveillance efforts near the Soviet coasts.

Both sides chose to acquire dedicated intelligence-collection ships to help their ocean-surveillance systems locate the adversary in distant waters and eavesdrop on communications. But the well-known episodes of attacks by Israeli and North Korean forces on the two U.S. intelligence ships USS *Liberty* and USS *Pueblo* in 1967 and 1968 led Washington to withdraw most of its dedicated spy ships, while Moscow continued to increase the number of its seagoing intelligence collectors.

The deployment of Soviet strategic submarines near the coasts of the "probable enemy" and the threat posed by nuclear-capable aircraft carrier task groups and Polaris submarines required more precise monitoring of U.S. naval ships. As Soviet navy captain Vladimir Kuzin explained, "The systematic reconnaissance of the forces of the probable enemy around the world was a prerequisite to ensure the high combat readiness of the Soviet navy."[3] After the widespread introduction of radio, radar, and sonar detection equipment in the Soviet fleet, radio and electronic emissions were a main source for information about the probable enemy.

To systematically collect these emissions in the waters where the United States and other foes were active, the Soviet navy created three categories of intelligence ships. These would be subordinated to the fleets' intelligence directorates or to the GRU, depending on the mission and the circumstances.

The expanding Soviet fishing fleet had provided Moscow with an opportunity to study the seas a full decade before its navy began distant ocean operations. Fishing was one of the Kremlin's answers to the food shortage then growing in the USSR. With an average monthly consumption of twenty pounds of edible fish per person, seafood constituted the greater part of the protein ration for the ordinary Soviet citizen. Another contribution of the fishing fleet was invaluable intelligence. Soviet trawlers engaged in coastal and naval observations whenever possible and often appeared in the middle of scheduled NATO exercises.[4] Employing echo-sounders and other sensors, Soviet whaling fleets could provide valuable information about ice movement; surface and subsurface current velocities, for use by submarines; antisubmarine weapons; and mine warfare. The massive Soviet fishing fleets could also study the distribution of terrestrial magnetism in distant regions to provide data applicable to radio communications, navigation, and degaussing (masking the magnetic signatures of steel warships to protect them against magnetic mines).Underwater television cameras that supported fishing could also be used to map the ocean bottom—an important requisite for submarine operations.[5]

In response to the West's growing Polaris strategic submarine fleet and nuclear-armed aircraft carriers, the Soviet navy needed dedicated platforms to conduct reconnaissance missions. The first intelligence ships used by the Soviet navy were converted fishing trawlers and survey ships. Displacing 1,200 tons, these little ships were seaworthy, were relatively fast, and had a long-range capability. They packed aboard radio, electronic, and hydroacoustic intelligence equipment to intercept signals and signatures emitted by U.S. naval ships. Within five years, the Soviet fleet of dedicated naval intelligence collectors (NATO term AGI) had grown from none to about ten units.

By 1975, twenty-five of these trawlers were in service. All of these ships flew the naval ensign and displayed an identifying pennant number on their hulls. They were not really in disguise, but their growing presence helped to build a detailed picture of NATO naval forces. The results of their close-range surveillance allowed the Soviet navy at home to study Western tactics, operational patterns, underway replenishment, and carrier flight operations. The trawlers were tasked by their respective fleet intelligence departments or by the GRU. The same intelligence trawlers could either trail NATO warships or stay on specific stations for a special purpose. For example, one Soviet navy intelligence trawler was stationed off the coast of Northern Ireland to monitor British army and Royal Ulster Constabulary communications and watch British and U.S. strategic submarines entering and leaving the Clyde River.[6]

These Soviet intelligence collectors shared common features: direction-finding loops on top of their masts to locate other ships by triangulation and extra boxes on the superstructures that contained working space for the monitoring specialists and their equipment. When exact locations of Western forces were in doubt, the trawlers requested surveillance-support Bear-D long-range reconnaissance aircraft or assistance by a reconnaissance satellite link. Often, they mounted small-caliber weapons for self-defense. Within the same class of ships, however, there was no standard fit, and the same unit could display different sensors from one mission to another. Most of the intelligence harvest was recorded and bought back for in-depth study.[7]

A typical example of the routine aboard an intelligence trawler provides a glimpse of what the crew members did during their long and often arduous patrols in heavy seas and extreme temperatures.

Onboard the *Vertical*

Soviet naval intelligence veteran Yuri A. Berkov recalled his assignment onboard the Northern Fleet's 1,200-ton radio technical intelligence ship

Vertical, based in Goriachiye Ruchyi near Polyarny.[8] Berkov, a junior officer who had just graduated from the prestigious Felix Dzerzhinsky Naval Academy, specialized in naval engineering. As Berkov explained, keeping track of the whereabouts of the U.S. Polaris submarines was the prime objective of his ship: "My first patrol was in the end of October. Our task was to observe the NATO's war games in the mid-Atlantic. The radio technical intelligence was assigned to detect all signals coming from coastal radio stations (most of them Norwegian). There was also a group of 'coastal' radio intelligence specialists who helped us. In the north Atlantic, we began to monitor NATO submarine radio station broadcasting via AN/BQQ-9. We were looking for patrol positions of American fleet ballistic missile submarines. My task was to locate the submarines and to avoid collisions with other ships."

In November, the *Vertical* was involved in the rescue of the *Leninsky Komsomol* K-3, the first Soviet nuclear submarine to experience a fire in the mid-Atlantic. Thirty-nine crew members died. Back in port, the *Vertical* underwent a yard period where Burkov took an active part: "When the ship was being repaired at the plant, I constructed wide-range ultrashort-wave antennae; those we had on the ship were unable to take the bearings of detected emanations, which made it difficult to pin down their source."

The *Vertical*, a first-generation Soviet intelligence collector that was built on the hull of a whale catcher.

In March 1967, the *Vertical* was again at sea, this time to monitor a U.S. carrier group: "Our task was to observe NATO's war games in the Middle Atlantic. At 90 kilometers from the games' area, I detected the U.S. antisubmarine aircraft carrier *Essex*. My new antenna helped me to do it. We approached the squadron. It was composed of seven ships—antisubmarine aircraft carrier *Essex*, frigate *Farragut*, and five destroyers."

On a previous cruise, the Soviet captain had sympathized with the commanding officer of the USS *Courtney*, one of the destroyers in the group. There ensued an unlikely and friendly exchange between the spy ship and her prey: "We offered vodka and cigarettes called 'Belomor.' As for them, they gave us beer, tins of pineapples, and pornographic magazines like *Playboy*. The exchange lasted about twenty minutes. . . . We observed the aircraft carrier for about a week; we analyzed emanations, filmed ships and antisubmarine planes—'Trackers,' trying to understand how they were able to detect well our submarines. As it soon became clear, there were two submarines: one of them was British, another one was Norwegian. The squadron was maneuvering, and we often disturbed their ships. They realized we were doing intelligence work and would not leave them alone. We managed to seize their radio message for the Pentagon in which they asked who were the people on the *Vertical*. Soon, they got the answer that the *Vertical* was the Russian intelligence ship (under Commander Leonid Shul'pin)."

"Once," Berkov remembered, "an antisubmarine helicopter hovered over us and began to take pictures. It was so close that we could see a photographer's face. My colleague Buturlin came out and showed him a fist. The American 'riposted' by throwing an orange. I stood close to Buturlin and threw the orange back. . . . In general, the British and the Americans showed no hostility; the Cold War was the governments' business. Simple people were just curious."

Holy Loch and Unconventional Submarine Detection

One of Berkov's most interesting comments referred to the nonacoustic and hydroacoustic devices used by the *Vertical* to attempt to locate Polaris submarines leaving their bases. Besides radio intercept equipment and sonar, the *Vertical* mounted a warmth detector that measured the slight variation in temperature that affected sea water after a nuclear-powered submarine passed by. Designated MI-110K in Soviet terminology, this equipment was top secret and unknown in the West at that time. It was being used by the Soviet navy in the immediate vicinity of the

U.S.-British Polaris submarine base of Holy Loch to detect departing strategic submarines. Berkov related, "In November, we had to go to the sea again. Our ship got a warmth detector MI-110K and a hydroacoustic device MG-409 with three series of mercury-zinc batteries. Ships *Bui* and *Giroskop* got the same radio technical armament. Now, our task was to search for patrol positions of American atomic missile submarines in the Northern Atlantic. MI-110K was the newest secret device for detecting warm submarines' traces. MG-409 was used to confirm a contact. By that time, the *Giroskop* had already come back and reported about some contacts with submarines. We've searched for submarines for two months. During that time, I learned how to deal with new devices and realized that searching submarines at sixteen knots speed was impossible. We had to zigzag crossing the same submarine's trace for many times. I invented my own theory proving that a ship had to move at twenty knots in order to detect a submarine's trace and 'catch' it. Besides, a good hydroacoustic station was needed in order to detect a submarine in the 'echo' regime. We celebrated the New Year 1968 at sea. Soon, we had to leave for Holy Loch, not far from Londonderry, Northern Ireland, where there was a British and American strategic submarine base. We detected British submarines when they were leaving the base and observed them until they were completely submerged."

These intelligence operations were often conducted in cooperation with a submarine. That same year a Soviet Project 633 Romeo class submarine was detected by Royal Navy aircraft while the submarine was on a surveillance mission inside British territorial waters near Holy Loch. After having detected the USS *Ethan Allen* and a pair of British and U.S. submarines communicating by using a new underwater system, the Romeo's radio intercept unit heard a message indicating that two warships were being readied to go after her. The submarine tried in vain to escape but with her batteries exhausted was forced to surface. She was then trailed all the way back to her Kola base by British, U.S., and Norwegian patrol maritime aircraft.[9]

In March 1968, the trawler *Vertical* engaged in an unusual mission: testing the Northern Fleet's vulnerability to foreign eavesdropping and intelligence gathering. For this mission, according to Berkov, the *Vertical* posed as an intruding intelligence ship: "We stayed in Goriachiye Ruchyi for a month. Then, our commander sent us to the White Sea to check whether our coastal sites and bases were well protected from audio and radio technical data-collecting devices. In March, we went to the sea; we went along the Kola Peninsula. We took down our flag and masked the name *Vertikal*: neither did we answer signals coming from border posts. We came close to the coast and recorded all transmissions and conversations on ultra short wave. As a result, we were able to pinpoint all Northern military infrastructure, air defense sites, coastal artillery, and submarine bases."

Berkov later became the captain of the *Vertical* and then returned to Leningrad for another assignment as his ship was being transferred to the Black Sea Fleet. The Goriachiye Ruchyi intelligence ships division had received a larger 3,000-ton *Khariton Laptev*, "stuffed with brand new data-collecting devices."

In 1970, the *Khariton Laptev* was tasked to monitor the launching of the new Poseidon ballistic missile, which was being tested from the nuclear-powered strategic submarine USS *James Madison* off Charleston's coast. On August 3, the *James Madison* exited its base, accompanied by the destroyer escort *Calcaterra* and the instrumentation vessel *Observation Island*, to perform the new missile's twenty-first launch since 1968. *Khariton Laptev* positioned herself to send her boats to recover debris from the missile silo membrane and U.S. telemetry buoys. U.S. aircraft tried to disturb the intruders by flying very low, but they were repelled with signal rockets, and the *Khariton Laptev* managed to escape with her bounty. Similar incidents took place over the years. On several occasions, Soviet intelligence collectors tried to cut the SOSUS (Sound Surveillance System) cables being laid by U.S. net-layers when they were still just below the surface.[10] Soviet veterans remain unwilling to discuss these occurrences for fear that they could be sued by the U.S. government while traveling to a Western country.

The Soviets were not the only side to employ floating intelligence collectors. The U.S. Navy and other Allied fleets deployed their own listening ships but with far less coverage than the large Soviet intelligence trawler force.

U.S. Surface Intelligence Gatherers

Tasked by the National Security Agency and manned by the navy, dedicated U.S. surface intelligence collectors were mainly focused on signal intelligence (SIGINT) to satisfy national, rather than tactical, navy requirements. Naval intelligence was normally a secondary mission for these ships. Seven ships had been commissioned in the early 1960s to eavesdrop on communications near the Soviet Union and other countries of interest. They were the technical research ships *Oxford* (AGTR-1), *Georgetown* (AGTR-2), *Jamestown* (AGTR-3), *Belmont* (AGTR-4), and *Liberty* (AGTR-5), and the miscellaneous auxiliaries *Pvt. Jose Valdez* (T-AG 169) and *Sgt. Joseph Muller* (T-AG 171).[11] All of these ships were active from 1964 to 1966, in the Carribean, the South Atlantic, and the South China Sea.[12]

Joint NSA/USN Naval Security Group detachments were also based on destroyers, submarine rescue ships (ASR), and other auxiliaries to conduct missions such as the DESOTO patrols. These operations provided more operational naval intelligence for the fleet commander. In 1963, an icebreaker had collected valuable information concerning the number, the class, and the characteristics of a Soviet naval convoy entering the Pacific after having followed the northeast passage all the way from Murmansk. A radar picket destroyer escort (DER) was kept in position to maintain close surveillance of the Soviet missile-range instrumentation ships throughout their intercontinental ballistic missile (ICBM) tracking operations in the mid-Pacific from November 1963 to April 1964. This inaugurated a policy of maintaining surveillance of such ships by surface craft. Ever since Communist China threatened to invade Taiwan in the late 1940s, the U.S. Navy kept a presence, usually destroyers, in the shallow and choppy Taiwan Strait. In later years, these destroyers often came close to the People's Republic of China (PRC) or North Korean coasts to provoke the Chinese or the North Koreans into using their radar and gunfire-control systems. The U.S. Navy would then collect and record the electronic results.[13]

Chasing Zulus

While en route from the South China Sea to its home port in Yokosuka, the USS *Orleck* (DD-886) was diverted to surveillance duty. In 1966, this was a plum assignment for a destroyer that had been equipped primarily for antisubmarine warfare and had spent long months tediously patrolling the Taiwan Strait or the Gulf of Tonkin. Its new job would consist of tracking eight Soviet diesel submarines on the surface, which were accompanied by a mothering Don class submarine tender.

This group of Soviet Pacific Fleet submarines—including long range Project 611 Zulu class boats—was en route to Vladivostok from training operations in the calm spring waters of the Philippine Sea. The crew members of the *Orleck* enthusiastically welcomed the assignment. On their antisubmarine-warfare ship, they seldom enjoyed the opportunity to match skill with real submarines.

The eight-ship Destroyer Squadron Three was home-ported in Yokosuka, Japan, with the Seventh Fleet flagship, the USS *Oklahoma City*

(CLG-5). Its crews included "Westpac sailors," descendants of the old Asiatic Fleet. These were senior enlisted men who remained permanently in Japan and transferred from one ship to the next as the squadrons rotated every three years to new home ports in California. As a result, some of the *Orleck*'s senior first-class and chief petty officers were experienced salts who had resided in Japan since the Korean War and in several cases since 1945. Many of these crewmen lived in houses in and around Yokosuka, in the picturesque Kanagawa Prefecture. Japan in the 1960s was economical and offered the navy men a pleasant respite between deployments, with few of the restraints of life in U.S. West Coast ports.

The *Orleck*'s wardroom mainly consisted of happy-go-lucky young officers. They were a spirited group with high morale, spiked by action in the South China Sea off the coast of North Vietnam. As they steamed at flank speed to take station on their quarry, they probably caused a great deal of curiosity because of their ship's three oversized battle ensigns flying from the mast head and two yard arms—as did the mysterious olive-drab van lashed to the ship's small helicopter flight deck aft. Three days earlier, the van had been hurriedly loaded in less than two hours in Subic Bay. With the van came a handful of communications technicians (CTs), and one rumpled and studious-looking lieutenant junior grade, who was curiously fluent in Russian. The presence of the van with strange antennae protruding from the top and the sudden arrival of the CTs, whose very existence was secret, helped make the surveillance assignment even more thrilling.

For a single destroyer equipped with only slightly updated World War II antisubmarine warfare (ASW) detection gear, it was difficult to maintain continuous surveillance on eight submarines—especially if some of them submerged. The *Orleck* was expected to hold contact with any submerging submarine and pass the contact off to a maritime patrol aircraft whenever possible, while at the same time not losing the main group, which consisted of the tender and the remaining submarines. It was initially an interesting and unusual assignment, until the second week, when the Soviet seamen showed no sign of tiring or returning to the Sea of Japan and their home port.

A young lieutenant was the senior watch officer, who was responsible for the training and performance of the officers' watch bill. He regularly stood watches on the bridge. Keeping track of the Soviets was a good way to pass a four-hour watch, and with a little luck, he might learn something interesting. Perhaps the rumpled lieutenant junior grade would emerge from his van, skulk to the bridge, and pass on some wisdom about what the Soviets were up to. It had not taken long for everyone to ascertain that this man, nicknamed Spooky, was in charge of an electronic eavesdropping team.

The officers standing the long watches welcomed Spooky's visits to the bridge. He usually came outside during the night watches, when the captain was not present to see his disreputable appearance—he always looked as if he had just climbed out of a bag filled with soiled laundry. Although the ship's company officers were aware that only three other officers had been cleared to know what Spooky was doing, the bright young lieutenant junior grade frequently gave out tidbits of advice, albeit poorly concealed, when something was about to happen with the Soviets. He called his nuggets of information "direct tactical support," and the *Orleck* officers all realized that it was coming from his SIGINT van. Initially, the ship's company officers were awed with the capability of the group, called a Naval Security Group Detachment, when one night during the mid-watch, Spooky materialized on the wing of the bridge in the warm tropical night. He said, "In a few minutes you're going to see a couple of surface contacts coming out of the north. They'll be a Soviet tanker with a Kotlin class destroyer in company to join us." He then vanished.

A half-hour later, two radar contacts emerged steaming from the north. They were picked up on surface search radar at about 35 miles and were heading in the destroyer's direction. They were exactly as Spooky had predicted. After that, the other officers listened intently to every word he said, fascinated by the power of vision he commanded in the van.

A favorite Soviet tactic was to steam at a slow 8 knots, with two of the Zulu submarines moored next to the tender on a long boat boom as their crews took advantage of the better facilities of the large, 7,000-ton tender. The remaining six submarines steamed a circle around the tender. The trial came as the submarines shifted positions, one of the two on the tender's boom slowly drifting away from the large mother ship, then submerging stealthily to avoid notice. They sometimes did this at night, when a slight sea made the surface search radar less capable of observing the maneuver, and when frequent rain squalls hampered visibility.

The *Orleck*'s reputation as an ASW ship hung on a fine line as a constant stream of situation reports were read first by the squadron commodore, then by the Seventh Fleet commander, and finally by the many lofty staff members, analysts, and soothsayers who followed the details of the Cold War ASW situation.

One morning on the bridge, with weather conditions deteriorating, the watch officer was unable to determine whether there were still two subs on the tender's boom. The captain's night orders stated that a minimum distance of 1,500 yards was to be maintained at all times at night to avoid generating a dangerous or embarrassing situation. The watch officer, however, was expected to give the alarm if any of the submarines submerged, for to lose one both visually and on sonar would result in professional embarrassment to the ship.

With this in mind and unable to determine whether one or two subs were by the tender, the concerned young watch officer gradually maneuvered *Orleck* astern of the tender, swinging gently from his port quarter to his starboard. This is normally a perfectly harmless maneuver, except that he had to cross the bow of one of the surfaced submarines 2,000 yards astern of the tender. As the U.S. destroyer approached the tender at the minimum distance prescribed by the captain's night orders, the watch officer crossed the submarine's bow by about 500 yards, which in his calculations seemed adequate. He inched the ship forward to the new position, still unable to discern whether one or two subs lay alongside the tender ahead.

Finally, after nearly an hour of uncertainty, and no new sonar echo indicating that one submarine might have submerged, the young officer, with growing confidence, threw caution to the wind and inched forward toward the tender, to 1,200, then to 1,000, and next to 800 yards. Still unable to count the submarines, he reduced the distance to about 600 yards and ordered the signal bridge to illuminate the nest on the starboard side of the tender. It was a good thing he did, for he saw clearly in the brief illumination of the red-lensed signal light that the outboard submarine had only her sail awash while she gradually slid away from the tender. The young American had caught them, he thought, and sighed with relief. Now all he had to do was to increase the separation back out to the required 1,500 yards and guide the sonar onto the submerging contact. The *Orleck* dropped back cautiously, ensuring that the submarine astern was clear. The officer of the deck, technically in violation of the standing night orders, was certain that the combat information center and the CIC watch officer, who was watching the situation from an enclosed tactical plot, as well as the rest of the enlisted watch, would not breathe a word to anyone that the officer of the deck (OOD) had brought the destroyer to a narrow 600 or so yards of the Soviets.

As daylight slowly arrived and the end of the watch drew near, the ship's sonar men maintained strong contact on the submerged submarine, and all the other submarines were in sight or on radar. The OOD felt elated by having returned stealthily to the prescribed limits of the station, when the captain appeared on the bridge. He appeared satisfied to see that all was in order. Suddenly, the Soviet tender began to signal the *Orleck* by flashing a signal light.

During the course of the surveillance, the two sides communicated by flashing light only when necessary, and usually when one side wished to impart some advance warning of a potentially dangerous maneuver. So when the other side's light flashed, the watch realized that a significant message was coming from the Soviet bridge. After a few minutes, the signalman arrived in the pilot house and said, "Flashing light message from the Russian sir, in English."

There was silence on the bridge as the captain read the signal. Gradually, the young lieutenant's elation over foiling the Soviet maneuver dissolved into remorse when the captain looked up with a frown. "When you are relieved of the deck, report to me in my cabin."

Shortly afterward, the lieutenant stood in the captain's sea cabin, while the latter held the signal in his hands and read it aloud in the Russian's halting English.

> Your officer on watch during the last hours of night is conducting himself in a dangerous and provocative manner by reducing the distance between ships to less than 500 meters, crossing the course of one Soviet ship, forcing it to maneuver to avoid danger. I demanding that your officer of the watch be severely punished and that such irresponsible actions be avoid in the future. I will report this gross display of poor seaman action to our fleet headquarters for appropriate diplomatic actions.
> The Soviet Commander

No further word was ever said of the incident.

Click Beetle

During April 1964, the U.S. submarine rescue vessel *Chanticleer* (ASR-7) had shadowed the Soviet Pacific Fleet's spring naval exercises with considerable success. This last mission against Soviet naval activities had convinced the commander in chief of the Pacific Fleet (CINCPACFLT) to propose to the chief of naval operations (CNO) in February 1965 to outfit new naval intelligence collectors to supplement the earlier ships. These had been operated by the NSA and were not specifically dedicated to ocean surveillance. CINCPACFLT explained the mission of the operations in a September 1965 message as "to determine Soviet reaction to a small unarmed naval vessel which is overtly a naval surveillance ship deployed in Soviet Navy operating areas." Thus, the surveillance ship was to be perceived as an observation platform wearing the "uniform" of the U.S. Navy, operating in places where it was entitled to be, and in no way disguising its nationality, appearance, or function.[14]

Under the code name "Click Beetle," the U.S. Navy initiated these new reconnaissance operations in the Western Pacific with a new converted small ship, the *Banner* (AGER-1). From August 1965 to December 1967, the *Banner* was often the target of Soviet and Chinese Communist harassment. As intelligence historian Captain Wayne Packard explained,

> The hostile actions included shouldering, closing to short range and maneuvering dangerously, one minor collision, closing with guns trained on *Banner*, surrounding the ship with trawlers, and signaling "Heave to or I will open fire." The *Banner* would be either sent away or supported with destroyers. In 1967, though, the commanding officer of *Banner*, Commander Charles R. Clark, perceived no threat from the Soviets because they had many surveillance ships operating and would have much to lose by seizing his ship. He remarked, however, that "The North Koreans and the Red Chinese were a very doubtful factor, because nobody knew what they would do, but the fact that we were in international waters, doing legal operations, was our greatest protection."[15]

The success of the *Banner*'s Click Beetle operations led to the fitting out of two additional small surface collectors: the *Pueblo* (AGER-2) and the *Palm Beach* (AGER-3).

The *Banner*'s operations had fulfilled high-priority SIGINT national intelligence objectives, while procuring naval intelligence on the Soviets. According to Packard, the concept of "a small ship acting singly and primarily as a naval surveillance and intelligence collection unit had been tested." The *Banner*'s proximity to the coast had allowed her to capture SIGINT signals that were not interceptable from shore-based sites or were not intercepted in sufficient depth by other mobile or shore facilities. A wealth of photographic, acoustic, and hydrographic intelligence had again been collected on targets of opportunity.[16]

The recipe was extended to the Atlantic. Between June 21 and September 14, 1966, the *Atakapa* (ATF-149), equipped with electronic intercept equipment, first conducted electronic surveillance of a Soviet fleet exercise off the coast of Norway. While waiting for the exercises to begin, the *Atakapa* sighted and trailed two Mediterranean-bound Zulu class submarines. Her successes proved once again the concept of employing small ships as intelligence collectors. Then the *Atakapa* entered the Baltic Sea, where she gathered considerable SIGINT data, this time of little value since much of it was available from other national or Allied sources. The Soviet reaction to the *Atakapa*'s cruises—in both the Norwegian and the Baltic seas—was benign.[17]

USS *Liberty* and the Komar Missile Boats

On June 8, 1967, while conducting reconnaissance off the Egyptian coast, the intelligence collector USS *Liberty* (AGTR-5) was attacked and severely damaged by Israeli aircraft and torpedo boats. She lost thirty crew members. After turning down an offer of assistance by a Soviet destroyer, the *Liberty* proceeded to Valletta, Malta, for immediate repairs but never returned to serve as an intelligence collector. The Israelis claimed to have mistaken the U.S. intelligence collector for a much smaller Egyptian transport and paid compensation to the victims. Many survivors, including the executive officer James Ennes, believed that the attack had been deliberate to prevent the United States from learning about an imminent Israeli offensive against Syria.[18] The documents related to U.S. foreign relations covering the Arab-Israeli War of 1967, published in 2004, give the latest U.S. government account of the incident. It maintains that the attack was one of mistaken identity. In a column published in the U.S. Naval Institute Proceedings, naval historian Norman Polmar draws the reader's attention to what may constitute a key piece of the puzzle.

> One of the most significant documents is a lengthy transcript of the EC-121 (U.S. spy plane) intercept of the communications between the Israeli helicopters (flying near the *Liberty* after the Israeli attack) and their controllers. One of the key transmissions to the helicopters after they had been told there were survivors in the water and to rescue them reads: "Pay attention: if any of them [survivors] are speaking, and if they are speaking Arabic, you take them to el Arish. If they are speaking English, not Egyptian, you take them to Lod [an airport near Tel Aviv]. Is this clear?" Again and again, ground controllers told the helicopters to ascertain the identity of the ship. The U.S. documents are highly critical of the Israelis for not making certain of the ship's identity before attacking, for (at one point) estimating the *Liberty* to be moving at 30 knots, and for a policy of automatically attacking any high-speed ship not determined to be Israeli. Although the documents, especially a lengthy telegram from the U.S. defense attaché to the White House, explain how these situations came about, the bottom line is that

Israeli "overeagerness" in a combat environment certainly contributed to the tragedy.[19]

Polmar pointed out that on the day of the attack, the U.S. embassy had sent a message to the secretary of state quoting the Israel Defense Force (IDF) intelligence chief, who had told the United States of Israeli intentions "to give Syria a blow to get more elbow room." Polmar stressed that the *Liberty* conspiracy theorists were proved wrong in thinking that the Israelis were attempting to keep their preparations to attack Syria a secret. That information was made available to both the U.S. government and the public on the day of the attack. After fourteen years of researching the incident, the author, judge, and former navy carrier pilot A. Jay Cristol also supports this position. He criticized the account of James Bamford, the author of a book describing the incident, who maintained that the attack was deliberate:[20]

> [Bamford] completely ignores that the United States had publicly announced to the world at the United Nations Security Council only two days before June 8, 1967, that it had no warships within hundreds of miles of the combat zone. The chains of reactions were started by an Israeli army report of explosions at El Arish. Since Israel controlled the air and the ground, they made the assumption that they were being shelled from the sea and a warship was in eye view. In view of the U.S. public announcement, it seems more logical for the Israelis to have assumed that a haze grey warship sailing within eye view of the ongoing combat was an enemy vessel rather than a U.S. ship.[21]

Cristol's remark on the explosions at El Arish is supported by Alexandre Sheldon-Duplaix's conversation with an Egyptian pilot. They were crossing the Suez Canal onboard the French destroyer *Montcalm* on April 20, 1988. A former naval officer, who in June 1967 was serving on a Komar missile boat, the canal pilot claimed that the Egyptian navy had launched several STYX missiles against the Israeli ground forces progressing near El Arish on June 8. He assumed that the Israeli attack on the *Liberty* had been caused by the Komar's swift action and withdrawal, which had left the Israelis at a loss to explain the origin of this bombing coming from the sea.

U.S. secretary of state Dean Rusk and Admiral Thomas Moorer, the commander in chief Atlantic, never accepted the Israeli explanation. To many survivors and to author James Bamford, the Israeli torpedo attack that followed the air raid on the *Liberty* could hardly be blamed

An Egyptian Komar missile boat. The swift missile attack by Egyptian Komar boats on El Arich may have provoked the Israeli response against the ill-fated USS *Liberty*.

on a mistaken identity. It vindicated their claim of a deliberate and cynical attack to finish off an undesirable witness. Cristol, however, contends that the Israeli air attack lasted about twelve minutes and was terminated as soon as the Israeli air force determined that the ship was not an Arab vessel. He further explained, "While the Air Force was initiating rescue operations, the torpedo boats approached, stopped, and began signaling to the *Liberty*. The response of the *Liberty* was to begin shooting at the torpedo boats, which thereupon began the torpedo attack. It lasted less than 15 minutes, during which time the navy torpedo boats believed they were facing an enemy who initiated the shooting at them." Confronted with the facts, Israeli naval intelligence presented the following account, summarized by Cristol:

A routine Israel Navy reconnaissance flight at dawn on June 8 sighted *Liberty* at about 6:00 a.m. steaming southeasterly and south more than 70 miles further west of El Arish. Positive identification was made and the information passed to Naval Intelligence Headquarters and the *Liberty* was marked on the battle control board at Naval Headquarters. Five hours later, the *Liberty* mark was considered old information and removed from the battle control board. At 11:00 a.m., shifts changed and the information about the *Liberty* was not known to the officer who assumed command. At about 1:00 p.m., when the presence of a ship steaming west, 14 miles off the coast of the Sinai and reported to be shelling Israel Army positions from the sea[,] became a tactical issue, the Navy Officer in command did not know about the dawn sight- ing of *Liberty* many miles to the west.

A CIA analysis concluded that the *Liberty* and the Egyptian trans- port looked alike from a distance and could have been mistaken for each other. More conclusive is Dr. Marvin Nowicki's statement. The U.S. Navy Hebrew linguist who had embarked onboard the NSA EC-121 spy plane wrote to Bamford: "We recorded most, if not all, of the attack. Further, our intercepts, never before made public, showed the attack to be an accident on the part of the Israelis."[22] Nowicki later published a letter in the May 16, 2001, edition of the *Wall Street Journal* stating that the Israeli attack was indeed a mistake: "My position, which is opposite of Mr. Bamford's, is the attack . . . was a gross error." Released intercepts from the NSA do support this position.[23]

The Egyptian pilot's claim that Komar missile boats had caused the El Arish's explosions, which in turn prompted the Israeli air force to look for a culprit off the coast, may provide an additional clue to exonerate Israel from a deliberate attack against her best ally.

Click Beetle Phase II and the USS *Pueblo*

After the *Liberty* incident, the vice chief of naval operations directed that reconnaissance ships be armed in the future. On December 14, 1967, the intelligence ships *Banner*, *Pueblo*, and *Atakapa* were ordered to mount a min- imum of two 50-caliber machine guns prior to their next missions. This would hardly have been an effective measure against the jet fighters and the torpedo boats that had attacked the *Liberty*.[24] Undeterred by the *Liberty* incident, the U.S. Navy continued its collection operations in the Mediterranean, the Black Sea, off the African coast, in Cuba, and in the Western Pacific. (The modified fleet tug *Atakapa* operated in the Mediterranean from June to October 1968.)

In 1967, the *Pueblo*, a twenty-five-year-old army coastal freighter, was refitted as a surface reconnaissance ship to participate in Operation Click Beetle Phase II. Commander U.S. Naval Forces Japan Operation Order 301–68 of January 3, 1968, read the following:

> I had successfully tested the operational feasibility and political implications of using one small trawler-type ship as a naval surveillance

and intelligence collection unit. Phase II expanded the program to use two ships (*Banner* and *Pueblo*) to provide continuous coverage of a selected area or operation. The program's objectives also included testing Soviet reaction to the continuous presence of a U.S. intelligence collection ship in Soviet naval operating areas. It was expected that the experience gained from Phase II, and the procedures and equipment developed therefrom, would lead to the implementation of Phase III—the employment of more AGERs.[25]

Click Beetle Phase II's instructions ordered the collection ships to remain no closer than 13 nautical miles from land and avoid any action that might be considered provocative. Weather conditions on the Soviet shoreline at that time of the year and the aggressive attitude of Red Chinese ships dictated the choice to send the unseasoned *Pueblo* for a first mission off the coast of Korea, "the land of the calm morning." The risk was considered minimal because her sister ship *Banner* and the larger *Oxford* had not been threatened during their previous operations in that area. The Foreign Broadcast Information Service had overheard Pyongyang's warnings against provocative acts in territorial waters by "spy boats" disguised as fishing boats, but the North Korean navy seemed to restrict its anger to South Korean fishermen and warships.

Commander Bobby R. Inman, the chief of the Current Intelligence Branch, CINCPACFLT, was not worried. He did not see a Director National Security Agency (DIRNSA) message of December 29, 1967, which noted increased activity by North Korean naval forces, including the sinking of a South Korean patrol escort, but he later doubted that he would have changed his assessment.[26] On January 11, 1968, the *Pueblo* left Japan. Twelve days later, she was surrounded by North Korean gunboats and fired upon while trying to escape, which resulted in the death of one sailor. Commander Lloyd Blucher surrendered his ship, rather than risk losing his men in icy waters. The remaining eighty-two crew members were taken captive to Wonsan. For the next eleven months, the North Koreans tortured their prisoners, coercing some to sign "confessions" of guilt and to make political radio broadcasts. President Lyndon Johnson turned down several military options and agreed to negotiate. On December 23, the crew was released. The navy did not court-martial the captain and his first officer.[27] Despite the crew's efforts, they had not been able to destroy all of the sensitive material that was necessary to accomplish its mission.[28]

The Future of the Spy Ship

During 1968, U.S. spy ships continued to conduct SIGINT and visual surveillance patrols in the north Atlantic. (The third AGER conversion, *Palm Beach,* patrolled in the Norwegian Sea and the eastern North Atlantic area during July and early August 1968.) Plans to acquire more AGERs were shelved. In June 1969, *Palm Beach* operated in the eastern Mediterranean. During August and early September, *Belmont* conducted surveillance of the Soviet helicopter-carrying cruiser *Moskva* in the eastern Mediterranean. And by the end of 1969, most of the surface reconnaissance ships had been decommissioned. They were to be replaced by faster and better-armed radar picket destroyer escorts. The *Thomas J. Gary* (DER-326) and the *Calcaterra* (DER-390) were reconfigured to act as naval tactical reconnaissance ships (NTRS) by the installation of a special electronic package (OICS Van) to meet the requirements for "multi-sensor collection of intelligence on maritime targets of opportunity or interest, for on-board processing of data and its preliminary evaluation, and for transmittal of tactical and significant information in near real-time." The *Calcaterra* operated in the eastern Mediterranean during the Jordan Crisis (September 17–25, 1970). But financial restrictions led to the cancellation of two more destroyer escorts for the Pacific fleet, and the United States abandoned the multisensor surface reconnaissance spy ships altogether.[29] The North Korean attacks on the *Pueblo* and a year later on an EC-121 spy plane had forced Washington to back down for fear of a second front in Asia.

Despite the October 3, 1969, and April 22, 1970, South Vietnamese attacks on Soviet intelligence trawlers reporting to Hanoi on incoming U.S. air strikes, the Soviet navy seemed to be more circumspect in the deployment of its spy ships. As a Soviet veteran later recalled, "We were not acting like the Americans. We would not have risked a ship like the *Pueblo.* We were more thoughtful and we were planning every move extremely carefully."[30] Ironically, as the value of spy ships was being debated in the United States in the aftermath of the *Pueblo*'s seizure, Defense Secretary Robert McNamara praised the usefulness of Soviet spy ships before a closed Senate committee: "It is important to us at times that the Soviets have these ships . . . near our forces. They had one associated with our Mediterranean task force during the Israeli-Egyptian conflict . . . which enabled the Soviets to know that Nasser and Hussein's statement that U.S. aircraft had attacked Egyptian forces was in error, and, it was extremely important to us under those circumstances that the Soviets know that."[31]

The Swedish intelligence collector *Orion*.

The United States did not completely abandon the concept of surface reconnaissance but instead reverted to the earlier practice of using an amphibious or an auxiliary platform. On the opposite side, the Soviet navy expanded its fleet of intelligence collectors, while the Soviet fishing fleet numbered some "villains": trawlers too neat to be engaged in regular fishing. Gulls usually ignored them, a sign that betrayed the absence of a fishy smell. In Britain since the late 1940s, Hull and Grimsby trawlers had been involved in collection operations in the Barents Sea under the code name Hornbeam. The loss of the *Gaul* in the Barents Sea in 1974 revived the rumors about the involvement of British fishing trawlers in secret intelligence work.[32] West Germany, Norway, and France, among other NATO members, also had intelligence-collection ships, as did neutral Sweden and Finland and Warsaw Pact members Poland and East Germany. Unlike the *Gaul* and the Soviet "villains," they all flew the naval ensign and were considered warships.

For their part, Soviet "naval" spyships or AGI successfully maintained permanent and semi permanent stations around the globe during the 1970s and the 1980s as part of their national ocean surveillance system. The Mirnyy whale catcher type and Mayak, Okean, and Alpinist trawler classes started to deploy in the Mediterranean in 1962 and to the South/East China seas in 1964 with the Vietnam War. The larger Nikolay Zubov

The *Kavkaz*, a more sophisticated Primorye class Soviet intelligence collector that was introduced in the 1970s.

and Moma were used for operations on the U.S. East and West coasts and near Hawaii, monitoring U.S. SSBN operations off Charleston and Puget Sound. The first patrol took place in 1965. After 1970, the much more capable Primorye and Balzam classes were used notably to monitor U.S. missiles tests in the Pacific. The Primorye and the Balzam were the most sophisticated surface collectors with extensive intercept and analysis capabilities. Some Soviet AGIs also had underwater detection sensors (sonobuoys, nonacoustic) and electro optic sensors.

The Soviet oceanographic fleets, both naval (AGOR) and civilian (AMGS) also participated in the surveillance effort. In wartime AGOR and AMGS could have been used as communications relays, as minelayers thanks to their precise navigation equipments or as acoustic decoys generating noises to confuse SOSUS and mask the deployment of Soviet submarines.

Soviet Space Events ships were another valuable category of spyships. Space operations control ships could control satellites while Space Events support ships were just monitoring space events. The first were often seen in the Caribbean, midway between the Soviet Crimean and Pacific ground based stations; the second usually sailed in the South Atlantic or in the Indian Ocean.[33]

12

A NAVAL INTELLIGENCE
REVOLUTION
1970s

The 1970s were characterized by a naval intelligence revolution. First, the deployment of the Soviet navy in distant waters prompted the Soviet Union to expand its fleet of spy ships. Second, we now know that the Soviet Union was able to read encrypted U.S. naval communications and anticipate nearly every move by the U.S. Navy for more than a decade. This advantage could have proved decisive in time of war. Third, the introduction of space-based surveillance systems allowed both the Soviet Union and the United States to improve their abilities to locate each other on the high seas. These new methods somewhat supplanted the more traditional HUMINT (human intelligence) methods of collecting that information. Finally, U.S. naval intelligence completely changed its appraisal and analysis of the Soviet maritime threat.

By January 1968, Soviet out-of-area submarine deployments showed surprising sophistication and improvement. One nuclear-powered submarine, which was believed to be a November class boat, on patrol from December 4, 1967, through January 19, 1968, puzzled U.S. naval intelligence. Operating out of her base of Petropavlosk in the Kamchatka Peninsula, the submarine transited the Aleutian chain, tested the SOSUS system off the U.S. West Coast in close coordination with a spy ship, and at high speed intercepted the aircraft carrier USS *Enterprise* (CVAN-65), which was deploying to the western Pacific. The nuclear-propelled frigate USS *Truxtun* (DLGN-35), which was escorting the *Enterprise*, diverted to the submarine, gained sonar contact, and directed a P-3 Orion maritime patrol aircraft to the submarine, which surprised her on the surface at dusk. Commenting on this operation, the admiral in charge of antisubmarine operations in the Pacific stated,

The extensive out of area patrol conducted by [the presumed November class submarine] revealed significant capabilities and patterns heretofore unobserved in the Pacific. . . . The extensive coordinated operations with the [intelligence collector] *Gavril Sarychev* (AGI) was the first operation of this nature observed in the Pacific. As such, it is indicative of very advanced mission planning and an effective submarine command and control and support broadcast system. The maneuvers of both [the presumed November submarine] and *Gavril Sarychev* imply at least a general knowledge of the West Coast SOSUS network. Whether the locations of the arrays are known with preciseness or whether their locations were deduced . . . is open to conjecture. The intercept of USS *Enterprise* (CVAN-65) provided a dramatic demonstration of the growing proficiency and confidence of both the crews of the Soviet's nuclear submarines and the command and control network directing them. . . . There are indications that *Enterprise*'s impending departure for Westpac prompted the Soviets to test the responsiveness of their at sea submarine resources by undertaking a multiple intercept. Although no conclusive evidence exists, it is possible that *Enterprise* was intercepted, or intercept attempted, by other out of area contacts . . . on 5 January 1968, and . . . on about 12 January 1968 as well as by [the presumed November submarine]. In addition the [intelligence collector] *Sarychev*, . . . may have contributed measurably from possible intelligence intercept in the Hawaiian area. Thus . . . the overall operation . . . required an extremely effective and responsive command and control system, highly competent and confident submarine crews and reliable submarine systems.[1]

Walk In

Just as the Soviet nuclear submarine that intercepted the USS *Enterprise* in the western Pacific was about to deploy, warrant officer John Walker allegedly strolled into the Soviet Embassy in Washington, D.C., and sold a radio cipher card for several thousand dollars. He negotiated an ongoing salary of $500 to $1,000 a week for passing more information from the submarine fleet message center, NAVCAMS, Atlantic Fleet, at Norfolk naval base, where he worked. After his arrest on May 20, 1985, Walker attempted to justify his actions by claiming that the classified navy communications data he had sold to the Soviet Union had already been compromised during the capture of the USS *Pueblo* on January 23, 1968.

The exact date of Walker's first contact with Soviet intelligence remains doubtful. John Walker's wife, who turned him in to the FBI, testified that the first "unexplained" money and high-priced gifts came in just before Christmas 1967. "Late in 1967," wrote Major General Oleg Kalugin, the former chief of KGB counterintelligence, in his controversial memoir. Kalugin gave an account of Walker's visit to the Soviet Embassy without specifying the month, which could have been November or even earlier:

> He had driven up from Norfolk, looked up the address of the Soviet Embassy in the phone book, and literally walked in the front door. Once inside, he told someone at the front desk: "I want to see the security officer or someone connected with intelligence." . . . One of our English speaking officers went down to see Walker. . . . He took Walker's material and came upstairs where [Solomatin, the Station's chief] and I were sitting. . . . We had never seen anything like this . . . and it was clear that what Walker was passing to us was real. In addition Solomatin, who had grown up in the Black Sea port of Odessa[,] was a navy buff. His eyes widened as he leafed through the Walker papers. "I want this," he cried. Our officer returned downstairs, where he gave Walker an advance of several thousands [*sic*] dollars and arranged a meeting with him later in the month at an Alexandria, Virginia, department store.[2]

Kalugin's statement sounds convincing. Yet in a 2001 thesis presented at the U.S. Army Command and General Staff College, Major Laura J. Heath, U.S. Army, showed that the FBI investigators had doubts about this version. They suggested that Walker might have started working for the Soviet Union at an earlier date, perhaps as an accomplice of his brother, Lieutenant Commander Arthur Walker. Arthur was suspected of having sold classified information to the Soviet Union in the early 1960s. FBI special investigator Robert H. Hunter explained in his published account of the Walker investigation that John Walker repeatedly failed polygraphs on the original, "overwhelming impulse" version of how he started to spy. Moreover, Walker's description of the Soviet Embassy included an iron fence that was added much later, in 1974. As Hunter pointed out, Walker did, however, pass a polygraph on the following statement:

> I know you believe I'm lying about the beginning of the operation because of [my failing] the polygraph and the fence story. Let me give you a scenario: [my brother] Arthur got in some type of financial trouble, probably in New York, and became involved with a number of New York loan sharks. Art needed money desperately, and I gave him some

classified documents to sell. After Art sold the documents and got some money, he became frightened and wanted to drop out. I saw no harm in selling the information to the Soviets, as the countries were not at war and would never go to war. I felt it was an easy chance to make some money, so I walked into a Soviet embassy somewhere in the world.[3]

John Walker stumbled on two polygraph questions: Was his brother, Arthur Walker, involved in any way in selling classified material to the Soviets prior to 1968? Was Arthur Walker involved in any way in initiating John's first contact with the Soviets?

Lieutenant Commander Arthur Walker had joined the navy in 1953 and served as a submarine officer and an instructor in antisubmarine warfare. His clearance gave him access to top secret information related to the sound surveillance (SOSUS) system. He might have compromised the information on the SOSUS that the Soviet Pacific Fleet apparently had when the *Enterprise* was successfully intercepted by a nuclear submarine in January 1968. John could also have been a source on SOSUS if he had started his betrayal during his 1965–1966 assignment onboard the Polaris submarine USS *Simon Bolivar* (SSBN-641). Heath believes that John could have walked into a Soviet embassy, at a port call "somewhere in the world," and initiated his spying before December 1967, giving Moscow more time to plan the seizure of the *Pueblo* through a second hand. Experts who grilled Walker and compared supporting evidence of his treason also suggested that Walker might have actually begun spying for the Soviets immediately after he reported to Norfolk in March 1967.[4]

Walker and the USS *Pueblo*

When John Walker gave the Soviet Union a copy of a key list for the KL-47/KL-7 encryption machines, the KL-7 was the most widely used encryption machine in the West. It was in service with all branches of the U.S. armed forces, the CIA, the State Department, and several NATO Allies. The KL-7 machines had first been introduced in the 1950s by the U.S. National Security Agency. Walker had accessed all of the cryptographic material related to the KL-7, including keys and classified manuals, while serving onboard the Polaris submarine USS *Simon Bolivar.* Similar but more complex than the famed German Enigma machine, the KL-7 was about the size of a teletype machine, with a three-row keyboard and shift keys for letters and figures. Each of the eight rotors had thirty-six contacts. To establish a new encryption setting, operators would select a rotor and place it in a plastic outer ring at the correct offset. This process would be repeated until all rotor positions were filled. Key settings were usually changed every day at midnight, Greenwich Mean Time. The KL-7 was aboard the USS *Pueblo*, and Soviet intelligence knew it would be there.[5]

Although Walker claimed that he had compromised only what had already been lost on the *Pueblo*, Walker's information may have prompted Moscow to ask Pyongyang to seize the *Pueblo*, as Major Heath explained,

> A close analysis of the details shows that John Walker probably did have a role in causing the USS *Pueblo* incident. From the Soviet standpoint, instigating the capture of a U.S. Navy ship, even second hand, would be a highly risky move, and would only be done if they thought that there was a correspondingly large payoff. Capturing one or more working copies of encryptors might fit the bill; but there is one problem: they could reasonably expect that the U.S. would respond to such an obvious compromise by making some changes to the encryptor, which would render the seized copies unusable. The U.S. did, in fact, do so; and the Soviets had John Walker giving them copies of the change orders that allowed them to modify the seized devices when the changes were made.[6]

All contingency procedures for security had failed the crew of the *Pueblo* on the day of her seizure. The ship had an incinerator, but it was located on the deck and consequently was exposed to North Korean gunfire. Weighted bags had proved to be too heavy to lift and could have been recovered easily in the shallow waters. The KL-47/KL-7 encryption machines had resisted the crew's attempt to destroy them with sledgehammers and fire axes.[7]

John Walker, allegedly nicknamed "number one" by Soviet KGB chairman and future premier Yuri Andropov, enabled the Soviet Union to win a decisive advantage over the U.S. Navy for more than a decade by providing the KGB with the U.S. Navy Fleet Broadcasting System (FBS) keys during at least the period 1967 to 1983. How was it possible that a single individual and his later associates could achieve this? Major Heath offered the following explanation:

> The evidence shows that FBS was designed in such a way that it was effectively impossible to detect or prevent rogue insiders from compromising the system. Personnel investigations were cursory, frequently

delayed, and based more on hunches than hard scientific criteria. Far too many people had access to the keys and sensitive materials, and the auditing methods were incapable, even in theory, of detecting illicit copying of classified materials. Responsibility for the security of the system was distributed between many different organizations, allowing numerous security gaps to develop. This has immediate implications for the design of future classified communications systems.

As Major Heath underlined, the protection of the FBS keys was flawed. The eight hundred ships in service during Walker's time of duty would have had several sailors with routine access to the keys as part of their daily tasks. Hundred of sailors worked in the communications centers in shore stations. Heath estimated their total number "in tens of thousands." And it could have been "above 100,000," if one added in "the personnel who might have gained access intermittently, either by fraud or deception."[8]

Heath pointed out that when John Walker was assigned to the supply ship USS *Niagara Falls* in 1971 as she prepared to deploy to the waters off the coast of Vietnam, the ship had an additional six months of "reserve on board" keys.[9] Since the vessel was a supply ship and had a communication security vault onboard, it was also used to ferry new keys and classified materials to other ships, including submarines and aircraft carriers, and held a large amount of keys to be used for the land forces in Vietnam as well. John Walker was assigned to the USS *Niagara Falls* for a total of three years (1971–1974) and was able to compromise the keys for the U.S. Navy worldwide during that period. All of the security procedures failed to catch him. As the classified materials custodian, he was the only person who was allowed to be in the code vault alone. John Walker routinely volunteered both to work as a courier and to keep materials for others. In this way, he was able to compromise the cryptographic keys for the ground forces' communications nets in Vietnam. The Soviet Union also gained access to the air force keys that were used to secure the transmission of orders to U.S. Air Force B-52 bombers in Guam. They were being stored in advance on the *Niagara Falls*, and Walker passed them to the KGB during port visits.[10] The quantity of material that he divulged was so large that during his debriefing, John Walker was unable to recall precisely all of the things he had compromised. His KGB Minox camera wore out, and he had to buy a new one to continue providing the keys to his handlers.

Walker retired voluntarily from the navy on July 31, 1976, apparently because he feared that his wife, an alcoholic, would commit an indiscretion. He had spent the last two years of his career supervising the distribution of classified material in Norfolk. After his retirement, Walker became a private investigator in Norfolk, Virginia, a front cover for a spy ring that included his brother, Arthur, now a defense contractor; a former trainee groomed and seduced by John, Jerry Whitworth; and later Walker's son, Michael. Whitworth and Michael served in the navy.

The KL-47 encryption machines were now being replaced by electronic systems such as the KW-26 ROMULUS and the KW-37 JASON.[11] Whitworth provided information on these systems and their modifications, helping the Soviets to continue reading most U.S. naval communications. From 1976 to 1978, Whitworth served onboard the aircraft carrier USS *Constellation*, where he was responsible for protecting the cryptographic material. He then held John Walker's previous position on the *Niagara Falls*, a supply ship that carried advance knowledge of fleet movements. As FBI investigator Hunter wrote, John and Whitworth stole "nearly 100% of crypto key and other valuable materials on the *Niagara Falls*."[12] Whitworth then worked in a communication center ashore before finishing his career as a chief radioman onboard the USS *Enterprise*. Michael Walker's involvement came in 1983, just as Whitworth was retiring. A yeoman in the USS *Nimitz*'s operation room, he had access to a wealth of classified instructions and messages.

Barbara Walker destroyed the spy ring when she reported her former husband to the FBI in November 1984. He had just recently stopped paying her. Former KGB agent Victor Cherkashin, however, claimed in his book *Spy Handler* that Walker was compromised by an FBI spy named Martynov, who overheard a conversation by chance in Moscow.[13]

One of the Greatest Espionage Successes in History

U.S. naval intelligence had noticed a growing number of worrying indicators of Soviet knowledge of U.S Navy secrets. Soviet submarines had developed an uncanny ability to stay just outside the effective range of U.S. sonobuoys—exactly as if the Soviets knew their performance. Soviet subs also appeared near U.S. bases just before American submarines were scheduled to deploy. And Soviet intelligence-collection ships were always on time and at the right place to monitor U.S. fleet exercises. "We now know that they did, in fact, have a copy of the Ops Plans—and everything else of any importance to the U.S. Navy, for a period of almost twenty years," explained Richard Haver, the deputy director of naval intelligence.[14] The Soviet navy knew in advance of the deployments of most, if not all, U.S. ships, including the strategic Polaris and Poseidon submarines, and apparently could monitor their communications. The Soviet navy also knew its own vulnerabilities and took successful measures to correct them.

In two affidavits submitted before the district court of Maryland, Rear Admiral William O. Studeman, then the director of naval intelligence, explained the damage inflicted by the Whitworth-Walker ring. He quoted the Soviet defector and counterintelligence agent Vitaly Yurchenko, who had been briefed by the KGB on the Walker case before defecting:

> The KGB considered the Walker/Whitworth operation to be the most important operation in the KGB's history. The information obtained by Walker enabled the KGB to decipher over one million messages. Averaged over John Walker's career, this equates to Soviet decryption of over 150 messages per day. This certainly ranks [the] Soviet intelligence operation as one of the greatest espionage successes in intelligence history. . . . Yurchenko related that a high KGB official informed him that the information obtained from Walker would have been "devastating" to the United States in time of war.

In his affidavit, Admiral Studeman described the information that Walker had been able to compromise. As a communication watch officer at the submarine forces, Atlantic Command, from 1967 to 1969, Walker had access not only to the communications received and transmitted during his watch, but also to a great variety of classified material that "[gave] the Soviets in-depth knowledge of U.S. national defense plans, policies, logistic data, forces levels and tactics." Studeman added that these disclosures endowed the Soviets with "an ability to make almost real-time tactical decisions because they knew the true strengths of our forces." Studeman concluded his affidavit on a depressing note: "The possibility exists that other personnel, identified by Walker as vulnerable to recruitment, could still be working for the Soviets. . . . His abuse . . . has jeopardized the backbone of this country's national defense and countless lives of military personnel. It is my judgment that the recovery from his traitorous activity will take years and many millions of taxpayer's dollars."[15]

Studeman also had to evaluate the damage caused by yeoman petty officer Michael Walker. Walker had stolen fifteen hundred classified documents, using his access to the aircraft carrier USS *Nimitz*'s storage room for keeping bags of classified waste paper called "burn bags." Father and son had discussed future assignments that would give Walker further access to classified information. Walker's secret was, per Studeman, his "ability to gain the confidence and trust of his superiors."[16] Over a period of twenty-one months, Walker had given the Soviets "technical details of the Navy's most modern weapons systems aboard the *Nimitz*." He had compromised the nuclear weapons' control procedures and information on the identification process of ships and aircraft, "which could help assist in developing countermeasures." Even more damaging, he had divulged details of the U.S. command and control procedures. Admiral Studeman summarized the significance of this information:

> The most dangerous time in any tactical situation is the transition from peace-time to combat operations. This transition process can involve several thresholds during which the "rules of engagement" or procedures governing the military force our commanders may use change in response to perceived threats from hostile forces. As the perceived threat increases, higher levels of authority are required to authorize more powerful responses. . . . The information given to the Soviets by Michael Walker helps them anticipate our actions and what force our commanders are allowed to use. Thus forewarned, they can deceive our forces or gauge our posture to achieve surprise. Of course, this same capacity is available to any client states or terrorist groups with whom the Soviets choose to share this information.[17]

The KGB handling of the Walker spy ring might have been an even greater success than the Allied ULTRA/Enigma victory during World War II. The Walker spy ring could have led the United States to lose a war against the USSR during the 1970s. As acknowledged by Yurchenko, "This knowledge could mean the difference between victory and defeat in war."[18]

All Source Ocean Surveillance

While the Walker spy ring secretly undermined all U.S. naval operations for more than a decade and a half, the growth of the Soviet navy after the collapse of the Cuban adventure triggered a concurrent expansion of Western intelligence and surveillance systems that were designed to keep tabs on the increasing Soviet forces. The United States had been accustomed to the security of having two wide oceans as buffer zones. Now it was suddenly faced with a belligerent power that was capable of breaching those buffers with an effective and modern blue-water navy.

The resulting effort to counter the new threat from the sea on both of America's flanks not only gave new purpose to the maintenance of a strong naval operating force, but spawned the development of the gigantic ocean surveillance information system, called OSIS, and a refurbished operational naval intelligence network. The reporting and locating systems, created by some very keen minds, soon became remarkably effective and grew capable of extraordinary feats, especially in the field of antisubmarine detection and surveillance. The expansion of the new ocean surveillance system into a workable fleet support network was spiced with intrigue and sometimes treachery and humor. Some stories of this vast organization are worth telling.

Once, in London, a lieutenant was summoned to brief a group of U.S. congressmen paying a visit to the ambassador. The lieutenant ran across Grosvenor Square from his headquarters to the U.S. Embassy, with three commanders and a captain in tow, in typical large staff fashion. The lieutenant was carrying a large folded plotting board. In the heat of the moment, a red tag dropped off the secret board as the group of wheezing staff officers crossed the crowded street. An elderly English lady tapped the lieutenant on the shoulder with her cane and whispered, "Young man, you just dropped a Russian submarine in the road," and handed him a piece of red tape emblazoned with the words "Soviet Charlie Class Submarine."

The Project 670 cruise missile nuclear submarine (NATO class Charlie) represented a significant improvement in Soviet capabilities. Unlike its predecessors from the Echo class, these submarines could fire their missiles while submerged and threatened U.S. carrier groups more effectively than could the surface ships that had previously been used for close-range shadowing. Up until then, the Soviets had been forced to rely on vulnerable surface ships to track U.S. carrier groups in a time of crisis.

In the frequently tense environment of the Mediterranean, beginning with the Six-Day War in 1967 through the Yom Kippur War in 1973, and the 1986 Gulf of Sidra strikes against Libya, the U.S. Sixth Fleet and the Soviet Naval Mediterranean Command operated cheek-by-jowl in near-wartime conditions. This gave them ample opportunity for close-range shadowing.

During the period when the late Admiral "Ike" Kidd commanded the Sixth Fleet in the 1970s, the "tattle-tale" tactic was incorporated into fleet doctrine in order to keep close check on Soviet surface combatants that were equipped with surface-to-surface missiles. These missile units were being used to continuously shadow and target U.S. carriers. Theoretically, when preparations for carrier air strikes were observed, the Soviet surveillance unit was authorized to neutralize the carrier by knocking out the flight deck. Consequently, Admiral Kidd's tattle-tale unit, which stayed glued to the Soviet surveillance ship, was in a position to strike the Soviet ship in advance. The equation grew more complex, though, as the Soviet navy increased the numbers and the multimission capabilities of its deployed ships and submarines.

Surveillance of the opponent's forces at sea is long accepted and continues today, yet with considerably less danger. In the early days of the Soviet fleet's expansion, however, close surveillance was fraught with the danger that a chance incident might escalate into a major confrontation. No Soviet-U.S. agreement yet existed to prevent incidents on the high seas; there was no special signal table, other than the International Code of Signals, for communicating intentions. Both sides operated only under the traditions of seamen's prudence and gentlemanly conduct.

Then in May 1965, Secretary of Defense Robert S. McNamara approved a recommendation that removed HUMINT collection from the navy's list of responsibilities. But a few months later this decision was reversed, and on December 7, 1965, navy secretary Paul Nitze assigned to the director of naval intelligence the task of implementing a clandestine intelligence-collection program in coordination with the Defense Intelligence Agency (DIA), its defense attaché system and the Naval Investigative Service:[19]

> The Navy has been largely inactive in the field of clandestine intelligence collection. The steadily emerging pattern of limited warfare engagement has, however, clearly indicated the need for the development of clandestine assets by the military services in advance of a limited warfare engagement. The Secretary of Defense has therefore established . . . a means for the increased participation by the Department of Defense in clandestine intelligence collection activities. DIA has authorized the individual services to engage in such activities.[20]

This instruction described the specific nature of a clandestine intelligence operation: "an activity . . . planned and executed . . . in such a way as to assure secrecy or concealment of the operation and its sponsor." Placed under the responsibility of the director of naval intelligence, clandestine intelligence collection excluded covert action. Similar collection operations were being set up by the army and the air force. The latter had its own clandestine intelligence team to deal with the recovery of Soviet, U.S., and other space objects that were considered to be of special importance under Project Moondust.[21]

Navy Task Force 157, housed in northern Virginia, was a secret intelligence unit with extraordinary powers. Hidden behind ten cover companies, it had a $5 million budget and comprised seventy-five operators, posted around the world. TF-157's mission was to keep tabs on Soviet shipping and naval capabilities by using pleasure yachts to watch over sensitive points such as the Strait of Gibraltar, the Panama Canal, and the Turkish Straits. Other missions included assessing Soviet armaments for the Strategic Arms Limitation Talks (SALT) and recovering downed U.S. and Soviet aircraft. The mere existence of a HUMINT capability within the navy was initially classified as secret, and the open reference to Task Force 157 was to be avoided.

In 1973, TF-157 scored a decisive intelligence coup when it reported that the Soviet Union had shipped nuclear weapons to Egypt during the October Arab-Israeli War. The production of the new navy clandestine service was impressive:

> Much of the information acquired from activity sources consisted of basic intelligence such as port and harbor data, information concerning foreign merchant ships, and routine naval order of battle reporting. This included the first high quality, ground level photographs of the new Soviet Mod-Kildin guided missile destroyers and Kara cruisers. TF157 also provided details of the Syrian Navy's actions in the 1973 War. A newly established unit based in the Persian Gulf had targeted Soviet and Iraqi naval activities in the area as well as seaborne arms shipments to Iraq.[22]

In 1974, TF-157 continued to provide a considerable flow of naval intelligence, including many high-quality photographs of communist shipping in the Turkish Straits and 110 movements by ships deemed suspicious by the Drug Enforcement Administration. TF-157's agents had conducted missions into the Leningrad area, Black Sea ports, the Kola inlet, and Soviet ports in the Far East. In the Middle East, their reports contained information from scores of missions into Egyptian, Syrian, and Somali ports.[23] In the Far East, sources had entered China and North Korea nearly one hundred times. One agent had spent nearly three months in a Communist Chinese port and kept a detailed log of all naval and air activity in the port during that period.[24]

Downfall

Although the navy's clandestine activity provided exceptional results, TF-157 was crippled by its disproportionate spending and internal strife.

One of its contract employees by the name of Edwin Wilson feared that his contract would be canceled, so he approached the director of naval intelligence, Rear Admiral Bobby Ray Inman, and offered to create a new intelligence unit that would take over TF-157's responsibilities. Horrified, Inman chose to dissolve "the activity" and declared that it was "out of control."[25]

On December 31, 1975, Captain Donald Nielsen, the commander of TF-157, appealed directly to Inman:

> Conflicts of view are never fun when they require one to tell his boss things he probably doesn't want to hear. . . . We are perhaps at least partly responsible for this, for the very physical and organizational separation that gives us much of our operational flexibility also keeps us out of sight—and probably out of mind. Without being here, you can have no appreciation of the volume, nature or variety of the work being done here on a routine basis. . . . Perhaps if I had made more appearances on your calendar you might have a different perception of us. . . . Whatever you think of this organization, I have a strong belief that what it is doing is right, efficient and good for the Navy. I believe to dismantle it in reaction to a cut . . . is wrong. Looking briefly at the big picture, [the] Navy would abdicate its position, recognized throughout the community, as the leader in Defense clandestine collection—the organization with the best program, clearest objectives, most efficient management, and most effective results. Navy participation in national clandestine HUMINT policy forums would cease to be pertinent. . . . The mistrust and stigma of leaving the field at this juncture, coupled with the always-Herculean staffing required with other agencies and within the Navy to develop such a capability, make it a matter of years, easily 5 to 10, even to re-enter the field, much less to re-create what we have now. You, another DNI, or any CNO may someday desire/require it. . . . I know your dislike for production figures, and we have carefully avoided "dog and pony shows" concerning our accomplishments. Suffice it to say that the take is significant and more worthy of your pride than your contempt.[26]

Nielsen enumerated the bilateral intelligence arrangements with various countries (Iran, Turkey, Greece, Germany, Denmark, the UK, Japan, and Italy) that would need to be modified or terminated. He also stressed that the navy would lose the capability "to move people, money, and equipment through the private sector in support of intelligence collection operations, and other intelligence undertakings, . . . a unique ability to effect such actions outside overt government channels, yet within full Navy control." Nielsen reminded Inman that TF-157 had moved $5 million for the secretary of the navy in 1974 and won the secretary's appraisal: "one bunch that did things right." Nielsen's appeal was to no avail. "The activity" was closed.

Among the many reasons for Task Force 157's poor image was that both army and navy HUMINT organizations had misbehaved during the period of anti-Vietnam protesters and civil rights campaigns. They had pursued strong evidence that Soviet intelligence was deeply involved in funding covert support of these movements. U.S. intelligence was thus availed of its best HUMINT experts.

During the administration of President Jimmy Carter, from 1976 through 1980, Admiral Stansfield Turner served as director of central intelligence. On presidential order, he promptly dismissed a core of experienced HUMINT personnel within the agency. Turner decided that the United States would put all of its marbles in the technical intelligence arena. Those purged were the most senior HUMINT experts, many with experience during World War II in Europe, the Pacific, and the Middle East.

The Turkish Straits

One delicate mission of navy HUMINT had been to monitor the traffic through the Turkish Straits. As the Soviet navy ventured out in force into the world's seas and the open oceans, its ships were forced by geography to pass through vital straits that surrounded its Pacific, Baltic, and Black Sea fleets. In the case of the vital Mediterranean, the Turkish Straits became the ultimate prize to be controlled in time of war. Until the end of the ancient rule of the Ottoman Empire, the Russian Black Sea Fleet was completely locked inside the Black Sea. Czar Alexander I once spoke of the Turkish Straits as the "key to the door of my house."

More than 4,000 miles of Russian inland waterways link the five inland seas of European Russia: the Baltic, White, Black, Azov and Caspian seas. Hence, the importance of access to the Turkish Straits. After the completion of the massive Volga-Don and Volga-Baltic canal systems, transfers between the seas were possible by large 5,000-ton ships and barges of the gigantic Soviet river-coastal fleet.

The West developed various devices and plans during the Cold War to defend the Turkish Straits in the event of war. NATO created the Mobile Force Concept and first exercised the notion in 1970 with amphibious landings in Greek Thrace. The object of the NATO Mobile Force was to thwart a Warsaw Pact thrust from the north, which was designed to split Greek and Turkish defenses west of the Bosporus, and to seize the straits. The international covenant regulating maritime passage of the straits assumed new importance in the Cold War naval campaign.

Deep concern for the defense of the Turkish Straits had begun in 1936, when the winds of war were threatening Europe for the second time in the century. The anxiety caused Turkey to send a diplomatic note drawing attention to the issue to the signatories of the Lausanne Straits Convention. The note proposed a new accord for the use of the straits. The Soviet Union, out of intense concern for the security of the Black Sea littoral against expanding fascists, initially supported Turkey. The resulting agreement was the Montreux Convention, which defined restrictions on warship use of the Turkish Straits in wartime and when war was deemed imminent by Turkey.

The Montreux Convention, still in force today, limits the total tonnage and transit time of all major combatants and submarines that are allowed in the straits. Black Sea powers are given an advantage over non–Black Sea powers in a provision of the convention that dictates that Black Sea powers must declare their naval warships that intend to transit the straits southbound into the Mediterranean five full days in advance. Warships of non–Black Sea powers are required to declare transit intentions eight days in advance. This declaration must be made by diplomatic note to the Turkish Foreign Ministry, in Ankara.

The Soviet navy needed to ensure that it could augment its Mediterranean force at will, so it resorted to a system of routinely declaring a group of reinforcing units, including amphibious ships with naval infantry on board. In this fashion the Soviets always had a force declared, and they could transit on any date, thus avoiding the delay of declaring a full five days in advance.

The Soviet government used the continuous contingency declarations and tricks with the wording and the nomenclature to ensure that they would be ready to increase their Mediterranean force in times of need. When tensions were high, they could keep a "fleet in being," albeit behind a major geographical choke point. This situation provided the background for a unique Western intelligence operation, which was begun in the late 1960s and worked extremely well until it was no longer needed.

For naval intelligence to gauge the Soviet Union's intention to augment its Mediterranean command, it was necessary to gain advance knowledge of the contents of Soviet diplomatic notes. But the notes were passed by hand, not electronically, and no amount of signal intelligence could read the declarations. Therefore, brilliant naval minds resorted to the oldest form of intelligence and wove an elaborate system to discreetly extract the required information from a human link.

A female courier was found to be carrying the transit declaration notes across Ankara from the Soviet Embassy to the Turkish Foreign Ministry. Western intelligence carefully planned to have her delicately compromised by a dashing Western naval officer, then seduced and recruited. As a backup, another young clerk was found to open the envelope and record the declared Soviet transit intentions. He was tactfully inducted into the club with the liberal application of expensive gifts, and the problem was solved. The desired transit declarations were suddenly regularly available at minimal cost from the willing accomplices.

Credit for properly devising the scheme goes to an imaginative naval attaché whose identity remains confidential. The routine worked for many years during the height of the most tense crises in the Middle East. A more elaborate intelligence collection organization was eventually formed in Istanbul to assist the naval attaché who lived in Ankara. Western intelligence formed a special unit in Istanbul to provide timely reporting and photo intelligence collection about Soviet warships as they passed through the narrow Bosporus.

As the Soviet fleet grew in size as well as sophistication, the U.S. and Allied ocean surveillance system grew proportionally, as did the requirement for timely reporting from Istanbul. To carry out the actual collection of information about the transiting ships, an attaché had organized a team of coast watchers that was initially made up of a mix of foreign and assistant naval attachés who resided in Istanbul. Upon receiving a tip that transiting ships were expected, the team members would fan out to take their photos. One member watched with high-powered binoculars from a strategic observation point overlooking the approaches to the Bosporus; others deployed in a small craft to the narrowest part of the strait, to obtain photos at close range and identify the ships. The unit shot the most remarkable quality of photographs, often even identifying senior Soviet officers aboard.

In general, the Soviet navy resorted to extraordinary means to camouflage the true identity of their deployed ships. Whereas most world navies openly display names and permanent hull numbers on their ships, Soviet cover-and-deception measures were intricate and included painting out-of-area pendant numbers on all of their deployed warships to confuse the observer. Although the tactic could be thwarted through the use of SIGINT correlations with high-frequency radio call signs, it was amusing to see the

Somebody's intelligence at work in the Turkish Straits.

extremes at which the Soviet mind worked. Often, the ship's side numbers were changed under cover of darkness, and occasionally one side of the ship displayed one hull number, while the other side had a different one.

It was crucial to be able to identify the ships not only by class but by name, to keep track of their inherent armament capabilities, whether they carried nuclear weapons, and whether they required in-port maintenance and overhaul—all valid intelligence objectives. The need to establish identifying features, in view of the Soviets' countermeasures, gave rise to the naval science of dentology. Although this was primarily a tool designed to identify surfaced submarines, dentology could also characterize surface combatants by fingerprinting them according to minor dents in their steel hulls. This grew into a precise science until the Soviet navy gave up the deception practice and in the 1980s began to display the true names of its ships.

The Istanbul reporting was extremely successful and the small collection craft so flagrant in its photo efforts that the Soviet sailors often countered by using large mirrors to reflect sunlight back against the cameras. The completion of the bridge spanning the Bosporus changed the task to one of merely walking out onto the bridge for excellent overhead coverage. In the meantime, the operational intelligence centers were improving their real-time support to the fleet with new methods of locating high-interest ships at sea.

Fusion

Within OPINTEL (operational intelligence), the U.S. Navy integrated sophisticated SIGINT, infrared tracking, passive acoustics, and other forms of electronic intelligence into a network of efficient correlation centers. From August 1960 onward, satellites began to provide limited photographic and signal intelligence. This capability, although crude in its infancy, could determine some positions of Soviet naval radar emissions. OPINTEL was also receiving regular acoustic input from SOSUS and info from the fleet weather central circuit and the naval movement report control center (a system devised in World War II to keep track of friendly ships).[27]

By the early 1970s, the system evolved and was renamed OSIS (ocean surveillance information system). It could detect aircraft preparing to take off from remote Soviet airfields and give advance warning of surface ships readying to leave their anchorages and submarines about to depart home ports. Western systems could track ships, submarines, and aircraft under and above the oceans and even pinpoint ships in congested areas and strategic ocean choke points, such as Malacca, the English Channel, and Gibraltar.

In 1971, the new OSIS system began to operate in the Mediterranean with the establishment of the first Fleet Ocean Surveillance Information Facility (FOSIF) in Rota, Spain. Rota was the first in a global system of centers to provide tailored, all-source tactical intelligence to U.S. and Allied fleets operating in contiguous ocean areas.[28] FOSIF Rota was the first such facility and was designed to support the Mediterranean-based U.S. Sixth Fleet. It was followed by the FOSIF in Kamiseya, Japan, for the Pacific Fleet. These facilities operated in parallel with the Fleet Ocean Surveillance Centers at the respective fleet headquarters in Norfolk, Pearl Harbor, and London.

During the first major worldwide Soviet navy exercise, called Okean 70, FOSIF Rota was the first center to evaluate Soviet operations in the context of the Sixth Fleet's own operations, giving due consideration to political and military activity in the littoral countries.[29] The ocean surveillance system that evolved was capable of sorting through merchant marine traffic in large areas of the oceans and singling out small Soviet intelligence trawlers hiding in fishing fleets and submarines steaming with sails awash. Sophisticated SIGINT combined precise communications analysis, called COMINT, with analysis of all forms of electromagnetic signals emanating from Soviet surface ships, submarines, and aircraft. Electronic fingerprinting by a combination of shore-based, shipboard,

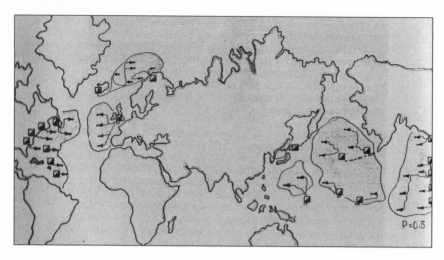

The Russian vision of the U.S. SOSUS hydrophone network.

and satellite platforms evolved to the point where individual Soviet naval units could be identified and located regardless of their position whenever they operated their radar for a brief moment. As Soviet naval communications shifted from the earlier cumbersome high-frequency bands to short-range UHF and VHF satellite systems, Western detection and exploitation capabilities grew to match Soviet advances.

U.S. passive acoustic detection and tracking systems, fixed and deployed on surface ships and submarines, made giant strides toward the goal of total ocean surveillance coverage. Airborne all-source tracking platforms such as the Grumman E2 Hawkeye became more capable of tracking multiple naval threats from all directions. The Grumman E2 Hawkeye was the navy's tactical version of the larger theater-level airborne early warning and air control system called AWACs. These ocean surveillance systems kept pace with and eventually surpassed the challenges posed by the burgeoning Soviet fleet.

The immense underwater detection and tracking system called SOSUS was built on the principal of arranging fixed passive underwater detection arrays. It reached its peak of efficiency when its upgraded computer systems grew capable of identifying and classifying submarines, not only by class and type, but by individual acoustic fingerprinting. The SOSUS system, together with the Lockheed long-range ASW patrol aircraft, P-3 Orion, and the Los Angeles class nuclear attack submarine, made it possible to keep pace with the growing number of Soviet ballistic missile submarines of the Yankee, Delta, and Typhoon classes. The latter were equipped with the longer-range, 8,000-nautical-mile, submerged-launched ballistic missiles, SLBMs. Although these missiles had an extended range that allowed them to be fired from submarines in Soviet home waters and still reach their targets in the continental United States, the American ASW system, combined with Los Angeles class killer submarines, grew capable of locating and destroying the launch platforms wherever they were.

The U.S. locating methods were based on high-frequency direction finding, called HF/DF, and exploiting shorter-range UHF and VHF satellite communications that used overhead SIGINT systems. These methods expanded greatly throughout the 1970s and became an integral part of the independent aircraft carrier battle group. The advance technology of SIGINT and acoustics, linked with overhead satellite tracking systems, virtually exposed the oceans to America's tactical and strategic vision.

As authors Christopher Ford and David Rosenberg wrote, "The most significant innovation during the 1970s was the arrival of [a] new ELINT collection system in the autumn of 1976. Analysts were now able to track all ships and other ELINT 'emitters' at sea on a near real time basis."[30]

Meanwhile, under the seas the United States already had one of the most extensive intelligence-gathering systems in the world. The daring operations of its special collection submarines *Halibut, Seawolf, Grayback*, and *Parche* were revealed in 1997.[31] These submarines had been modified and since the late 1960s were successfully used with remotely piloted deep submergence vehicles, DSVs, to observe, photograph, and recover invaluable intelligence from the ocean depths. They mounted special devices attached to their periscopes to shoot infrared photographs and collect information about the cruise missiles tested in the Okhotsk Sea. They recovered parts of Soviet long-range strategic missiles from the bottom of the ocean. In 1967 and 1976, U.S. submarines also "stole" two mines near Russkiy Island and recovered two atomic bombs from a Soviet plane that had crashed into the sea.[32]

In 1981, the former NSA employee Ronald Pelton walked in the Soviet Embassy in Washington and volunteered intelligence documents in exchange for money. Pelton was asked to shave his beard and was given worker's clothes to exit the Soviet Embassy. He was later invited to Vienna, where he met with top Directorate K officers. The KGB veteran Cherkashin remembered the traitor's most startling revelation: "Most of Pelton's information concerned general NSA activities. There was only a little top secret intelligence until he dropped his first bombshell during a trip abroad. . . . It was then that he handed over documents about the NSA operations called Ivy Bells . . . the code name for a highly ingenious, wildly expensive and risky U.S. Navy operation that tapped into a secret Soviet communications cable at the bottom of the Sea of Okhotsk, which separated the Russian mainland from the Kamchatka peninsula."[33]

The Soviet Pacific Fleet knew there was a security leak because U.S. aircraft would appear when missiles were tested. Under the code name Operation Combala, the Soviets attempted to locate the U.S. listening device described by Pelton. On October 20, the survey ship *Tavda* discovered two seven-ton cylinders 6 kilometers off the coast of Kamchatka at a depth of 84 meters. There, the Soviet communications cable stretched between the fleet headquarters in Vladivostok and the submarine base of Petropavlosk. Attached to the cable's 120 wires were 120 tape recorders. The U.S. apparatus worked on a battery of uranium 238 and had the capacity to transmit information to a satellite via an antenna attached to a buoy that could be lifted to the surface. In the Barents Sea, the Soviets found traces of similar equipment that had been removed.[34] As Cherkashin remarked, an embittered NSA employee had sold a multimillion-dollar intelligence operation for only $35,000.[35]

Despite this setback, the U.S. ocean surveillance system was born with drama and excitement, and its effectiveness increased with the growth of the Soviet fleet. In the midst of the preparations to launch the new intelligence centers in Rota and London in 1970, U.S. intelligence officers were interrupted by repeated crises in the Middle East. Each time the situation grew critical, scores of Soviet naval ships, some with naval infantry (their marines) aboard, positioned themselves tactically between U.S. naval forces and the crisis area. As Ford and Rosenberg explained, "All in all, the various advances and innovations that came together during the 1970s helped produce a sophisticated, worldwide ocean surveillance system capable of providing Navy commanders anywhere on the planet with an integrated, 'God's eye view' of their operating environment. As a system of aerial, surface, and subsurface systems continuously providing locating data on maritime activity," OSIS was soon to be regarded as an effective intelligence system, one of the strongest aspects of naval intelligence at the time.[36]

Beard Bite

One intricate operation that was designed to assess the effectiveness of Soviet ocean surveillance in 1975 was to use a routine transit by a nuclear-powered aircraft carrier. The scheme was named Beard Bite. Picture the vast Pearl Harbor Ford Island command center of the U.S. Third Fleet, which on the early morning of the operation stood empty. A young watch officer peered up at a gigantic chart of the Pacific basin that covered the entire wall. Within hours, the wall map erupted with activity and became the focus of the entire staff, from the three-star Third Fleet commander down to the lowest watch standers. The Pacific Fleet commander in his Makalapa headquarters, appropriately located in the hills above Pearl Harbor, also joined in the commotion.

The activity fixed on the blue plastic track line laid out neatly in an arc on the charts between Pearl Harbor and Subic Bay. The chart paper along that line was frayed from the constant movement of colored magnetic ship forms transiting in great arcs across the wall plot. One end of the chart, showing western Australia, hung above the coffee pot sitting on a table. There, the cleaning woman had left her mop, still in a scrub bucket, with its handle leaning on New Zealand and Tasmania. The other end of the chart, depicting Baja California, was behind the coatrack, where the command duty officer hung his black raincoat during the Hawaii monsoon season. The khaki cap perched above the coat had tarnished gold leaf or "scrambled eggs" on the visor, revealing the seniority of its owner. Most of the Third Fleet's command duty officers were career-worn commanders.

The enormous chart stretched from floor to ceiling. Along the bottom, the chart portrayed Mururora Atoll and Antarctica. At the top of the map, the name Provedinija was bent at the corner where the wall met the ceiling, symbolizing that northern port's extreme isolation. Colored forms cluttered the broad Pacific chart between the Hawaiian Islands and west to Japan and the Philippines. On the extreme left, figures of red surface ships, submarines, and aircraft sat clustered adjacent to the black outline of the Soviet Union's land mass. Columns of brilliant red submarine forms were arrayed to the right of Petropavlovsk-Kamchatka, depicting the large Soviet submarine order of battle. In the Sea of Japan, another blur of red forms was surrounded by narrow choke points leading to the broad Pacific: La Perouse in the north, Tsugaru in the center, and Tsushima and the Korea Strait to the south. The names were barely legible in faded ink; they listed the constraining passages deplored by the Soviets. In the calm before the storm, the watch officer gazed at the track of the carrier that was poised to depart Pearl Harbor in a few minutes, then head for Subic Bay in the Philippines. The aircraft carrier's mysterious caper would strike fear in the bowels of Soviet naval staff officers in Vladivostok and the Moscow naval high command on Griboyedova Street.

To the initiated in the Ford Island command center, the aircraft carrier's track showed an odd jog at a position marked with a date and a time of the following morning. At that point, the carrier would begin a diversionary feint and go electronically silent. This maneuver was intended to activate the Soviet ocean surveillance reconnaissance system: SIGINT and radar satellites, SIGINT ground sites, and over-the-horizon radar. The Soviets would mobilize all of their assets to obtain the carrier's position. This would allow them to station an anticarrier nuclear attack submarine, which was lurking in mid-Pacific, to intercept and trail the large carrier.

The unusual electronic silence of the carrier, the USS *Enterprise*, would bring out the "Bears." These giant Tu-95D long-range reconnaissance aircraft with owl-like wings could take off from bases within the Soviet Union (shown on the wall to the extreme left of the command center chart) and cover the Pacific in an arc that encompassed the Hawaiian to the Fiji islands, the Coral Sea, and even the Philippines. During the Cold War, the Soviets frequently flew their Bears over or near U.S. carriers during mid-Pacific transit or when the carriers and their formations ranged into threatening proximity to the USSR. The Bears' mission was as much political as it was tactical. It was staged to demonstrate the Soviets' ability to track the carrier and their readiness to substitute air-to-surface-missile-carrying aircraft that had a similar range in place of the reconnaissance Bears.

The appearance of the giant bombers near the carriers was a routine display of Soviet defensive preparedness. When the carrier reacted by sending combat air patrol fighters to intercept these huge lumbering airplanes, it was a similar statement of readiness. But both sides held back on revealing some measure of their capability. Perhaps the U.S. fighters intentionally intercepted the Bears at less than their best range, to reserve some surprise. Likewise, the Soviets' aircraft might have deliberately missed the carrier by several hundred miles to hide their hand.

This sophisticated game of tactical poker played on for years. Both sides applied their military forces with sufficient skill to justify generating additional funds at home, so that they could expand their advantages with more capable gadgets. In this case, rather than trying to hide the carrier, the United States was using electronic silence to provoke the Soviet side into displaying more of its reconnaissance capability—to find the carrier without using purely passive means. Thus, the stage was set for a new hand in the nautical poker game.

During the transit, the *Enterprise* appeared to prove that a U.S. carrier could effectively evade the Soviet ocean reconnaissance system, hide

A Soviet Tu-95 Bear D reconnaissance flight over the USS *Enterprise* in the Pacific Ocean in 1978.

in the broad ocean like a needle in a haystack, and move undetected to any spot in the Pacific to carry out a mission—in total electronic silence. As the 90,000-ton behemoth departed Pearl Harbor on her initial transit leg, a tiny Soviet SIGINT trawler, bristling with antennae, lurked just outside the sea buoy. The trawler slowly fell in astern, and the two proceeded toward the jog in the blue wall track. Before reaching the turn, the nuclear-powered *Enterprise* suddenly surged to 30-plus knots and quickly left the trawler in her wake. The course change for the USS *Enterprise*, which took her due south of the normal carrier transit lanes, coincided with initiating complete electronic silence. The rest of the show would be pure pleasure to watch, as the Soviets were compelled to search for the carrier.

The operation had been conceived by an imaginative intelligence officer at Third Fleet Headquarters, Commander Peter Trentwell White, who was considered by many of his seniors to be a loose cannon. Yet his success in gaining the authority to tamper with the western Pacific transit of the sacrosanct *Enterprise* was a feat that demonstrated his remarkable tenacity.

The drama began as the *Enterprise* achieved and maintained her assigned silence—no mean task for a carrier transporting an air wing that was inclined to favor the comforting sound of its pilots jabbering on VHF circuits. Within hours, two pairs of Soviet Tu-95 Bears took off in rapid succession and ranged out to the Pacific to search. The first attempt missed the carrier, and the second apparently failed by a wide margin, even after the *Enterprise* began to transmit gradually on designated frequencies. The U.S. operation was a total intelligence success; the Third Fleet officers watched as the Soviets activated one surveillance system after another, to no avail. To the carrier advocates, the operation was proof that the Soviets weren't really as good as assumed. But then again, the Soviet submarine that was suspected to be waiting to intercept in mid-Pacific was never positively located, and perhaps she had not been there at all.

As the transit continued, the Americans enjoyed the performance. It was especially satisfying for them to imagine that in Vladivostok, located on the left side of their wall chart, a naval officer speaking excitedly on the phone with another, probably his senior in Moscow, as he described his version of the action unfolding in the broad Pacific.

This was one of many Cold War operations that pitted fleet operations against the vast ocean surveillance systems of both sides. Later in Moscow, during years of frequent contact with senior Soviet naval officers, U.S. naval attachés had many opportunities to ask senior Soviet officers about their opinion of the U.S. Navy's ability to hide aircraft carriers in the broad ocean. In each case, the Soviet officers gave the impression that the Soviets had been amused by the idea that U.S. commanders believed it had been proved possible. General of the Army Yakov Moiseev, the former chief of Soviet general staff, claimed in a November 1994 interview that in the grand tradition of Russian chess abilities, they sometimes chose to give the U.S Navy the impression that the Americans had indeed managed to hide an aircraft carrier. In fact, said Moiseev, "we never lost track of American carriers, merely used your deception operations as beneficial training exercises for our ocean surveillance systems."[37]

The Pony Express

Pony Express was the operational name of the effort that was designed to collect data from the Soviet Union's tests of intercontinental ballistic missiles (ICBMs) and their attached warheads. The United States needed to know exactly what ICBMs the USSR had and how they flew. To determine their capability and characteristics, it was necessary for the U.S. military to place a collection platform down range to see what was being tested. The United States could then gather data, debris, and telemetry in order to analyze exactly what the tests said about the Soviet weapons.

Strategic missile–collection operations enjoyed the highest priority, mainly because of the strategic arms limitation talks that were under way and the need to know what the United States had to counter as a real threat. In the mid-1970s, this information could not be gleaned solely by satellite. The United States therefore used specially configured platforms to gather valuable data directly from the down-range missile-reentry areas over the Kamchatka Peninsula and Pacific Ocean impact areas.

The concept was simple: put a platform down range in a Soviet test area and try to catch the warhead before it impacted—Operation Pony Express. The Soviet Union stretched more than ten time zones from west to east. The Soviet Strategic Rocket Forces could test their intercontinental ballistic missiles by launching them from sites in the European side of the USSR to land on the volcano-studded and sparsely inhabited Kamchatka Peninsula on the Pacific side. The United States could station specially configured ships and aircraft outside Soviet territorial waters and airspace near the Kamchatka coast to observe the incoming test warheads on radar and by telemetry as they completed their trajectories, reentered the Earth's atmosphere, and impacted in the remote Kamchatka hills. From these observations, it was possible to determine the accuracy and the number of warheads that were intended to ride on each missile and to intercept other valuable data.

The surface platform designed to carry out this collection was called the ARIS, or Advanced Range Instrumentation Ship. The United States had two of these vessels, the reconfigured World War II troop ships USNS *Vandenberg* and USNS *Arnold*. They had been fitted with large antennae for onboard high-resolution C-band radar to detect extremely long-range targets. The two ships were initially tasked with carrying out an identical mission for U.S. ICBM testing in the Pacific Missile Range, which stretched from California's Point Magu to Vandenberg Air Force Base.

The ARIS could study the accuracy and the behavior of U.S. ICBMs and warheads, while concurrently evaluating the antimissile defense system against the inbound ICBMs. The two ships were manned by civilian crews and operated by the U.S. Navy's Military Sealift Command (MSC). During Soviet missile testing on the Kamchatka Peninsula, the two ARIS's were placed under the operational command of Commander Third Fleet in Pearl Harbor. This made it possible for one commander and his staff to coordinate the ships and the special aircraft that collected the data.

The U.S. naval ship *Vanderberg* monitored Soviet tests within 20 miles of the Kamchatka Peninsula.

In addition to the C-band radar, there was sensitive monitoring equipment aboard the *Arnold* and the *Vandenberg*, which made the two ships and their personnel extremely precious. The command structure and the contingency plans governing Pony Express operations were formed with the experience of the USS *Pueblo* fresh in mind. With similar risks to consider, the Third Fleet staff drafted elaborate plans to ensure that tactical armed air support was available in case an attempt was made against Pony Express ships. The precautions were designed specifically against Soviet forces that might attempt to capture USNS *Arnold* or *Vandenberg* while they observed missile tests within 20 miles of the remote Kamchatka coast.

The specially configured aircraft participating in these operations were RC-135s, flown by the U.S. Air Force out of bases in Alaska, and U.S. Navy P-3 and special EP-3 SIGINT and ELINT Orion patrol aircraft from Naval Air Station Adak. U.S. Air Force tactical support units, consisting of F4 Phantom fighter-bombers, were deployed to the advance base of Shemya to cover all contingencies. The operations called for exact timing and precise planning because the distances were long and the fighter aircraft necessarily relied on tankers to reach the ARIS station if protection of the participants was required.

No personnel were lost during these operations, but there were close calls. One occurred when USNS *Arnold* lost all power in heavy seas and drifted helplessly for several hours before regaining electrical and main propulsion. A real nightmare scenario would have been one of these defenseless ships being driven in stormy seas against the rocky cliffs of Kamchatka, while the Third Fleet Headquarters in Pearl Harbor stood helplessly by. The two ships were old World War II Victory hulls and had engineering plants that were built in the 1940s. The whole operation was choreographed like a dramatic ballet and called for split-second timing to ensure that all participants were in place and ready for the Soviet missile tests. The usually stormy and frigid northern Pacific waters left little room for mistakes.

Pony Express operations were often mounted on short notice, since they used a combination of intelligence intercepts and the Soviet Union's unpublished test schedules. Tip-offs were generally based on communications preparation activity noted at the normal launch sites of Kapustin Yar, Shary Shagan, and Plesetsk in the eastern Soviet Union. As the Third Fleet intelligence coordinator for Pony Express operations, Peter Huchthausen often flew to remote stations to brief the participants: at Kwajelain Atoll, Adak, and even Papeete in French Tahiti.

The Soviet Strategic Rocket Forces began to test their intercontinental ballistic missiles by firing into the open Pacific Ocean in the early 1970s, especially for the new longer-range missiles like the SS-18 and SS-19. In addition to research and development purposes, the Soviets primarily tested to prove to the world that they indeed had long-range missile capability. Much of the testing was pure showmanship to let the West know that they possessed an ICBM with multiple warheads and a range of more than 6,000 miles. The missile shots into the open ocean areas were always announced in the press well ahead of time. The United States placed ships on the scene to verify that the tests took place and to gain important evidence of Soviet capabilities. This complex operation served both the Soviets' and Western purposes. The broad ocean area testing involved even more players than did the missiles shot at the Kamchatka Peninsula.

The U.S. Navy assigned two small modified Claude Jones class destroyer escorts to the operation, to provide special collection capabilities in addition to the giant radar of the ARIS. The destroyers would wait down range in the anticipated impact area and would deploy specially designed flotation rings. When the rings were thrown on the water, they opened and a netlike device fell from the floating doughnut. These devices were designed to retrieve precious debris from the missile-reentry vehicle. The invaluable retrieved material allowed the U.S. Navy to analyze the warheads, as well as the parent missile and bus that carried the multiple warheads.

In one harrowing action, an incoming Soviet ICBM reentry vehicle split into its several dummy warheads, which splashed into the Pacific and bracketed the two U.S. destroyers at distances of less than 100 yards. To complicate the operation, the Soviets deployed their own range instrumentation ships, which had been built for down-range operations. These were the Sibir class space event–support ships, each carrying a helicopter. The presence of so many surface ships and aircraft often made the impact area resemble a carnival.

Operating in close quarters at the event would be three Soviet special-range instrumentation ships, the American ARIS, and the two U.S. destroyer escorts, the USS *Claude Jones* and *McMorris*. All of these ships positioned themselves in a circular area that was declared hazardous to navigation by the Soviet Union in an international "Notice to Mariners and Airmen." These announcements were issued several days in advance to warn innocent ships against wandering into the area and getting clobbered by reentry vehicles plummeting earthward from test ICBMs. To this scene would be added three Soviet Ka-25 Hormone helicopters

A Soviet Sibir class instrumentation-range vessel that was used to track Soviet and U.S. missile trajectories.

flying in a circular pattern, at least one U.S. P-3 Orion, possibly an E-P3 for tip-off, an air force RC-135, and sometimes an additional C-130 Hercules, each trying to get as near as possible to the reentry vehicles without getting in the way.

When the Soviet sailors realized the purpose of the floating nets deployed by the Americans, they complicated matters by ensuring that each one carried a full load of their ship's trash and garbage, jettisoned with great skill before the anticipated impact time of the reentry vehicle. The U.S. sailors awaited the results of the debris analysis with great interest after the operation's conclusion—one Stolichnaya vodka bottle, a heel of stale brown bread, unidentified cloth, possibly a sock, and so on. During one of these open ocean tests, a Soviet helicopter accidentally ditched in the sea. The U.S. forces' ships joined in the search and rescue and actually recovered one injured crew member, who was transferred to the Soviet ship after receiving first-aid treatment. Despite the humorous aspects of these operations, the overall atmosphere was often foreboding.

The down-range collection effort proved that the Soviets had the capability to reach the designed distance and achieve relative accuracy within their expected target area. But the missiles fired for such exhibits were usually from fixed sites and less often from actual submerged submarines. Later, the Soviets tested from submarines that launched from the northern Barents Sea areas, traveled across the entire USSR, and landed on the Kamchatka Peninsula range, where the United States could observe their accuracy. The West learned that they worked but also found out later that the unstable liquid fuel used in missiles deployed aboard Soviet submarines well into the late 1980s caused frequent and costly accidents and rendered the ballistic missiles of overall questionable reliability. Western intelligence has learned from data revealed since 1991 that there were many explosions and fires caused by the liquid missile fuel aboard Soviet submarines. On at least two documented occasions, ballistic missiles with their nuclear warheads were ejected uncontrolled from their submarine launch tubes as a result of accidental liquid fuel explosions.[38] Despite the success of the U.S. collection operations, the Soviets continued to expand their shaky strategic arsenals.

Mutiny

An extraordinary event occurred in 1975. It became a harbinger of the traumatic changes in store for the USSR, still ten years beyond the horizon. Although at the time some news appeared in the world press, the incident generally passed with little fanfare, due to the long-standing blackout on information about Soviet military affairs.

The incident was an actual mutiny carried out aboard the most modern antisubmarine ship then in the Baltic Fleet. This sleek 4,000-ton, Project 1135, multipurpose frigate (NATO class Krivak I) was named *Storozhevoy* ("sentry"). The insurrection was conceived and led by a deranged political officer, Captain Third Rank Valeri Sablin. The mutiny ultimately provided the idea for the popular novel *Hunt for Red October* by Tom Clancy. Unlike the Clancy novel, where the plot centers around a disillusioned anticommunist officer seeking refuge in the West, the real mutineers were radical Marxist-Leninists planning to "purge the Brezhnev regime of nepotism, bribery, careerism and contempt for the people."

In Stockholm's naval intelligence center, a young lieutenant sat at his radio receiver listening to Soviet naval communications. He was alone as duty officer and heard the intercepts of the Soviet aircraft in hot pursuit of the *Storozhevoy*. Flabbergasted, the lieutenant called for assistance, and still no one believed him.[39] During the final phase of the episode, Western SIGINT operators also sat in disbelief, listening in on radio nets that controlled the Soviet Tu-16 Badger medium bombers pursuing the ship. They heard senior Soviet commanders pleading with the reluctant bomber pilots to open fire against their own ship. The bombers fired on the frigate after Sablin ignored warnings to stop. Sobered by the sudden use of force, the band of revolting crewmen released the original commanding officer, who then led an assault on the bridge with a group of loyal officers. In the struggle, Sablin was shot in the leg and capitulated. Although he was supported by practically every member of the crew during the public protest against corruption, Sablin was turned on by his fellow conspirators. He was later tried and executed for treason against the motherland. He never recanted his radical reform views.[40]

Soviet Naval Intelligence Excellence

To monitor an enemy fleet approaching its coasts, Soviet naval intelligence first relied on HF intercepts and on land-based, over-the-horizon radar. In the 1960s, long-range reconnaissance aviation and surface reconnaissance ships extended the Soviets' capabilities to monitor U.S. and Western naval activities. The USSR's total number of intelligence collectors increased from less than ten in 1962 to sixty in 1979. Fitted with powerful radar, a modified intercontinental bomber, the Tu-95RTS BEAR D, entered service in 1964. A total of forty were produced.

Airborne radar could detect surface targets at ranges of up to 200 nautical miles and could provide submarines and surface ships with targeting information to launch their attacks over the horizon.[41] Flying above 10,000 feet, BEAR D would generally detect a carrier task force at 150 nautical miles and relay that information to a submarine 100 nautical miles behind it, more than 250 nautical miles from the carrier. Taking advantage of the "duct" effect, the Soviet Union also developed ship-borne, over-the-horizon radar, which used the properties of troposphere propagation to achieve detection ranges of up to 200 nautical miles in favorable conditions.[42]

After 1970, space reconnaissance systems helped the GRU and the fleet intelligence centers to create a global picture of "probable enemy" naval activities. The reconnaissance service was tasked to provide targeting information to the fleet and had a direct operational responsibility. Designated Morya-1 (Sea-1) and Legend, the Soviet maritime space reconnaissance constellation comprised both active and passive satellite systems to detect a group of ships and identify their radar signatures and nationality. Flying in a low orbit, the US-A active radar satellites could reposition a target every twenty or thirty minutes and provide information on the speed and the direction of the ships. Passive US-P satellites then intercepted the ships' electromagnetic signatures for identification. At least six ocean reconnaissance satellites were in orbit at any given time. Given their low orbits, thirty to thirty-five satellites had to be launched every year.[43]

The defector Victor Suvorov, in his unreliable book on the GRU, described the Soviet naval reconnaissance system and the coexistence of two "cosmic" reconnaissance organizations:

> The Soviet Union possesses two independent cosmic intelligence organizations, one within GRU and the Navy's cosmic intelligence organization. Although naval cosmic intelligence serves the interests of the Soviet Navy High Command, all information is shared with the GRU. The co-operation between the two cosmic services is co-coordinated by the chief of the General Staff. Should a very serious situation arise, the same task may be set at the same time for both services and the results can be collated and compared.[44]

The Soviet space system remained vulnerable in wartime and had to be supplemented by other means of detection and monitoring. These included intercepting U.S. fleet broadcasts, which enabled Moscow to track every U.S. ship during the 1970s.

The Soviet Union never built a global acoustic detection system comparable to the U.S. SOSUS. A series of hydrophones, however, were deployed off the coast of the USSR to detect approaching submarines and surface ships.

Highly centralized, the Soviet system was comparable to the U.S. OSIS system. And like the U.S. system, it reached a peak of excellence in the late 1970s and in the 1980s, to reposition the "probable enemy's" carrier groups every half hour.

Changing National Strategy

During the summer of 1981, Rear Admiral Sumner Shapiro, the director of naval intelligence, convened senior U.S. naval intelligence officers for a brainstorming meeting. The participants included analysts from the ONI, the CIA, the DIA, the Naval War College, and the Center for Naval Analysis. This navy think tank had come up with a revolutionary analysis. So far, the strategy of the Soviet navy had been seen as aggressive, and it was anticipated that the Soviet Union would fight another "Battle of the Atlantic" against Western sea lines of communications (SLOCs) in the Atlantic. It fit with U.S. intellectual maritime traditions. Captain Alfred Thayer Mahan's sea-power theories focused on control of the sea lanes and decisive engagements. In 1978, an unclassified U.S. Navy brochure titled "Understanding Soviet Naval Developments" was promulgated throughout the fleet to help operational forces better understand their potential adversary. U.S. Naval Intelligence viewed the Soviet navy as "an aggressive force capable of challenging the United States in all aspects of maritime activity"; "The Soviets are firm believers in the old adage that the best defense is a good offense"; "The Soviets are employing their Navy in much the same way as the United States and Great Britain . . . and can now perform most of the functions of a naval power in waters distant from the Soviet Union."

Information received from deep-penetration U.S. sources placed within the highest Soviet circles, combined with intellectual debate among U.S. experts on the Soviet navy, led to a complete reassessment of past beliefs. It had become clear that even if a war with the United States did not involve the actual use of nuclear weapons, the Soviet Union would focus on protecting its strategic submarines in order not to alter the nuclear balance in its disfavor. The Soviet navy had no strategy of its own. It remained focused on the concept of "nuclear correlation of forces," which would decide the issue of a war. From a Soviet perspective, there was no point in attacking Western sea lines of communications or seeking a decisive "Mahanian" battle. It was more important to keep its nuclear deterrent alive. The naval strategy of the Carter administration had focused on protecting sea lines of communications, and it was now proved to be all wrong.[45]

As Captains Vladimir P. Kuzin and Vladislas I. Nikolskiy explained, this debate had already taken place in the Soviet Union when the "young school," supported by Premier Khrushchev and Marshal G. K. Zhukov, who advocated attacking Western sea lines of communications, had given this mission to the Soviet navy. From a Soviet perspective, fighting another battle of the Atlantic was both too costly and of little use since Moscow intended to conquer Western Europe in less than three weeks, the time that the first convoy would need to reach Europe: "If Zhukov had known that for the real destruction of the communication lines of the probable enemy (making attrition greater than replacement) at that time it was necessary to sink over two million tons of merchant ships each month, which would have required by the most modest calculations about one thousand Soviet submarines and the monthly production of no less than fifteen thousand torpedoes. Given those figures, it is most likely that Zhukov would have posed this task for the Soviet Navy alone."[46]

Revolutionary Estimate

As a result of these intelligence and analytical breakthroughs, U.S. intelligence completely changed its appreciation of Moscow's naval priorities. The "revolutionary" National Intelligence Estimate (NIE) of 1982 now listed Soviet naval priorities as follows:

- Providing combat stability (protection and support) for Soviet SSBNs, principally through the creation and maintenance of submarine safe havens, or bastions, in Soviet SSBN deployment areas
- Defending the USSR and its allies from NATO sea-based strike forces (aircraft carriers and Western SSBNs)
- Supporting ground forces involved in land combat against NATO in Europe or elsewhere
- Interdicting some Western SLOCs

The new analysis determined that the Soviet navy's objective in wartime would be to seize control of and defend the Kara Sea, the Barents Sea, the northern portions of the Norwegian and Greenland seas, the Sea of Japan, the Sea of Okhotsk, and the Northwest Pacific basin. Its top priority would be to provide strategic submarines with "combat stability" by protecting them against Western attacks.[47]

As Ford and Rosenberg wrote:

> The late 1970s and the early 1980s represent one of the few instances in history where the acquisition of intelligence helped lead a nation to completely revise its concept of military operations. In the words of two participants, this period thus stands as an example of how good intelligence, well analyzed and well applied by teams of intelligence officers and line officers working together, enabled the U.S. Navy to devise a strategy and a set of war plans which would have helped ensure victory, should we have had to fight a war with the USSR."[48]

From now on, the United States would focus on a new aggressive maritime strategy to threaten Soviet strategic submarines in their home waters and divert resources previously invested in the defense of the sea lines of communications.

13

WAR SCARE AND PSYOPS
1981–1987

The U.S.-Soviet imbroglio has the "familiar characteristics, the unfailing characteristics, of a march toward war—that and nothing else."

George Kennan

Never, perhaps, in the postwar decades was the situation in the world as explosive and hence, more difficult and unfavorable, as in the first half of the 1980s.

Mikhail Gorbachev, February 1986

Doomed to Strike First

With the information recently published or made public about secrets previously held by both sides of the U.S.-Soviet schism, it is now possible to look back and understand the frightening period of 1981 to 1987, when tensions were at their highest since the 1962 Cuban missile crisis.

During the 1970s, the Soviet Union had achieved strategic military parity with the United States. The navy that was created primarily by the efforts of Sergei Gorshkov, the fleet admiral of the Soviet Union, was second to none in its number of warships and submarines. The United States still surpassed the USSR in the number of warheads on ballistic-missile nuclear submarines, but, unlike Washington, Moscow did not consider the missile submarines a main component of the strategic triad: namely, air, ground, and sea-launched missile forces. According to Soviet thinking at the time, strategic submarines were meant to participate in the destruction of the military and administrative centers of the probable enemy, alongside the other strategic forces. They were not officially considered a deterrent force of a special nature. The Soviet navy's main concern remained the West's aircraft carriers and ballistic missile–carrying nuclear submarines.

237

Accordingly, tracking these forces at close range was vital to prevent them from carrying out a nuclear strike on the Soviet Union.[1] For Moscow, the main endeavor was to prevent the enemy from seizing the initiative. Hence, a "combat service" was established to perform the tracking function. Based on its calculation of the enemy's intentions, the Soviet navy would be forced to make a preemptive strike. Failing to do so, in their minds, would mean losing the war.

It was for these reasons that in the 1950s and the 1960s, light gun cruisers of the Sverdlov class tracked U.S. carriers during times of increased tension. Should the tracking ship observe the carrier to be preparing a strike, which was a pretty difficult matter to conceal, the Soviet unit would simply open fire to destroy the carrier's flight deck. Whether one believed these large behemoths were really unsinkable or not, it would be easy to put the vessel's flight deck out of commission with a few well-directed rounds from a cruiser's main battery or missile. These Soviet gun platforms were later replaced by a whole group of longer-range submarine-launched, air-launched, and surface-launched cruise missiles.

During the 1970s and the 1980s, the theoretical number of antiship cruise missiles that the Soviet navy could fire in a salvo against a Western aircraft carrier group grew from fifty to about one hundred, with ranges of up to 300 miles. This theoretical number of one hundred was considered adequate to either destroy (with nuclear warheads) or neutralize (with conventional warheads) a Western carrier group. To fulfill that critical task, detection systems were of necessity greatly improved. Soviet naval aircraft could detect and track (with the detection radar YEN-D and Rubin) an opposing surface fleet from an altitude of 10,000 meters and a distance of 180 to 240 miles. The Soviet naval TU-95RTS Bear long-range bomber, introduced in 1966, and the Ka-25RTS (Hormone) helicopter both utilized the powerful Uspekh-U target designator. They could transmit targeting data to missile-shooting surface ships of the Kynda, Kresta I, Kirov, Slava, and Sovremenny types and, likewise, missile-equipped, nuclear-powered submarines of the Charlie and Oscar I classes. Other aircraft and helicopters mounted less powerful radar with a range of 30 miles for data transmissions to the surface and submarine shooting forces (onboard the Be-12 seaplane and the Ka-25 and Mi-14 helicopters).[2]

Taking advantage of a wave-propagation property known as the ducting effect, which was not exploited at the time by Western navies, the Soviets had developed an over-the-horizon target-acquisition capability to strike enemy vessels beyond the normal radar-detection range. But the most revolutionary Soviet systems were the automated information

The main concern of America's and NATO's navies in the 1980s: an Oscar I cruise missile nuclear submarine SSGN (top) and a Victor III nuclear attack submarine SSN (bottom).

receivers that the Soviets used to process the targeting data provided by intelligence satellites that collected radar images and radioelectronic intercepts of Western battle groups. (Korvet became operational in 1975, followed by Korall-B and Kasatka in 1981–1982.) The Soviet soft space intelligence satellites, however, were highly vulnerable to U.S. antisatellite weapons, which could be launched in the first critical moments of action.[3]

In the mid-1980s, the Soviet navy relied more on large antisubmarine ships equipped with torpedoes or missiles of the Kashin, Kresta II, and Udaloy types. The Soviets also relied on long-range, shore-based antisubmarine aircraft, the TU-142 or ASW Bear and the IL-38 May. Later, they would use nuclear-powered submarines of the Victor and Akula types to attack enemy ballistic-missile nuclear submarines. In the 1980s, the number of antisubmarine platforms dedicated to anti–ballistic missile submarines increased tenfold so the Soviets possessed roughly three maritime patrol aircraft, three antisubmarine destroyers, and two antisubmarine nuclear submarines to each U.S. strategic-missile submarine. The West's large Trident- and Poseidon-carrying missile submarines were then patrolling the far reaches of the Atlantic, Indian, and Pacific oceans, far out of Soviet weapon ranges.

To counter these longer-range, submarine-launched ballistic missiles, Soviet Admiral Gorshkov decided to construct the Soviets' first large aircraft-carrying cruisers of the Kiev type, which could provide air defense to submarines and surface ships that were going after Western submarines. Soviet naval science also created powerful detection equipment: sonar antennae and the world's original nonacoustic sensors to detect submarine wakes and electromagnetic signatures. Yet despite the efforts of Joseph Barr, the late atom bomb spy Julius Rosenberg's former fellow spy, who was now at work as a Gorshkov protégé, Soviet military electronics still lagged behind those of the West.[4]

Nonetheless, the emphasis on a decisive missile strike implied that the Soviet navy intended to be ready at any time to anticipate an attack by the West. As Captain First Rank Lev Vtorygin, a former GRU officer, explained, "All our submarines were prepared for the first strike; just for the first strike; most of them would not survive; our men knew it; they were the people of the first strike; what will be after that does not matter; we did not have factories and plants to repair these submarines anyway nor the bases to support them, nor the good living facilities of the West. We were sure that NATO and the U.S. would start the war; we were living every day in anticipation and preparation for that war."[5]

Restricted Access to Foreign Ports

The Soviet blue-water fleet required a fully maneuverable logistics force and naval bases overseas. Polish, East German, Yugoslav, Finnish, and Swedish yards were beginning to provide the necessary support for the Soviets, especially the use of floating dry docks. But solving their lack of overseas bases required shifting to a policy that contradicted the political declaration that the USSR did not seek military ports in foreign countries because it did not threaten anyone. This premise had been conveniently set aside when the Soviet navy gained access to foreign ports in Syria, Guinea Algeria, Libya, and later Vietnam. But as former commander in chief of the Northern Fleet Admiral Ivan Kapitanets explained in his book on naval strategy, the Soviet navy never had the ports it needed:

> The most difficult task for the Soviet Navy was to create a system of bases. The USSR had no bases overseas. Floating workshops and auxiliary ships could not resolve this problem. Through diplomatic channels, the USSR got the permission to enter the ports of Syria, Egypt (1967), Algeria (1969, renewed in 1978) and Cuba (1970). In 1971, according to

the request of [the] Guinean government, the USSR was granted access to the port of Conakry. In 1972, Soviet ships began to enter the ports of Somalia, in 1977—those of Benin. In 1978, they were granted access to the ports of San-Tome and Principe. In 1979, Vietnam granted its permission to use the naval base of Cam-Ranh. . . . The American government tried to restrict Soviet ships' access to foreign ports, usually through diplomatic pressure. This policy led to consequences: in the beginning of the 70s, Egypt denied our ships access to its territorial waters. Our visits to Syrian and Algerian harbors were more complicated; later, Somalia and Ethiopia also closed their ports.[6]

As Captain Vtorygin stated, the Soviet navy never enjoyed and would never enjoy the privileged status of the U.S. Navy: "The U.S. Navy is on duty all over the world, but we don't have such possibilities to keep our ships permanently at sea; the difference is that the U.S. Navy is a separate war organization; just because of that they have a separate minister; they can't imagine the existence of the United States without a strong navy; in our country, there were some military experts who even expressed the ideas that the Soviet Union did not need a Navy; it's foolish but it was said."[7]

Gorshkov's Golden Fish and an Ailing Economy

From Moscow's perspective, "The United States was conducting an arms race with the main purpose of exhausting the Soviet economy."[8] As Vtorygin recalled, "All the time, we were trying to catch up. They have it; we need to have it too."[9] Enormous resources spent by Gorshkov were, in the opinion of competing branches of the armed forces, disproportionate with the navy's contribution to the defense of the Soviet state. The dispute over the use of aircraft carriers between Admiral Gorshkov and former Pacific Fleet commander in chief Admiral Amelko—then a naval deputy to the commander in chief of the armed forces—became legendary. Gorshkov managed to have a secretary reinstate an aircraft carrier project that Amelko had deleted from a naval procurement recommendation to be submitted to the defense minister.[10]

The powerful Georgi Arbatov, the director of the Institute for U.S. and Canada Studies, also opposed Gorshkov's aircraft carrier. He emphasized the economic dimension of the Cold War: "Are we going to follow the Americans if they jump from a sky-scraper?" Also opposing Gorshkov's carrier program was the Central Committee of the Communist Party. According to Vtorygin, "They will rejoice if we waste all that money. The Cold War is essentially about economics. America's gross domestic product is twice ours . . . and our fiercest adversaries want to bleed us so that we would collapse politically and socially."[11]

Despite internal Communist Party Central Committee opposition, the Soviet navy continued to spawn other expensive programs such as the titanium-hulled submarines (Alfa, Papa, Sierra). These highly effective submarines were dubbed "golden fish" because of their high cost. Another example of Gorshkov's extravagances was his project for a naval laser gun. The laser, an acronym for "light amplification by stimulated emission of radiation," could be a potentially dangerous weapon. The first working laser was demonstrated in May 1960 by Theodore Maiman at Hughes Research Laboratories in California. Until the Soviets began research to develop a weapons-grade laser, lasers had not been powerful enough to destroy hardware. The "death ray" announced by the Serb genius Nikola Tesla was still a long way from realization.

Under the project name AYDR, however, the Soviet navy converted the Vytegrales class cargo ship *Dixon* into a test platform for a laser gun produced at the port of Kaluga. Three aircraft engines provided the 50 megawatts of electrical power required by the laser gun. Under the code name Voroshilov, the *Dixon* fired her first laser shot during the summer of 1980 in the Black Sea. For a laser gunshot having a duration of 0.9 seconds, the preparations for the firings took more than 22 hours. The laser beam was intended to remain undetected. The land target was hit at a distance of 2.5 miles, the actual hit being recorded by a rise of temperature on the target. The beam was capable of piercing an aircraft fuselage. The system was initially designed for spaceships, where there would be an absolute vacuum and no humidity. Gorshkov had hoped to install the laser weapon system on major surface combatants. The laser gun required too much energy and space, however, for effective naval use.[12]

As with similar expensive projects that were intended to outrange the enemy, economic realities forced the Soviet navy to seek more modest goals. Laser directors for conventional guns were mounted on several cruisers and on the Sovremenny class destroyers commissioned in 1980 and were used successfully on occasion to blind Western pilots flying near the ships.[13]

Reagan's Psychological Operations

In Washington, many people interpreted Gorshkov's expansion of the navy's global capability to be an attempt to thwart Western interests. As CIA National Estimates NIE 11–15–82 explained:

A once top-secret laser test platform, the converted cargo ship *Dixon*. Soviet lasers were occasionally used to blind Western pilots who flew too close.

The out of area operations of the Soviet Navy continue to reflect the Soviets' interests in strengthening their position in the Third World (especially in areas of potential Western vulnerability). . . . Soviet naval forces have demonstrated support for friendly nations and sought to inhibit the use of hostile naval forces against Soviet allies.[14]

As Professor John Hattendorff underlined in his book *The Evolution of the U.S. Maritime Strategy*, the plan to beef up the U.S. Navy and extend its related operations to meet the Soviet naval challenge was devised during the Carter administration by Admiral John Holloway, the chief of naval operations (1974–1978):

Admiral Holloway focused on the 600-ship goal as a general objective. The Department of Defense reported to Congress in January 1977 that over the following 15 years it would need almost $90 billion more than the amount funded. With a force reaching 568 ships by 1985, increasing to 638 in 1990, the navy could maintain both sea control and presence, but in the Indian Ocean this could not be done simultaneously with the other theaters. In reaching this conclusion, the report stressed the basic issue in relating force level budget decisions to strategy.[15]

This new "Maritime Strategy," elaborated in the late 1970s, and its "600 ships" corollary became the well-publicized objectives of the Reagan presidency and of the flamboyant navy secretary John Lehman. Then, in 1980, the new Republican administration started a secret campaign of naval psychological warfare against the Soviet Union. This was President Reagan's response to Leonid Brezhnev's doctrine of exporting socialist revolutions, such as in Afghanistan, Angola, and Nicaragua.

In February 1981, President Ronald Reagan authorized the resumption of secret naval and air operations along the Soviet periphery, thus reactivating earlier electronic intelligence or "ferret" missions. During the 1960s and the 1970s, more than twenty thousand reconnaissance missions had been flown along the Soviet and Chinese borders. Some of them deliberately penetrated into Soviet airspace at a time when the Soviet Union had no early warning radar. In 1970, the United States had abandoned these missions after the North Koreans shot down a navy EC-121 reconnaissance aircraft off the coast of North Korea.[16] The Reagan administration would now repeat these operations, which aimed at locating gaps in Soviet radar and air defense installations. His administration also allowed the U.S. Navy to launch simulated strikes against the USSR. As former undersecretary of defense Fred Ickle explained, "It was very sensitive. . . . Nothing was written down about it, so there would be no paper trail."[17] CIA historian Benjamin B. Fischer gave the following account of the naval side of this psychological warfare campaign:

> According to published accounts, the U.S. Navy played a key role in the PSYOP program after President Reagan authorized it in March 1981 to operate and exercise near maritime approaches to the USSR, in places where U.S. warships had never gone before. Fleet exercises conducted in 1981 and 1983 near the far northern and far eastern regions of the Soviet Union demonstrated US ability to deploy aircraft-carrier battle groups close to sensitive military and industrial sites, apparently without being detected or challenged early on.

These exercises reportedly included secret operations that simulated surprise naval air attacks on Soviet targets.

In the August to September 1981 exercise, an armada of eighty-three U.S., British, Canadian, and Norwegian ships led by the carrier *Eisenhower* managed to transit the Greenland-Iceland-United Kingdom (GIUK) Gap undetected, using a variety of carefully crafted and previously rehearsed concealment and deception measures. A combination of passive measures (maintaining radio silence and operating under emissions control conditions) and active measures (radar-jamming and transmission of false radar signals) turned the Allied force into something resembling a stealth fleet, which even managed to elude a Soviet low-orbit, active-radar satellite that was launched to locate it.

As the warships came within operating areas of Soviet long-range reconnaissance planes, the Soviets were initially able to identify but not track them. Meanwhile, U.S. Navy fighters flew at low altitude to avoid detection by Soviet shore-based radar sites and conducted an unprecedented simulated attack on the Soviet planes as they refueled in-flight.

In the second phase of this exercise, a cruiser and three other ships left the carrier battle group and sailed north through the Norwegian Sea and then east around Norway's Cape North and into the Barents Sea. They then sailed near the militarily important Kola Peninsula and remained there for nine days before rejoining the main group.

In April to May 1983, the U.S. Pacific Fleet held its largest exercises to date in the northwest Pacific. Forty ships, including three aircraft carrier battle groups, participated along with AWACS-equipped B-52s. At one point, the fleet sailed within 450 miles of the Kamchatka Peninsula and Petropavlosk, the only Soviet naval base with direct access to open seas. U.S. attack submarines and antisubmarine aircraft conducted operations in protected areas ("bastions") where the Soviet navy had stationed a large number of its nuclear-powered ballistic missile submarines (SSBNs). U.S. Navy aircraft from the carriers *Midway* and *Enterprise* carried out a simulated bombing run over a military installation on the small Soviet-occupied island of Zelenny in the Kuril Island chain. In addition to these exercises, according to published accounts, the U.S. Navy applied a full-court press against the Soviets in various forward areas. Warships began to operate in the Baltic and Black seas and routinely sailed past Cape North and into the Barents Sea. Intelligence ships were positioned off the Crimean coast. Aircraft carriers with submarine escorts were anchored in Norwegian fjords. U.S. attack submarines practiced assaults on Soviet SSBNs stationed beneath the polar ice cap.[18]

From the U.S. perspective, the first series of these naval and air operations was a complete success. Both the navy and the air force estimated that they could elude Soviet ocean surveillance and early warning systems and penetrate Soviet air defenses. In 1983, the chief of naval operations noted that the Soviets "are as naked as a jaybird there [on the Kamchatka Peninsula], and they know it." The same appreciation was given for the northern maritime border. But the U.S. Navy also had another purpose in mind. Some senior officers were convinced that Soviet intelligence was reading U.S. communications. They were hoping to get confirmation from the Soviet reaction to their fleet approach.[19] But from the Soviet perspective, U.S. and NATO forces exercising in the protected waters of the Norwegian fjords or the Sea of Japan would be very difficult to neutralize. The rationale for initiating preemptive strikes before these forces could reach their protected zones of operations grew even stronger.

In September 1982, four Backfire long-range jet bombers approached the carriers USS *Midway* (CV-41) and *Enterprise* (CVN-65) as they were conducting fleet exercises near the Aleutian Islands. At a distance of 120 miles from the carriers, the Soviet bombers simulated release of their cruise missiles. The following day, four more Backfire bombers approached the fleet in another mock air strike.[20] For Moscow, these aggressive U.S. naval deployments toward the Soviet mainland seemed to foretell a sinister outcome: an attack on the Soviet Union.

Despite the tremendous advantage the Soviet navy had during the 1970s, when it had enjoyed—thanks to the Walker spy ring—access to U.S. naval communications, the Soviets continued to fall behind. Fischer thinks that in the early 1980s, Soviet intelligence could no longer read U.S. naval traffic, which increased the Soviets' level of anxiety while they tried to keep track of the U.S. fleet.[21]

Initiating Operation Ryan

In May 1981, Brezhnev briefed a closed session of the KGB on his concerns about the new aggressive U.S. policies. KGB chief Yuri Andropov stated that he felt the United States was preparing for a nuclear strike on the USSR and invited the KGB and the GRU to collect specific indicators of U.S. war preparations, under the code name Operation Ryan. RYAN is the Russian-language acronym for "nuclear war." The KGB officer in charge of coordinating Soviet intelligence collection was sent to Washington, where he was assigned to prepare daily briefs for the Politburo. Benjamin Fischer believes that Andropov had deliberately chosen a worst-case scenario so that he would not repeat the Soviet analytical intelligence failure of not reacting to intelligence that predicted the German invasion of 1942. But like Stalin—who distrusted clandestine intelligence, which was susceptible to manipulation by the enemy for psychological purposes—Andropov preferred to survey indicators that could not be twisted. Forty years earlier, Stalin had focused on mutton prices in Nazi-held territory, assuming that the Wehrmacht would order warm clothing for winter operations. He didn't realize that Hitler had hoped to defeat the Soviet army before winter. Likewise, Andropov required the London's KGB *rezidentatura* to look for signs of increased quantities of blood in blood banks and meat in cold storage.[22] As KGB veteran Boris Grigoriev recalled, "I was getting requirements which were outside of my responsibilities; nobody said anything to me but I was worried because I was to look for the kind of intelligence that would signal war."[23]

Soviet leaders seemed to genuinely believe that the United States had already decided to go to war. At a meeting with Warsaw Pact chiefs of staff at Minsk in September 1982, Marshal Nikolai Ogarkov, the first deputy defense minister and the chief of the General Staff, described the recent U.S. naval exercises as "the material preparations for war." He further qualified the current maneuvers of the NATO states as "dead serious." The Warsaw Pact's supreme commander, Marshal Viktor Kulikov, estimated that the enemy was capable of launching a surprise attack in all parts of Europe simultaneously.[24] The rise to power of the KGB chairman Yuri Andropov would not alter this judgment. A dogmatic communist, the ailing Andropov had never been to the United States. He was convinced that Reagan's anti-Soviet rhetoric reflected a growing danger for the motherland.

Star Wars: The Ultimate Challenge to the "Evil Empire"

Andropov's fears were soon confirmed. On March 8, 1983, President Reagan denounced the Soviet Union as an "evil empire." This low-key speech at an evangelical convention gained wide media attention and struck the Soviet Union as another bellicose statement prompting Soviet defense minister Marshal Nikolai Ogarkov to demand a sharp rise in the defense budget. Less than two weeks later, on March 23, Reagan called on the scientific community "to give us the means of rendering these nuclear weapons impotent and obsolete." He ordered that ground-based and space-based systems be developed to protect the United States from attack by nuclear ballistic missiles. The "Strategic Defense Initiative" was received in Moscow as a breach to the strategic parity that had led to détente. The Soviets saw it as the signal for a new arms race. Andropov publicly accused the United States of "inventing new plans on how to unleash a nuclear war with the hope of winning it." Marshal Ogarkov acknowledged to a U.S. journalist that Moscow's Achilles' heel was its lapses in military electronics: "Modern military power is based on technology, and technology is based on computers. In the U.S., small children play with computers. . . . Here, we don't even have computers in every office of the defense ministry."[25]

As a GRU analyst in the 1980s, Captain Lev Vtorygin remembered that from the first release of the Strategic Defense Initiative's anti–ballistic missile shield, his organization considered it to be unfeasible. The GRU assumed that the purpose of the SDI was solely to force the Soviet Union to create countermeasures and further exhaust its resources.[26]

Vtorygin further supports the claim made by the Russian writers A. Kopakidiy and D. Prokorov that the June 10, 1984, U.S. Air Force intercept over the Pacific of a Minuteman reentry vehicle by a kinetic kill vehicle had been faked.[27] "The intention was to oblige the USSR to spend more," he said.[28] Soviet intelligence allegedly discovered that it was indeed disinformation. Several Soviet ships were sent to collect data on the test. They intercepted a signal coming from the Minuteman reentry vehicle to assist its successful interception by the kinetic kill vehicle. As Vtorygin recalled, "It became clear that the test was specially prepared to fool the Soviet Union. We told our leaders. But they would not trust us."[29] Indeed, the Soviet leadership took the Strategic Defense Initiative very seriously and felt compelled to react.

On October 22, 1985, in a secret speech to the Bulgarian government in Sofia, the new Soviet premier and Communist Party chairman Mikhail Gorbachev acknowledged that the primary challenge was that the U.S. industrial complex would establish absolute technological dominance over the rest of the world:

> The idea of the SDI is broader. Its purpose is to secure the permanent technological superiority of the United States, not only over the social-ist community, but over [the U.S.] allies as well. . . . Apparently they want to put their military-industrial complex . . . into a situation that no American administration would be able to reverse. If this were to hap-pen, we would have to make our own military-technological decisions, although this would be the last thing we would want to do."[30]

Since the 1960s, the Soviets had built ships dedicated to monitoring space events, national and foreign, civilian or military. Some ships were subordinated to the Academy of Sciences, even though they could have military use.[31] Other ships with similar space-tracking capabilities were especially built for military purposes with modern radioelectronic intelligence equipment and sonar to serve both the Soviet navy and the

SSV33 *Ural*, an early warning ballistic missile–detection platform and command post. The very concept of this ship illustrated the paranoid fear of a nuclear first strike on the Soviet Union.

Strategic Rocket Forces.[32] The GRU had requested two dedicated intelligence ships with practically year-round autonomy, but it was allocated only one. To answer the challenge, the Soviets constructed a nuclear-propelled, space control–monitoring ship of 35,000 tons that was capable of 21 knots. Her trajectory-telemetry measuring equipment and sensors differed little from those of an earlier space event–support vessel—the SSV33 *Ural*, commissioned in 1988. Like the two Marshal Nedelin–class space-support ships, the *Ural* deployed to the Pacific, where the United States was conducting Star Wars experiments.[33]

"Farewell": Revenge of the Dupes

Evidence suggests that the U.S. plan to derail the Soviet economy with the Strategic Defense Initiative may have been partly inspired by information gained from the Soviet defector and former KGB agent Vladimir Vetrov, also known as "Farewell." Working at the KGB Moscow headquarters during the early 1980s, Vetrov evaluated the intelligence collected by Directorate T, which was dedicated to Western technologies. Disenchanted with the Soviet system and his own future, Vetrov—who had previously been stationed in Paris—volunteered his services to the French counterintelligence organization the Direction de la Sureté du Territoire (DST) through a French contact who had once helped him. Vetrov was lucky to have escaped detection by his peers. He was handled at first in the most risky fashion by an inexperienced traveler. This handler was later replaced by a tall French military attaché, whose size made him easy to follow on Moscow's streets. The improbable handler managed however to evade successfully Soviet surveillance.[34]

From March 1981 to February 1982, Vetrov reportedly supplied his contact with more than four thousand secret documents and allegedly revealed the names of more than two hundred KGB Line X officers stationed in the ten Western cities under regular diplomatic covers. Vetrov's stolen documents were of extreme value. They listed the current and future military and naval programs and the corresponding technologies acquired or required from specific Western companies. They referred to various Soviet projects and detailed the theft of certain foreign weapons system blueprints, for example, a French naval gun that later inspired the development of a similar Russian gun. They revealed Soviet industry's reliance on Western technologies and the efficiency of the "Military Technical Committee" as an interface between industry and the intelligence services.

As Captain Vtorygin explained, this technical committee helped the Soviet economy to make substantial savings in research and development: "We had very close relations with our industry; their requirements were our requirements; our information would be provided to them; the Military Technical Committee collected the needs from our industry and passed it on to the intelligence services; this system worked out very well, and one of its main purposes was to save money."[35]

Vetrov's information helped the newly elected French socialist President Francois Mitterrand establish his credibility as a trustworthy U.S. ally when, during the July 1981 Ottawa summit, Mitterrand presented Agent Farewell's intelligence to Ronald Reagan.[36] Former special assistant to the secretary of defense Gus W. Weiss recalled the Americans' bewilderment:

> Upon receipt of the documents (the Farewell Dossier, as labeled by French Intelligence) [the] CIA arranged for my access. Reading the material caused my worst nightmares to come true. Since 1970, Line X had obtained thousands of documents and sample products, in such quantity that it appeared that the Soviet military and civil sectors were in large measure running their research on that of the West, particularly the United States. Our science was supporting their national defense. Losses were in radar, computers, machine tools, and semiconductors. Line X had fulfilled two-thirds to three-fourths of its collection requirements— an impressive performance.[37]

Instead of arresting or expelling the identified Soviet agents, the CIA devised a more Machiavellian scheme. It would feed the agents the pieces of equipment and the semiconductors they were seeking. Former special assistant to the secretary of defense Gus W. Weiss explained,

> I met with Director of Central Intelligence William Casey on an afternoon in January 1982. I proposed using the Farewell material to feed or play back the products sought by Line X, but these would come from our own sources and would have been "improved," that is, designed so that on arrival in the Soviet Union they would appear genuine but would later fail. US intelligence would match Line X requirements supplied through Vetrov with our version of those items, ones that would hardly meet the expectations of that vast Soviet apparatus deployed to collect them. . . . Contrived computer chips found their way into Soviet military equipment, flawed turbines were installed on a gas pipeline, and defective plans disrupted the output of chemical plants and a tractor factory. The Pentagon introduced misleading information pertinent to stealth aircraft, space defense, and tactical aircraft. The Soviet Space Shuttle was a rejected NASA design. When Casey told President Reagan of the undertaking, the latter was enthusiastic. . . . In a further use of the Farewell product, Casey sent the Deputy Director of Central Intelligence to Europe to tell NATO governments and intelligence services of the Line X threat. These meetings led to the expulsion or compromise of about 200 Soviet intelligence officers and their sources, causing the collapse of Line X operations in Europe. . . . The heart of Soviet technology collection crumbled and would not recover. This mortal blow came just at the beginning of Reagan's defense buildup, his Strategic Defense Initiative (SDI), and the introduction of stealth aircraft into US forces.[38]

Former U.S. Air Force secretary Thomas C. Reed, then serving in the National Security Council, explained how the CIA took advantage of the circumstance to provide the Soviet agents with bogus software and sabotage a new Soviet pipeline to Western Europe, which denied Moscow much-wanted hard currency earnings: "The pipeline software that was to run the pumps, turbines and valves was programmed to go haywire, after a decent interval, to reset pump speeds and valve settings to produce pressures far beyond those acceptable to pipeline joints and welds. . . . The result was the most monumental non-nuclear explosion and fire ever seen from space [in the summer of 1982]."[39]

Vetrov paid the highest price for his treason. On February 22, 1982, he attempted to murder his Soviet mistress, who had tried to blackmail him with a threat to expose his espionage, but by mistake he killed a man who had interfered. Vetrov was sentenced to twelve years for what appeared to be a love crime. Then, a year later, French president Mitterrand expelled forty-seven Soviet diplomats after the discovery of a KGB tap unrelated to Farewell.

In 1997, Russian journalist Serguei Kostin claimed that on March 23, 1983, the staff director of the French foreign affairs minister showed the Soviet ambassador copies of a document provided by Farewell to prove the reality of KGB espionage in France and justify the expulsions. With the ambassador's description, the KGB would have had little trouble identifying the document, checking its distribution list, and finding Vetrov. Assuming that—against normal rules—the French counterintelligence service had shared Farewell's identity with the United States, Kostin also suggested that the leak could have come from CIA defector Edward Lee Howard, whose treason had forced William Casey to resign. Whatever the source of his exposure, Vetrov was executed in the Soviet Union on January 23, 1985.[40]

Korea Airlines' Erratic Flying

The most severe war scare occurred in 1983 with the downing of a South Korean airliner by Soviet fighter aircraft. The story begins in April 1978.

South Korean Airlines KAL Flight 902 left Paris on a course to Anchorage, Alaska, where it was to refuel and proceed on to Seoul. Carrying 109 passengers and crew members, the Boeing 707 flew near the North Pole and then suddenly changed course. It headed south toward Murmansk and the heart of the Soviet Northern Fleet, in the opposite direction from its initial destination. On reaching the Kola Peninsula, Flight 902 was intercepted by a Su-15 Flagon piloted by Captain A. Bosov. After initially failing to make contact with the intruder, Bosov was ordered to shoot it down. He fired two missiles, which damaged the plane's left wing and punctured the fuselage, causing the airliner to descend rapidly. Forty minutes later, the aircraft crash-landed on the frozen Korpijärvi Lake, 250 miles south of Murmansk and just 20 miles from the Finnish border. Two passengers were killed; 107 survived and were rescued by Soviet helicopters. The passengers were released two days later. That flight had not been fitted with an Inertial Navigation System (INS), and, according to Korean officials, the pilots used the wrong sign of magnetic declination in their calculations. The pilots also failed to note the position of the sun, almost 180 degrees off from where it should have been. Aviation experts were perplexed by errors of that magnitude.

Five years and five months later, in the middle of the new Cold War, another South Korean airliner flew off course. Coming from New York, Korean Airlines Flight 007 refueled at Anchorage and at 1:00 p.m. Greenwich Mean Time on September 1, 1983, departed for Seoul with 240 passengers and 29 crew members. The Boeing 747 flew much farther west than planned, cutting across the Soviet Kamchatka Peninsula and then over the Sea of Okhotsk toward Sakhalin Island. Soviet air defense radar tracked the aircraft for more than an hour over Kamchatka. The order to shoot down the airliner was given just as it was about to leave Soviet airspace for the second time after flying over Sakhalin. At 3:26 a.m. Tokyo time, a SU-15 Flagon piloted by Lieutenant Colonel G. Osipovitch shot down the airliner, killing all 269 people onboard. President Reagan was prompt to react, calling it "an act of barbarism, born of a society which wantonly disregards individual rights and the value of human life and seeks constantly to expand and dominate other nations."

Moscow took six days to acknowledge the shoot-down and three more days to organize a press conference, in which Marshal Ogarkov, the chief of general staff and first deputy defense minister at that time, put the blame on the United States and Japan for using a passenger airliner to mount an intelligence mission: "It has been proved irrefutably that the intrusion of the plane of the South Korean Airlines into Soviet airspace was a deliberately, thoroughly planned intelligence operation. It was directed from certain centers in the territory of the United States and Japan. A civilian plane was chosen for it deliberately, disregarding or, possibly, counting on loss of human life."[41]

This public statement reflected the strong suspicions of Western intentions within the Politburo. CIA historian Benjamin Fischer quoted a December 1983 classified memorandum written by the defense minister and the KGB chairman after the analysis of KAL 007's black boxes and released by Russian president Boris Yeltsin in 1992:

> Assessment of the factual data obtained by the analysis of the flight recorders and the stand the U.S. Administration adopted after the airplane was shot down confirm that we encountered a thoroughly planned high scale political provocation by the U.S. Intelligence Community which had two objectives. In the first place incursion of an intruder

plane into the airspace of the USSR was to create a situation favorable for gathering intelligence on our Air Defense installations in the Far East by use of various systems including the Ferret spy satellite. If the intruder plane could fly through Soviet airspace unpunished, the U.S. would be likely to launch a campaign to stress inefficiency of our Air Defense in the Far East. In the second place, if the flight was terminated, a global anti-Soviet campaign was ready to be started in order to condemn the Soviet Union.[42]

As former GRU analyst Lev Vtorygin remembered, there was no doubt in his mind either as to the purpose of this second Korean airliner's airspace violation: "The flight of the Korean airliner into the Soviet Union was a deliberate action to activate air defense systems and collect intelligence. The plane flew above Kamchatka and nothing happened. The Soviet high command was unable to take the responsibility to shoot it down right away. And then a decision was made in Moscow to shoot down the plane. We wanted to avoid another humiliation [after earlier airspace violations] in the face of our people that we were not able to defend our airspace. I felt some satisfaction that we had done it."[43]

In fact, the Soviet military was under great strain. The previous exercise conducted by the U.S. Navy off Kamchatka during the spring had raised the state of alert in the Pacific Far East region. Moscow had formally protested the violation of its airspace by U.S. aircraft allegedly flying several times some 20 miles into Soviet airspace.[44] As Fischer explained, "The Soviet air defense command was put on alert for the rest of the spring and summer—and possibly longer—and some senior officers were transferred, reprimanded, or dismissed."[45]

Missing Bodies and Giant Crabs

Meanwhile, KAL 007 was still missing. South Korea had designated the United States and Japan as its representatives for the search-and-salvage operations, giving them the right to oppose any attempt by the Soviet Union to retrieve the aircraft in international waters. A U.S. and Japanese joint expedition of six survey vessels—protected by Task Force 71, consisting of a cruiser, two destroyers, five frigates, and two support ships—unsuccessfully searched an area to the north and the northwest of Moneron Island.[46] At the same time, Soviet ships apparently concentrated their efforts within their own territorial waters.

Admiral Vladimir Vasilyevich Sidorov, the commander of the Soviet Pacific Fleet, directed the concurrent Soviet joint military and civilian

Soviet salvage operations. Soviet naval divers were the first in the area, followed by civilian divers coming from Sevastopol after September 15. In his report, the commander of the U.S. task force described what he could see of the Soviet search-and-salvage operations:[47]

> Within six days of the downing of KAL 007 [on September 7], the Soviets had deployed six ships to the general crash site area. Over the next 8 weeks of observation by U.S. naval units this number grew to a daily average of 19 Soviet naval, naval-associated and commercial (but undoubtedly naval subordinated) ships in the Search and Salvage (SAS) area. The number of Soviet ships in the SAS area over this period ranged from a minimum of six to a maximum of thirty-two and included at least forty-eight different ships comprising forty different ship classes.[48]

Russia tried desperately to improve its image over the case, and on October 16, 1992, the Russian newspaper *Izvestia* published five formerly top secret documents concerning the KAL 007 shooting, which were released by President Yeltsin. They acknowledged that the Soviet interceptors had made no attempt to contact KAL 007 or to fire warning shots; they admitted that the Soviet Union had secretly recovered the black boxes after having located the wreckage on October 20, in international waters, about 5 miles outside of Soviet territorial waters at a depth of 180 meters. The documents also revealed that Soviet ships had performed "Imitation search efforts in the Sea of Japan" to mislead the United States and Japan as to the true location of the plane.[49] On this occasion, Soviet ships had harassed U.S. ships during the initial phase when the Soviet navy had not yet located the wreckage.[50] In 1991, the Republican Staff Study Group of the Congressional Committee on Foreign Relations gave the following description of these intimidations:

> The Soviet Navy and auxiliary vessels committed many serious violations of the 1972 Incident at Sea Agreement . . . such as attempted ramming of several U.S. and allied ships, presenting false flag and fake light signals, locking on the radar guidance of their weapons . . . sending an armed boarding party to threaten to board a Japanese auxiliary vessel chartered by the U.S. They engaged in a naval live-firing exercise northwest of Moneron Island, and sent Backfire bombers armed with air-to-surface nuclear-armed missiles to threaten the U.S. Navy search task force . . . move[d] U.S. sonar markers . . . manipulated the U.S. Navy search efforts into searching for decoy "pingers" on the sea bottom in very deep, international waters.[51]

In his book *Rescue 007: The Untold Story of KAL 007 and Its Survivors*, author Bert Schlossberg claimed that most of the crew and the passengers must have survived the crash because neither their bodies nor their luggage were found in or near the submerged wreckage. He suggested that the plane might have crash-landed on the water, and the KGB border guards had rescued the survivors and picked up the floating suitcases. This claim seems quite extraordinary. It is hard to believe that the Soviet authorities would have chosen to keep the passengers of KAL 007 as prisoners when they had released the passengers of KAL 902 five years earlier. It seems even more unbelievable that such a secret could have been preserved after the collapse of the Soviet Union. Soviet commentators first suggested that KAL 007 had no passengers onboard, an opinion that is still held by Rear Admiral Ivanov, a former counterintelligence officer.[52] But then, what happened to the very real people who boarded that plane?

Captain Vtorygin explained that the myth of the empty plane was circulated in the Soviet Union to calm public consternation over the killing of innocent people.[53] The second explanation, that the bodies were pulverized, was immediately dismissed by crash specialists. The fact that the bodies had remained nearly fifty days underwater before the wreckage was discovered may support the third theory: that they had been eaten by giant crabs and sharks. As a former navy diver involved in this operation, Captain First Rank Oleg Malov maintained that he saw body parts in the wreckage, badly decomposed after nearly two months underwater.[54]

Schlossberg objected that skeletons and bones should have been found in great numbers. *Izvestia* correspondents Shalnev and Illesh interviewed Mikhail Igorevich, the captain of the Soviet *Tinro 2* submersible, who proposed a fourth explanation: that most passengers could have been sucked out of the aircraft and out of their clothes by decompression in midair and spread out over a much larger area.[55] A combination of the third and the fourth explanations may be closest to the truth.

The Verdict of the Black Boxes

Shortly after the 1983 downing of the airliner, a group of experts consisting of specialists representing the Defense Ministry, the KGB, and the Ministry of the Aerospace Industry worked on the black boxes that were secretly recovered in late October and forwarded to the Air Force Scientific Research Institute in Moscow. In the first memorandum to Premier Andropov, Marshall Ustinov, the then defense minister, and V. Chebrikov, the KGB chairman, noted that the secrecy of the black boxes' recovery seemed to have been preserved: "At present there is no data at our disposal indicating that any information on the salvage and dispatch of the said equipment to Moscow has been available to the Intelligence Community of the U.S. and Japan."[56] These internal documents, released by Yeltsin in 1992, further reveal the Soviet experts' absolute conviction that the airliner's intrusion was deliberate.

On November 28, 1983, the experts submitted their analysis of the black boxes to Yuri Andropov. Their reports noted that there was no attempt by the crew to take command of the airplane in order to correct its flight path, "in spite of having all necessary instrumental data indicating a considerable deviation from the international airway." Furthermore, the crew had been using the Inertial Navigation System information to make "false reports to air traffic controllers units on the airplane's position in regards to the international airways." Their analysis stressed that "neither single error nor error combination at programming the flight path by the crew [might have resulted] in the flight with constant magnetic heading under control by INS." Their memorandum concluded, "It seems reasonable to assume that the South Korean airplane followed a pre-planned flight track and its crew was aware of the actual position of the airplane within the full time of the flight."[57]

In December, Marshall Ustinov and V. Chebrikov further stressed to the Soviet premier that "The Inertial Navigation System was not connected to the autopilot and was used only for reference of the position that the airplane should have had on the international airway at any time during the flight." Ustinov and Chebrikov concluded, "Taking into account . . . the high professional skills of the flight crew in question and high reliability of the navigation equipment of the airplane it should be assumed, as established beyond any doubt, that the incursion of the South Korean airplane into the airspace of the USSR was intentional." Following the recommendations of the expert group, however, the defense minister and the KGB chairman suggested that flight recorders not be transferred to the International Civil Aviation Organization (ICAO) for fear that their data could be misinterpreted and used in an "anti-Soviet hysteria campaign."[58]

Able Archer 83: A Cold War Turning Point

To the Soviet leadership, the KAL 007 incident was a U.S. conspiracy orchestrated in cold blood to ostracize the Soviet Union. The subsequent anti-Soviet propaganda campaign at the United Nations and in other forums came as a confirmation. Benjamin Fisher quoted Andropov's September 29, 1983, declaration on U.S.-Soviet relations: "If anyone had any illusion about the possibility of an evolution for the better in the policy of the present American administration, recent events have dispelled them completely."[59]

Within four weeks, the United States put an end to a Cuban takeover of the Caribbean island—and British Commonwealth member—of Grenada. A week later, on November 2, NATO began a scheduled ten-day command-post exercise code-named Able Archer 83. The exercise featured an escalation, a simulated setting of Defense Condition One, the highest state of alert, followed by a simulated nuclear strike. Former KGB resident in London Oleg Gordievsky claimed that Moscow mistook Exercise Able Archer 83 for a ruse to obscure preparations for a genuine nuclear first strike. On the night of November 8, the KGB advised—incorrectly—its residencies that U.S. forces in Europe had been mobilized, perhaps signaling the beginning of a countdown to a surprise nuclear attack. The various KGB residencies were asked for immediate confirmation.

On that night, Peter Huchthausen, the serving U.S. naval attaché in Yugoslavia, received a surprise telephone call from his neighbor, Soviet Army Attaché Colonel Vladimir Zhuk. Although Zhuk lived less than 1,000 yards from the U.S. naval attaché's residence, Huchthausen saw little of the Soviet attaché. Since the Soviet invasion of Afghanistan in December 1979, most Western attachés were forbidden to mingle or socialize with their opposite Soviet numbers. But a late-night phone call was truly unheard of. Colonel Zhuk expressed concern and asked if Huchthausen would stop over at his residence for a talk. Since the Defense attaché, Huchthausen's senior, was out of town, Huchthausen agreed to go to see Colonel Zhuk. When the author arrived, he was escorted into a bleak room with dark red drapes covering the walls and the stern grimace of Vladimir Lenin peering down. "Are you aware of any NATO war preparations for attacking the USSR?" The colonel asked in perfect Serbo-Croatian. Taken aback, Huchthausen thought, then replied, "Colonel Zhuk, I am certain that if NATO were to launch a surprise attack against your country they surely would not inform me here in Belgrade in advance. I know of no such plans." Zhuk then pulled out a bottle of vodka and the two men drank a toast to peace.

An earlier burst of secret communications between Washington and London, which were actually related to the U.S. invasion of Grenada, seemed to vindicate Moscow's assessment. In response, the Soviets readied their nuclear forces and placed nuclear-capable aircraft—and probably other strategic forces—on alert.[60] Moscow, perhaps well advised by its agents, aborted its alert when the NATO exercise terminated on November 11. Historian Vojtech Mastny believes that East German intelligence may have reassured Moscow at a critical moment.[61] President Reagan was reportedly deeply moved when he heard about the Soviet fears. Upon learning of Operation Ryan, he had declared to his adviser Richard MacFarlane, "I don't see how they could believe that—but it's something to think about."[62] At the time, U.S. intelligence dismissed Gordievskiy's warning as propaganda. The following year, a May 1984 CIA estimate stated, "Soviet leaders [did] not perceive a genuine danger of imminent conflict."[63]

Vice Admiral Yuri Kviatkovskiy, the last chief of Soviet naval intelligence, could not remember a higher level of tension in November 1983.[64] Historian Beth B. Fischer believes that Able Archer 83 prompted President Reagan to change his policy toward the Soviet Union and seek a rapprochement with Moscow.[65] Reagan, in his memoirs, acknowledged that he had been surprised to realize that "Many people at the top of the Soviet hierarchy were genuinely afraid of America and Americans."[66]

Planning the Unthinkable

The product of miscalculations, a Soviet "preemptive" nuclear strike was perhaps avoided on November 9–10, 1983. President Reagan did not even want to think about a nuclear war. In his memoirs, he very much doubted that anyone would act rationally in such circumstances" "We had many contingency plans for responding to a nuclear attack. But everything would happen so fast that I wondered how much planning or reason could be applied in such a crisis. . . . Six minutes to decide how to respond to a blip on a radar scope and decide whether to unleash Armageddon! How could anyone apply reason at a time like that?"[67]

Nevertheless, during the 1970s and the 1980s, war plans envisioned a conventional war with "limited" nuclear strikes. The apparition of weapons like the Tomahawk cruise missile that were capable of destroying selective targets with conventional munitions also changed the threat perception. These weapons would perhaps allow the United States to contemplate victory against the Soviet Union without having to go nuclear.

At the turn of the 1980s, Reagan's new Maritime Strategy increased the stress on the Soviet decision makers. The Soviet navy resumed a partnership with the Soviet air force to target NATO and U.S. naval forces in the coastal areas, where the missiles salvo could be rendered inefficient by the land masses.[68] Again, by its emphasis on missiles, Gorshkov's navy was by definition a first-strike navy, totally dependent on intelligence and critical assessments on the "probable enemy's" intentions. Former GRU analyst Vtorygin did not perceive the Reagan administration's Maritime Strategy as something actually new:

> I did not see much difference between the American strategy of the 1970s and their strategy of the 1980s. I thought that the public declarations were just a cover. The main aim of the American navy was to eliminate the possibilities for the Soviet Navy to play an active role in the war. In the Pacific, they would first destroy our submarine bases; second they would destroy all our command posts and communications facilities so as to leave Soviet naval forces without means of communications, incapable to act against American forces; third, they would destroy our economic infrastructures. The Soviet goal was to make everything to limit American possibilities to destroy our defenses; we had no intention to attack Japan or some other American facilities in the Pacific; our strategic aviation would just take care of their communication centers; that was the main aim of the Soviet forces; we did not have enough forces to fight the US Navy in the vast areas of the Pacific; we would focus on how to defend our naval bases; not in a passive form, but in a somewhat active form. We would work more actively on the whole Atlantic area, especially in the Northern Atlantic and in the Mediterranean; but in the middle Atlantic, we could only use submarines to cut American lines of communications and follow and destroy carrier battle groups approaching the areas where they could use their aircrafts.[69]

Admiral I. Kapitanets, the commander in chief of the Soviet Northern Fleet, further explained the importance of the North Atlantic theater:

> The Northern Atlantic could be used by our naval opponents to target the main military and industrial sites of the Socialist countries and in support of their troops. On this theatre, the USA and their NATO allies were ready to concentrate their main forces to achieve maritime supremacy. . . . They would deploy their battle fleet consisting of 4 aircraft carriers, 10–12 multi-purpose nuclear submarines and 10–12 strategic nuclear submarines belonging to the USA, the UK and France as well as anti-submarine and amphibious expeditionary forces. The war actions in [the] North-East Atlantic would be the most intensive. . . .

NATO's strategic naval forces including submarine groups armed
with "Tomahawk" missiles . . . were considered as the main threat to
the region."[70]

Cold War historian Vojtech Mastny has conducted a pioneering
work on Warsaw Pact war planning in Europe during the Cold War,
using declassified archives from ex-Pact members Poland, former East
Germany, Hungary, former Czechoslovakia, Romania, and Bulgaria.[71]
Warsaw Pact war plans were "all offensive after 1975," remembered a
Cold War veteran from the Polish navy who served on the Warsaw Pact's
staffs. But he added that Admiral Gorshkov himself did not believe
that they would ever be used.[72] Western and Warsaw Pact war plans all
included an initial attack by the other side, which would be repelled and
followed by a counteroffensive. Captain Vtorygin explained why the
Soviet Union had to occupy Europe in the shortest possible time and
what role the Soviet navy would play:

> Europe is the main route to invade Russia; remember Napoleon and
> Hitler; the Soviet Union was not accessible from the North; the Far East
> was too far away; Europe was an ideal route; infrastructure were [*sic*]
> good for troop transportation; good railways, good roads; the popula-
> tion was well prepared to meet Americans and sympathetic to them;
> they would have the support from the rear. The Americans had already
> built up huge reserves in Europe. Everything was in place: weapons,
> ammunitions; spare parts; it was well planned. Their intention was to
> sealift their troops to Europe right away to man these weapons. When
> you think about our rationale, the main aim for the USSR was to seize
> Europe immediately not to permit American troops to land on that
> territory. Our airborne divisions would move quickly across Norway,
> France, Spain, Portugal and capture these territories and not give any
> possibilities for the Americans to send reinforcements; the Soviet fleet
> at that time was on a par with NATO naval forces; so it could only be
> used to make attacks from the flanks to support the flanks of our land
> armies and use submarines to prevent the American sealift. The main
> theatre for the expected war would be in Europe; and I am sorry to say
> that France would have suffered because the French territory was well
> situated and well prepared to meet American troops.[73]

Archives in Poland and Bulgaria have revealed that these two
countries had been given important responsibilities to lead the offensive
and seize the Danish and Turkish straits. Both Poland and Bulgaria had

The USNS *Stalwart* was a new kind of U.S. intelligence vessel. The T-AGOS would tow very-low-frequency passive antennae off the Soviet coast to detect departing and incoming submarines.

sizable amphibious forces and would have been rewarded with the operational command of the straits. In the case of Poland, ever since the 1960s, nuclear weapons under Polish control would have been used to support the offensive and to convince the Danish population and armed forces to cease combat after nuclear strikes were made on Esbjerg and Roskilde (Zealand Island).[74] As Graham H. Turbiville Jr., a former director of foreign military studies at Fort Leavenworth, explained, the Soviet Union was also monitoring key infrastructures in the continental United States to possibly disrupt the preparation and deployment of the U.S. military to Europe: "Soviet intelligence personnel . . . had for years closely studied and systematized U.S. and allied newspapers, journals, and other materials to identify and understand the critical war-supporting assets upon which the United States relied for mobilization, deployment, and war sustainability."[75]

Atrina

The last critical moment of the psychological confrontation may have been Operation Atrina, a Soviet effort to deceive U.S. underwater surveillance. Admiral Kostev gives the following account of this three-month-long exercise:

After 1984, the Chernavin era began. The most significant event was the operation called Atrina. This operation was the beginning of the end for the Soviet Navy. The North fleet prepared five multipurpose nuclear submarines. They had to go to the Western hemisphere into the Bermuda Triangle. Secrecy was kept until the very last moment. Even submarine commanders knew nothing until the day of their departure. Admiral Chernavin called the operation "underwater curtain." Other participants were two Kolguiev ships with flexible antennae and a naval aviation division. Admiral Kapitanets was in charge of coordination of all types of forces. In March 1987, the first submarine left Zapadnaia Litsa. After that four submarines left the ports one after the other. Soon they formed a kind of a curtain moving towards the west. Later Chernavin reminded that the goal of the operation was to evaluate the situation in the Atlantic Ocean, something which was difficult to achieve using other means of reconnaissance. During the operation Soviet submarines had several contacts with American and British submarines. It seems that the latter had also detected our atomic vessels. The American government began to worry. US naval forces started to hunt down Soviet submarines. However their attempts seemed to be unsuccessful. Five submarines reached the American coast. The US thought that Soviet multipurpose submarines were missile carriers. This fact aggravated the situation. According to Chernavin, that was the reason why the Americans sent three more maritime air patrol squadrons, three more reconnaissance ships and their escorting forces to find the submarines. Six US nuclear attack submarines were also dispatched.[76]

Atrina lasted for three months. According to Kostev, the U.S. efforts to track the Atrina participants were done in vain. Soviet submarines managed to hide successfully in the boundary layers of the ocean. While evaluating the operation's results the Soviet naval command concluded that the U.S. Navy had not enough forces to confront large-scale Soviet submarine operations in the Atlantic or Pacific oceans.[77]

14

SWEDISH WATERS
1980–1990s

During the last decade of the Cold War, the tension between the West and the Soviet Union was further exacerbated by a series of alarming reports of unidentified underwater incursions into neutral Sweden's territorial waters. These incidents went unexplained for years, but now new light is shed on the culprits involved.

The Soviet Union was the usual suspect. The USSR had traditionally been Sweden's main security concern, and since the early days of the Cold War, Stockholm was the tacit partner of the Atlantic alliance, despite Sweden's proclaimed neutrality. Its waters could be used as a safe haven by both sides, and its territory could serve as a passage for the Red Army to attack NATO member Norway. Safeguarding the thousands of islands, islets, and rocks that constitute Sweden's Baltic shoreline is nearly impossible by the country's modest navy, coast guard, and police force. Most of the islands are uninhabited.[1]

The intrusions continued for a decade after 1980, with additional trespasses in 1992 and 1993. The Swedish government convened three commissions, in 1983, 1995, and 2001, to assess and reevaluate the problem.

Caught in the Act

In 1980, the Swedish navy believed, from acoustic intelligence, that it had detected diesel-electric attack submarines of the Soviet-built Project 613 (NATO named Whiskey) operated by both the Soviet and the Polish navies. For a period of several weeks in March that year, the Swedish navy hunted what seemed to be two unidentified submarines operating in the country's inner territorial waters near the naval bases of Karlskrona and Muskö.[2] After much speculation, hand wringing, and false contacts, the next year the Soviet navy was caught in the act.

264

The grounding of the Whiskey submarine S-363 in Karlskrona appeared to be indisputable proof of Soviet submarine penetration in Swedish waters.

A Whiskey class submarine S-363 (wearing tactical pennant number 137 on its sail) was found by a Swedish fisherman stranded on a rock about a mile southeast of the main Swedish naval base of Karlskrona. United Press International reported that the submarine's commander blamed "faulty radar and bad weather" for his embarrassing plight. The Swedish government, however, quickly dismissed that excuse as unacceptable. Maritime experts claimed that sailing so deep into Swedish waters called for precise navigation, and they demanded an explanation.[3]

On October 30, Swedish prime minister Torbjörn Fälldin ordered his military forces to keep the area clear and to stop the unilateral Soviet effort to free the submarine. A navy rescue force was approaching. A Swedish force consisting of three Spica class torpedo boats turned away a Soviet salvage tug that had already moved inside Sweden's 12-mile territorial limits. In the meantime, a nearby coastal artillery emplacement locked its gun directors and aimed its 120 and 77 mm turrets at two Soviet destroyers escorting the tug. The Soviet force, under the command of Vice Admiral Kalinin, briskly retreated.

During the evening of October 30, the Soviet ambassador to Sweden visited the Swedish foreign ministry to express Moscow's regrets over the incident.[4] Both sides agreed that the submarine's commander, Captain Third Rank Pyotr Gushin, could be interviewed on a Swedish torpedo boat in the presence of two Soviet diplomats, and that the Swedish navy would get access to the submarine's navigation log. Swedish navy Rear Admiral Emil Svensson and Captain Erland Sonnerstedt remembered that the submarine's crew members were allowed four days to "correct" their log book and come up with a legitimate story. On November 5, after lengthy interviews with the submarine's commander, Sweden agreed, weather permitting, to release the submarine.

In the meantime, the Swedish Defense Research Institute conducted a series of environmental radiation measurements from inside the hull of a Swedish coast guard vessel moored alongside the Soviet submarine.[5] Twenty-two pounds of uranium 238 were allegedly detected, revealing the likely presence of nuclear-tipped torpedoes. This was not a good omen because Sweden, the Soviet Union, Finland, Norway, and Denmark were all parties to the Nordic Nuclear Free-Zone. Finally, a week later, the submarine was freed and joined the Soviet flotilla off the Swedish coast. The foreign minister stated that as far as his country was concerned, the incident was closed and that "there [would] be no further protests." Moscow challenged Stockholm's contention that the submarine might have been armed with nuclear-tipped torpedoes, while the U.S. government issued a statement confirming that it was standard practice for Soviet combatants to carry nuclear weapons.[6]

The Soviet Account

In a personal memoir of his naval career, Soviet captain first rank Sergei Aprelev provided a Russian insight, based partly on speculation, into the Whiskey class S-363 grounding. As a submariner himself, Aprelev should be qualified to present this version of what he called the beginning of the "periscope syndrome":

> That night onboard S-363 [tactical number 137] nothing could foretell what would happen. Just another day on patrol; still one week and a half to go before returning to base. The submarine's submerged operation area sector was located in the center of the Baltic. The monotony of the navigation was interrupted from time to time by the noises of the large ferries and by fishing craft that shuttled along the shores. The submarine nearly got entangled in a trawl. After a while they managed to free themselves without major damage, except for the bending of a gonio

navigation antenna the navigator used to take the acoustic bearings. This was the beginning of a series of errors that compounded their misadventures. First the navigator: he was a young ensign, but experienced and well trusted by the commanding officer, himself a torpedo specialist by training. Whatever the reason, it is the commander's sacred duty to check on his navigator who works alone on diesel-electric submarines of that size. . . . [Another] complicating factor was the presence on board of a more senior captain, the flotilla commander. . . . Statistically accidents tend to happen when there are two captains onboard. . . .

The flotilla commander had a reputation of being a clever man. . . . He was an experienced navigator, but nevertheless trusted the commanding officer. . . . And the commanding officer had two options: trust his navigator or constantly be on his back. It seems that he choose the first solution to boost his navigator's self-confidence. But the navigator, Boris Korostov, did not like the sextant. . . . [Instead he used] the Decca radio navigation system that covers the Baltic. . . .

To avoid confusing his patrol sectors, Korostov took bearings with the, now bent, gonio acoustic antenna. The navigator began to doubt their actual position. But he knew that their speed of three knots would not take them far. The young navigator rationalized that they would soon surface to recharge the batteries enabling him to clarify their position. The navigator did not share his concern for their shaky position doubts with the Captain. When the submarine surfaced, [Korostov] just assumed that the current might have drawn them slightly off course. Meanwhile the captain ordered a 030 degree true degree heading, accepting that his submarine was where his trusted navigator had said, in the middle of the Baltic. The officer of the watch was also confident in the navigator's position. The officer of the watch and lookouts busied themselves with their main task [which] was to detect the "probable adversary" . . . in time to order the submarine to dive. . . .

When the political officer, *zampolit* Vassiliy Bessedine, relieved the officer of the deck, the situation seemed normal: cold and dark as it should be at 8 pm during the month of October. . . . The *zampolit* skillfully maneuvered among small rocks that he mistook in the poor visibility for oil spills. Without his skillful handling the submarine would have run aground much earlier. Unfortunately [137] struck the reefs. . . . A rock penetrated the external hull, removing any hope to disengage the submarine under her own power. . . . The first thing to do was to report to the Baltic Fleet's headquarters. This was done immediately.[7]

Captain Aprelev explained that at dawn, a Swedish tug approached the stranded submarine with Captain Karl Anderson, the Karlskrona naval base chief of staff, aboard. Aprelev contended that Anderson had apparently accepted the Soviet explanation and in the following proceedings treated the Soviet crew courteously.

In reaction to Captain Aprelev's written account, Rear Admiral Svensson pointed out that at the time, the Swedish navy had not noticed any damage to the 137's gonio antenna and "could not accept the alleged Soviet confusion between rocks and oil slicks even in the poor visibility."[8] Aprelev's account stresses that the submarine was surfaced when grounding. Again, the Swedish navy objected, stating that the submarine had not been detected by its coastal radar and that the submarine should have seen a 60-foot-high lighthouse on its way in and not, as alleged, mistaken it for a fisherman. The Swedish Defense Staff stated that secret torpedo tests had been underway in the vicinity of Karlskrona. This, they suspected, had been the real motive behind the Soviet submarine's presence.[9]

A Navigation Error?

On three occasions, the government of Sweden reassessed its position on the Soviet Whiskey incident. In 1985, the Swedish navy disclosed that the Soviet submarine's log entries and gyrocompass course headings for the last twenty minutes of her track had been altered and that Swedish officers had found the radio navigation equipment to be in order. They also claimed that they had intercepted a radio communication from the commander of the Soviet rescue force, ordering the submarine to tell the Swedes that the submarine had made a navigational error.[10]

Ten years later, the Swedish Submarine Commission reiterated, "In view of the navigation and visibility conditions at the time of the incident, the violation must have been deliberate."[11] Then in 2001, a commission headed by an ambassador reappraised the incident and expressed its doubts over the Soviets' real intentions. He noted that the Soviet navy had been searching for the submarine in a totally different area of the Baltic: "The unusually clumsy nature of the intrusion and the submarine's unusual behavior raised questions as to whether or not the intrusion was deliberate. It has come to the knowledge of the inquiry that the Soviet side searched for the submarine for six hours in its designated patrolling area, east of the island of Bornholm, before realizing that the submarine was on Swedish territory."[12] After all, what seemed obvious was less certain.

The Harsfjarden Incident

In June 1982, just one month after Moscow had agreed to pay Sweden compensation for the "Whiskey on the rocks" incident, the Swedish navy and air force sealed off the Gulf of Bothnia in the northern Baltic Sea to prevent the escape of an apparent unidentified submarine.[13] Officers of three separate lighthouses had spotted a submarine in Sweden's territorial waters sailing north toward a 60-foot-deep channel at the southern end of the Gulf of Bothnia.[14] On June 9, Swedish navy helicopters dropped depth charges, while the Swedish authorities claimed that a second submarine had probably fled its territorial waters.[15]

In August of the same year, repeated submarine contacts were gained near Landsort. Two months later, helicopters and ships were again searching for a suspected foreign submarine, this time near Stockholm. These contacts were reported just after the elections, which were won by the Social Democrat Olof Palme. Then on September 25 to 27, a U.S. Navy force consisting of the cruiser USS *Belknap* (CG-26), the frigate USS *Elmer Montgomery* (FF-1082), and the supply ship USS *Monongahela* (AO-178) called in the Swedish capital. In the early afternoon on September 26, passengers aboard a city ferry spotted a periscope alongside the U.S. ships and a few hundred yards from the royal palace. A local couple who reported the sighting sketched the 15-inch-high apparent periscope.[16] Three days later, another couple reported seeing a small dark gray submarine sail—3 feet high and 4 feet wide—this time just outside of Stockholm harbor.[17] Then on October 1, two more observers witnessed two dark pipes about 3 to 4 feet from each other moving in the vicinity of the secret Muskö Naval Base, south of Stockholm.[18] Local newspapers revealed that a Swedish submarine crew had also reported an unknown submarine again 300 yards away from Muskö.[19]

On October 6, the navy informed the press that it had trapped what it suspected to be a Soviet submarine and that the navy was dropping depth charges to force the vessel to the surface. A military spokesman said that the submarine was trapped inside Harsfjarden Bay, 10 miles from the open sea, and could escape only through two narrow passages between the rocks where the navy had laid powerful mines.[20] While forty navy ships, including a submarine salvage craft, and ten helicopters were searching the zone, they dropped seven depth charges.[21] Then two days later, the Swedish military informed the press that the submarine that it had been hunting since the previous Friday had probably escaped. The military also suggested that it had detected a second submarine. Helicopters dropped six more depth charges on contacts and then met up with Swedish navy ships 20 miles south of Muskö to join the hunt.[22] More depth charges were exploded on the following day when the submarine apparently tried to slip out through the channel.[23]

On October 10, Swedish authorities revealed that a Soviet Ilyushin Il-38 maritime patrol aircraft was criss-crossing the Baltic, displaying "unusual interest" in the Harsfjarden area and possibly trying to make contact with the mysterious intruder. The source reported detecting radio frequency jamming that was intended to prevent communications between the Soviet aircraft and the suspected submarine.[24] At noon on October 11, the static magnetic anomaly sensors in the Harsfjarden area detected a submarine. A 600-kilogram mine was detonated.[25] The following day, Swedish general Lennart Ljung, the commander of the armed forces, suggested that the submarine might have escaped.[26] Then on October 12, the Swedish navy set off a powerful mine after noticing indications of a possible submarine in Hors Bay, an area farther to the south.[27] However, official documents quoted by Ola Tunander indicated that during the night of October 11–12, the bottom-fixed sonar system at Mälsten registered metallic sounds, apparently related to hammering work being performed on a presumably damaged submarine.[28] The following afternoon at 6:00 P.M., the Swedish Navy tape-recorded "a certain submarine" moving past the hydrophones out to the open sea.[29]

But the most disturbing events took place on the evening of October 13. First, Vice Admiral Bror Stefenson, the chief of defense staff, ordered Lieutenant Colonel Sven-Olof Kviman, commanding the Mälsten area, to observe a five-hour ceasefire. Then an attack on a submarine contact by three patrol boats and a helicopter was interrupted by an order from the naval base restricting to two the number of charges to be dropped.[30] Lieutenant Colonel Kviman was very upset, because these restrictions had enabled the submarine to escape.[31] Rear Admiral Emil Svenson and Captain Erland Sonnerstedt explained that this decision was made to protect the lives of innocent fishermen and sailors who could have entered the area during the hours of darkness.[32] However, these restrictions were imposed just on that single evening of October 13 and not during the previous or following nights. Finally, on October 27, one year after the Whiskey incident, the Swedish Navy called off its three-week search.[33]

Angered and embarrassed by what it believed had been repeated incursions by foreign submarines, the Swedish government ordered its forces to change tactics and to seek to capture and humiliate future intruders. Meanwhile, the navy would purchase new antisubmarine weapons.[34]

Following a six-month investigation, an official Swedish commission concluded that up to six Warsaw Pact submarines had been involved in a bold intrusion into the waters near Sweden's secret underground base of Muskö. The fleet was said to have included Whiskey-type submarines and three Soviet advanced miniature submarines. At least one submarine mounted tanklike treads to crawl along the sea floor. The tracks resembled those made by a Soviet scientific submarine that was active in the Azores, allegedly searching for the "lost continent of Atlantis," and also coincidentally exploring near U.S. SOSUS arrays. This minisub, the report disclosed, "may have crept fifty miles to the north into a submerged channel running to the center of Stockholm." The commission concluded "that the violations at Harsfjarden, and other violations during 1980–82 were by submarines belonging to the Warsaw Pact."[35] As researcher Ola Tunander explained,

> The Commission used a Defense Staff Report from 18 April 1983. It stated, *firstly*, that all visual observations had been interpreted as submarines from the Warsaw Pact. *Secondly*, two acoustic observations were made. In both cases, the conclusion was submarine[s] from the Warsaw Pact. *Thirdly*, the results of signal intelligence cannot be made public for security reasons. Signal intelligence proved definitely that there were Warsaw Pact submarines. *Fourthly,* the existence of tracks from midget submarines supported the conclusion that the Warsaw Pact was responsible for the intrusions. It would, according to the Defense Staff Report, be "almost impossible to keep such systems secret in the West."[36]

Experts suspected that the Soviet Union could have been gathering intelligence to plan an invasion of Sweden and Norway, in order to gain control of the vital northern Atlantic sea lanes in the event of war.[37] Olof Palme, the new prime minister, who was suspected of being more partial toward Moscow, seemed absolutely infuriated and convinced that the Soviets were responsible. He announced a new "shoot to kill" rule of engagement: "Violators in the future can count on the Swedish government to order the military to sink the intruder at once."[38]

Periscopitis and Dead Frogman

In November 1982, barely a month after the Harsfjarden incident, new bottom-crawler tracks were found in the northern part of Mysigen. At that time, the Swedish navy's antisubmarine equipment was totally inadequate. Sweden's last remaining destroyer, *Halland*, fitted with a variable depth sonar (a sonar transducer tethered to the ship by a long data cable), had just been decommissioned. Corvettes and minesweepers fitted with modern sensors were on order. Then in April and May of the following year, coinciding with Moscow's furious denials, the intrusions suddenly multiplied.

From May 1 to May 10, 1983, two submarines and one or more midget submarines were sighted at Sundsvall. A Swedish navy spokesman stated that helicopters had failed to find the midget. "But that does not mean that it is not lying on the bottom," he added.[39] It was finally accepted that the intruders had escaped under a Soviet or Finnish tanker and ferries transiting the area.[40] Meanwhile, Oslo reported a submarine incursion in Norway's Hardangerfjord. The Norwegian navy also seemed powerless. Carl Bildt, a young defense specialist within the conservative opposition and future prime minister, noted that "The Norwegian Navy with its frigates, German built submarines, U.S.-supplied P-3 Orions and access to Sosus data has so far not been more successful than its self reliant, neutral Swedish counterpart in handling this new and very complicated threat."[41]

On July 1, new regulations went into effect, giving the Swedish navy more scope to use weapons against intruding foreign submarines.[42] They were welcomed with a joke relating that a suspected submarine periscope spotted on Sweden's northeast coast turned out to be a sewage pipe placed in the water by pranksters. "It is not very funny," said the navy spokesman.[43] Nor was it funny to the Soviet press, which was still infuriated with Swedish allegations against Moscow. On June 3, an *Izvestia* article asked, "Do the participants in the practical joke have the slightest understanding of where the boundary between reality and fantasy lies?"[44]

But there was no fantasy on the Swedish side. In August and September, the navy again dropped depth charges after spotting what was believed to be a foreign submarine stalking naval maneuvers near the Karlskrona naval base.[45] At this point, there were more than forty violations in 1982 of Swedish waters by foreign submarines, all of them suspected of having come from the Soviet bloc, as well as twenty-five certain and thirty-eight probable submarine intrusions in 1983.[46]

In late April 1983, Anders Ferm, the Swedish ambassador to the United Nations in New York, transmitted to Michael Milstein, a retired GRU lieutenant general, and Georgi Arbatov, the director of the U.S.A. and Canada Institute of the USSR's Academy of Sciences, a note demanding respect for Sweden's borders, respect for Sweden's neutrality, and that the Soviet Union cease intentional submarine intrusions. The following month, Olof Palme sent a similar note to Moscow through the Soviet ambassador in Stockholm and temporarily recalled Sweden's ambassador from Moscow. He also announced that official visits between the two countries would be cut back sharply.[47]

The Soviet press agency *Tass* vehemently denied the Swedish charges that Soviet submarines had infiltrated Swedish waters: "It is essential to stress most categorically the total groundlessness of the allegations."[48] Then, reacting officially on May 6, Moscow again rejected Stockholm's formal protest: "According to precise and carefully verified information from the relevant Soviet authorities, Soviet submarines were not in Swedish territorial waters at the time given in the note; nor were any located within twenty miles of these waters. For this very reason they could not have carried out the activities described in the Swedish government note."[49]

The year 1984 started with a flurry of new incidents involving submarines and divers. From February 9 to early March 1985, the naval base of Karlskrona was again the scene of a fruitless chase. Bottom-mounted magnetic anomaly sensors detected three submarines and midget submarines and small motorized diver vehicles. Swedish forces detonated one mine, twenty-two depth charges, and twenty-eight sub-strength warning munitions. Quite strangely, however, the orders given by the Swedish high command did not call for the destruction of the intruders. The area was sealed off, but quite unexpectedly Vice Admiral Bror Stefenson again gave an order to lift the barrage to allow the passage of a merchant ship. Many assumed that the intruder had for the second time been permitted to escape by the commander in chief himself.[50]

The Soviet military daily *Krasnaya Zvezda* promptly dismissed the repeated accusations: "It is the desire of the Swedish right wing forces to see alien submarines off the coast of the country that stands behind periscope disease. These forces are interested in arms build-up, close links with NATO, and in promoting mistrust in the USSR's foreign policy. They want to spoil the traditional good-neighborly relations between the USSR and Sweden and for this purpose they incite anti-Soviet submarine hallucinations."[51]

The hallucinations did not go away. On March 5, a military spokesman said that military forces hunting another submarine had fired at three frogmen who were spotted emerging from the sea on the small island of Almoe, where a lone diver had been fired on the previous Wednesday night.[52] Soldiers uncovered a food cache and intensified their search, even stopping a funeral procession to open the casket. But nothing was found.

For more than a month and a half, Sweden had deployed its antisubmarine arsenal in vain to capture the elusive midgets and frogmen. The chief spokesman for Sweden's Supreme Command, expressing his frustration, explained, "We have tried and tried to explain the evidence by natural phenomenon but in the end there are reports that can't be accounted for that way."[53] In total, sixty incidents, of which twenty were either certain or probable, were reported for 1984.[54]

There was a slight decrease of incidents in 1985. On April 12 of that year, the navy fired antisubmarine grenades at a suspected submarine near the entrance to Karlshamn Harbor on Sweden's southeast coast and finally gave up after a thirteen-hour search.[55] In late 1985, the Swedish military came close to recovering the body of a frogman drowned by a fisherman's net. The fishermen had illegally laid their nets in a mined area. A diver was retrieved in the net, apparently trapped and drowned. The panicked fishermen dropped the body back into the water.[56] More incursions in 1985 were reported in five separate sectors.[57]

Again in May 1986, major incidents occurred, with a lengthy submarine search taking place in the Stockholm archipelago. More than thirty depth charges and sixty grenades were used against suspected submarines and midget submarines that had been reported. More contacts were chased in six other areas from June to October.[58]

Then in 1987, the Swedish government decided to cease publishing specific statistics on alleged intrusions. But despite that decision, the press continued to publicize incidents that took place again in the Stockholm archipelago in May and June 1988 and in January and February 1989. In the first case, the intruder was detected visually and acoustically in the middle of a Swedish antisubmarine exercise, provoking an avalanche of between one hundred and two hundred depth charges. According to the commander of the Coastal Fleet, the Swedes had never been so close to hitting a foreign submarine. Several sets of underwater breathing apparatus were recovered, but their ownership could not be determined.

In 1987, an analyst with the "Stockholm International Peace Research Institute" (SIPRI) published a major study on *Soviet Submarine Operations in Swedish Waters, 1980–1986*. Three years later, the Rand Corporation issued a report titled *Stranger Than Fiction: Soviet Submarine Operations in Swedish Waters*. Both authors had no doubt that the intruders were from the Soviet Union. The former study concluded that the Soviet Union had conducted submarine operations in Swedish waters "almost continuously since World War II."

Moscow's Imprint?

In June 1988, the Swedish newspaper *Aftonbladet* revealed that "a Soviet government official" had acknowledged that the Soviet Union had been responsible for sending submarines into Swedish inner waters from the late 1970s until 1985, only stopping the intrusions after Mikhail Gorbachev had come into office. The source claimed that the Soviet planners did not trust Swedish neutrality in the case of a general war and that they had "a need to know where they could hide from the Swedish Navy and from Swedish electronic equipment."[59] This "revelation" provoked a rash of denials.

The Soviet ambassador to Stockholm stated categorically, and with great certainty, "that there have never been any Soviet submarines in the Swedish archipelagoes or territorial waters." The chief spokesman for the Soviet ministry of foreign affairs wondered whether the newspaper *Aftonbladet* had not made up the whole story.

Admiral Vitaly Ivanov, the former commander in chief of the Baltic Fleet, stated, "No Soviet submarine [had] ever approached Sweden's territorial waters closer than 50 to 70 kilometers." Marshal Sergei Akhromeyev, the chief of the Soviet General Staff in 1988, declared that Soviet submarine commanders were under strict orders not to violate Swedish territorial seas and that any claims to the contrary were therefore groundless and incomprehensible.[60]

After the collapse of the Soviet Union, former Soviet Baltic Fleet submariners—some of Baltic descent—claimed to have participated in operations inside foreign territorial waters, generally to put ashore agents.[61] Such accounts were dismissed by submariner Captain Aprelev in his book, but these operations have been common for submarines in all nations, and it would be hard to believe that the Soviet navy had not engaged in such activities. Generally, the Soviet Union was more cautious than the West in its maritime and aerial intelligence-gathering operations, and the Karlskrona incident was certainly a navigational error at some stage of "137" operations.

Nevertheless, as Rear Admiral Emil Svensson, the Swedish expert on submarine intrusions, explained, the Soviet *spetsnatz* units based in Paldiskiy had conducted reconnaissance on the Swedish coasts using SIRENA and TRITON midgets supported by trawlers. Later, they had used the larger TRITON-2 midgets, but they were always dependent on a support ship. The very existence of this dedicated Soviet elite combat swimmer unit just across the Baltic was compelling circumstantial evidence in itself. The GRU also supported two then top secret underwater units nicknamed "Crab" and now known by their numbers 40056 and 45707. They had been created respectively in 1965 and 1976 to operate against the U.S. SOSUS hydrophone network in the Atlantic and the Pacific. They were not based in the Baltic but could have operated there from a mother submarine.

There was, however, other evidence pointing to Warsaw Pact activity in Sweden. In 1986, Swedish security services discovered that Polish intelligence personnel posing as Polish art dealers had visited the homes of some 120 Swedish pilots, with the purpose of identifying them and presumably of gunning them down at the outbreak of a conflict.[62] Similarly, it was assumed that Baltic Fleet *spetsnatz* teams inserted by submarines would neutralize Swedish naval bases, ports, radar sites, communications facilities, and coastal batteries far more effectively than a bombing campaign would.[63]

U.S. naval analyst Robert Weinland suggested that the USSR would want to occupy and use airfields in southern Norway early in a war and in the aftermath of the social unrest fostered by Solidarity, which had weakened the reliability of the Polish ally to fulfill its assigned mission and seize the Danish straits. According to Weinland, the surge in submarine intrusions could have been related to the internal situation in Poland.[64]

In 1995, a new Swedish commission on submarine intrusions questioned the Soviet involvement and the work of the 1983 commission. The Swedish signal intelligence agency FRA stated in a letter to the defense minister that it had no information on Soviet communications during the hunt for the Harsfjarden intruders. Moreover, signals emanating from Swedish waters, although they could not be positively identified, "were believed to originate from the west."[65] Both the minister of justice and the foreign affairs minister doubted the conclusions of the 1983 Submarine Defense Commission and had objected to issuing a formal protest to the Soviet Union.[66]

Confessions

In the year 2000, former U.S. defense secretary Caspar Weinberger gave a most remarkable interview on the Swedish TV program *Strip Tease*:

> Swedish TV: Did NATO have a motive for making intrusion into Swedish waters?
>
> Weinberger: It would certainly be part of NATO's activities as a defensive alliance to ensure that there were defenses in all parts of the area against Soviet submarine attacks. And there were undoubtedly instances where NATO was trying to ensure that there were adequate coastal defenses maintained as part of the whole alliance defensive line. I noticed the former Prime Minister Carlsson made some statements about NATO preparing for war. This is quite wrong. NATO was preparing to defend against an attack being launched. NATO had no offensive capabilities and no offensive intensions.

Weinberger: [Operations in Swedish waters were] part of a routine regular scheduled series of defense testing that NATO did and indeed had to do to be responsible and liable. [The Soviet Whiskey submarine in 1981] was a clear violation, and submarines can get in where they are not wanted, and that is exactly why we made this defensive testing and these defensive maneuvers to ensure that they would not be able to do that without being detected. . . . The [Navy-to-Navy] consultations and discussions we had were designed—with all countries not just Sweden—to assure that NATO was able to perform this mission and had ample opportunities to test through manoeuvres and other activities as to whether the defenses were adequate and whether or not the Soviets were acquiring any new capabilities that would require any changes in their defenses or anything of that kind. So, the result of all that I think was very satisfactory. Besides from that one intrusion of the Whiskey class submarine, there were no violations, no capabilities of the Soviets to make an attack that could not be defended against. . . . The point was that it was necessary to test frequently the capabilities of all countries, not only in the Baltic [Sea]—which is very strategic, of course—but in the Mediterranean and Asiatic waters and all the rest. . . .

Swedish TV: How frequently was it done in Sweden?

Weinberger: I don't know. Enough to comply to the military requirements for making sure that they were up to date. We would know when the Soviets required a new kind of submarine. We would then have to see if our defenses were adequate against that. And all this was done on a regular basis, and on an agreed upon basis.[67]

Swedish prime minister Göran Persson reacted the next day with a statement to the Swedish Parliament: "A former secretary of defense, a U.S. secretary of defense, in a long interview, in a clear wording has presented a rationale for what, according to his view, NATO apparently did in our waters."[68]

Revisiting the October 1982 Harsfjarden Incident

In 2001, another investigative body, the third submarine commission headed by Ambassador Rolf Ekéus, reexamined the facts. During this process, Swedish defense minister Björn von Sydow gave researcher Ola Tunander—a member of the commission—access to the previously classified reports on the Harsfjarden incident. Tunander's examination of these documents brought to light a number of disturbing facts.

The dark gray submarine sail seen by a couple in late September was a meter high and 1.5 meters wide and squarelike.[69] The submarine's sail spotted on October 4 at Sadhamm by the local Coastal Defense commander was square-shaped and about 10 meters high. As Rear Admiral Svensson explained, "In the dark, he observed the navigational lights (red and green) from an object coming at first nearer and then passing the look-out behind an islet. He estimated its height to be 10 meters. However it was impossible to define the angle of observation when he estimated the size of the object. You can either conclude that he observed a sail or a possible submarine. In 1982, I connected this observation to a submarine because a Fast Patrol Boat had detected a submerged object in the same area shortly after the visual observation. I am not sure I would have made the same interpretation today."[70] According to the October 5 Naval Base War Diary, the sonar reading showed a small submarine about 35 to 40 meters long. Two days later, a 3-by-4-meter dark square sail was observed at Berga.[71] By showing their sails, however, the submarines had deliberately attracted attention, a most unusual behavior for a Soviet submarine. The descriptions of the two square sails—one large, one small—did not match Soviet-type submarines. The sighting lasted several minutes, so identification should not have been a problem, especially if the submarines belonged to the well-known Soviet submarine classes.[72]

On the day of the sightings, the Swedish defense minister had agreed to use force to bring the submarines to the surface.[73] One hour after the explosion of a powerful mine on October 11, a bright yellow chemical substance rose to the surface. A patrol boat collected samples.[74] Yellow-green dye was a visual distress signal for submarines that was primarily used by the U.S. Navy and was detectable from satellite, both visually and by radar.[75]

Later in 2001, Ambassador Rolf Ekéus reported that no results had been found from the analysis of the samples and that the photographs of the yellow-green dye were missing.[76] The chief of naval operations, Eastern Military District, claimed that he was never informed about the yellow-green patch.[77] Rear Admiral Emil Svensson was also not aware of this incident and doubted its reality.[78]

During the night of October 11–12, the bottom-fixed sonar system at Mälsten registered metallic sounds, apparently related to hammering work being performed on a presumably damaged submarine.[79] Then, documents released to researcher Ola Tunander revealed that at 6 p.m. on October 12, the Swedish navy tape-recorded "a certain submarine" moving past the hydrophones out to the open sea.[80] The researcher pointed out that the original reports referred to the tape recording of a slow propeller rotation of about 30 to 40 rpm, which is characteristic of a Western submarine.

The first report from 1983, based on the two tape recordings (October 12 and October 13–14), made in the evening on October 14, stated, "The subsequent analysis made of the tapes recorded on 12–14 October shows that the classification 'submarine' ['certain submarine'] that was done on 12 and 13[–14] October is confirmed. . . . The conclusion . . . is that this submarine has a damaged propeller shaft, or one of the blades is broken."[81]

Rear Admiral Emil Svensson has argued on Swedish television that a three-minute, forty-seven-second tape recording of a propeller sound captured on October 12, 1982, most likely revealed a Soviet Whiskey class. Admiral Jan Ingebrigtsen, the chief of Norwegian Military Intelligence in 1982, told the same program that his people listened to the tape soon after the incident took place in October 1982 and denied that it could have been a Soviet submarine. The Swedish Navy Sound Analysis (MUSAC) and the Swedish Defence Research Institute Division for sound analysis (FOI) reexamined the recordings. The former said that it might have been a surface vessel, while the latter concluded in May 2008 that the tape-recorded sound most likely originated from the taxi boat *Amalia*, which had brought some journalists to the area that very day.

Ambassador Ekéus, during his 2001 investigation, found that all of the pages covering this incident in the Defense Staff War Diary— from the afternoon of October 13 to the early morning of October 14—were missing.[82]

In its reappraisal, the Ekéus Commission stigmatized the handling of the Harsfjarden incident and the loss of crucial evidence during the investigation: "The handling of tapes and other evidence, in particular, must be criticized. The tapes recorded during important events were not carefully stored, the results of water samples and oil analysis, as well as photographs, were not preserved, and data were found to be missing from the Supreme Commander's military diary." Military archivist Per Clason believes that the actors had not realized the importance of the notes they were making and had failed to keep clean records. He suggested that for this reason, the records would not stand in court should there be a legal proceeding.[83]

Rear Admiral Svensson rejected the possibility that the tapes could have been falsified:

> The tapes were recorded on a 7-channel Racal tape-recorder. There was no erasing head on the tape recorder so it was impossible to erase.

The recorded tapes were transported by helicopter and car directly to the Armed Forces Ministry. The investigation team included people from the Armed Forces research institute, from military procurement and from the anti-submarine school. Every detected noise was examined and entered in the minutes. The minutes are preserved; they were compared with the tapes. No differences were found. An independent expert has been asked to re-examine the tapes in 2008. He absolutely denies the possibilities to falsify the tapes in 1982. It would be difficult even today.[84]

The commission further blamed the Swedish navy for "an obvious lack of tactical leadership and central intervention from the military leadership" as a cause for "a number of rumors and speculative hypotheses." One was that growing public opinion considered "that some or all the intrusions were made by NATO submarines with the tacit consent of the Swedish Navy, and that the chiefs of staff deliberately allowed submarines to violate Swedish territorial waters, and further that the submarines were released with the tacit consent of the Government, etc."

The Ekéus Commission rejected both the indictment of the former Soviet Union and that of NATO: "The inquiry [had] not found any factual basis for these rumors. Today, the evidence advanced for identifying the Soviet Union as responsible for the violations does not seem conclusive. The Commission's chairman, Sven Andersson, and its members made a political rather than an objective judgment. They considered the identification necessary regardless of the evidence." The commission then added, "The same goes for the Government's decision to endorse the Commission's conclusion, although, in view of the prevailing political situation, it seems unreasonable to criticize this decision." Ambassador Ekéus's board of inquiry had acknowledged that Sweden's diplomatic protest to the Soviet Union was not based on hard evidence, although it dismissed the rumors of a NATO involvement. The origin of the intruders was becoming more mysterious.

In September 2002, Rolf Ekéus, the president of this most recent Swedish board of inquiries on submarine intrusions, interviewed former Polish president, prime minister, defense minister, and commander in chief Wojciech Jaruzelski. General Jaruzelski explained that Sweden had never been a Warsaw Pact objective:

From 1965, I attended every Warsaw Pact meeting, and not once did Sweden come up as a problem per se. Swedish neutrality was accepted as something obvious and natural. I talked to Grechko and Ustinov a few times about Scandinavia and the Nordic countries. Regarding

Finland they were convinced about where the Finns stood. The border to Norway was difficult and complicated. Norway was of interest only because the coastline was a base area for NATO's fleet and air forces. But as I understood it, no offensive activities were planned against Norway. That would have been difficult to implement. If the advance towards the west on the Continent was successful, the matter of Norway would solve itself. Sweden was always spoken of with respect, considering the up to date Swedish defense and the stable Swedish economy. Sweden would not depart from the position it havd, to risk becoming involved in a conflict. The Soviet Union was interested in Swedish neutrality being maintained. So its marshals did not foresee any action that could provide grounds to intervene. However there was concern that situations could develop where NATO drew in Sweden, perhaps not directly but to create a rear front.

With regard to Sweden and Scandinavia/the Nordic countries, in Poland we did not have access to Soviet plans which extended beyond the joint Warsaw Pact planning. For us it was self evident that the Soviet Union, as the only Warsaw Pact country with nuclear weapons, something which gave them special privileges, planned certain things by themselves and kept it under lock and key. We were enlisted in the total context, but did not have all of its parts. What we had was our own planning, and in terms of the Nordic countries it was focused on Denmark as it was a NATO country, and a landing on Zealand. The resources we had at our disposal gave us little opportunity to realize such plans, but the philosophy was that we would demonstrate our power. In the vicinity of Sweden there was Bornholm, which could possibly be taken, but that was further in the future, not in the first days of the war. Jutland was not a target in itself, but a way of securing the northern flank. Øresund interested us, as did Kattegat. There were reconnaissance missions there, but our navy was small.

When asked about submarine intrusions, General Jaruzelski also denied having known about them: "Both the Soviet Union and Poland had said that they did not have anything to do with it. And I have no reason to hide anything now. . . . What would be the objectives of such operations? Perhaps to test the reactions of the Swedish Navy, to investigate certain waters. It could also be a matter of a navigational error. . . . Midget submarines? In that case, it was something that the Russians kept secret from us."[85]

In fact, no large midget submarines corresponding to the general descriptions given by Sweden were known to be in service in the

A recently released Russian rendering of a Soviet Uniform class special submarine at work. Unknown predecessors might have been used for activities in Swedish waters. The two Soviet units for reconnaissance and special tasks submarines were based in the Northern and Pacific fleets.

Soviet Baltic Fleet during the period 1980 to 1988 when the intrusions were intense. The Project 865 Piranha (NATO name Losos) class small submarines came too late. Only two were built, out of the twelve originally planned: the MS-520 and the MS-521, launched respectively in 1986 and 1990. The Piranha were small submarines (28.2 meters long by 4.8 meters beam by 5.1 meters draft), ideally suited to operate in the Swedish archipelago. They carried divers for special operations and sabotage missions. Their titanium hulls made them immune to magnetic mines. Both submarines were first stationed in Liepaya and painted white and red to disguise their real purpose. They were hidden from prying eyes and satellites inside a concrete tunnel. When the Soviet Union collapsed, the submarines were transferred to Kronstadt, reactivated during 1995 to 1997, and finally decommissioned and broken up. The known midget submarines operated by the then top secret "Crab" units in the Northern and Pacific Fleets entered service after 1983. They were designated by NATO X-Ray, Paltus, and Uniform, and the first two classes operated from larger mother submarines. Neither category had ever been reported by the

West as having deployed in the vicinity of the Baltic. If smaller midgets had been built and deployed prior to 1983, the secret of their existence would have resisted the close scrutiny of Soviet ports by U.S. satellites and the wave of indiscretions that followed the collapse of the USSR.

Even if one can assume that some intrusions must have been intentionally conducted by the Soviet Union and its elite *spetsnatz* units, this explanation was far from satisfactory to address the Swedish "periscope syndrome."

In effect, after 1982, Swedish citizens had ceased to characterize the intrusions as "Soviet." Referring to the tracks discovered on the sea bottom during the Harsfjarden incident, Rear Admiral Ivan Ivanov, a former Soviet counterintelligence officer with previous assignments in the Northern, Pacific, and Baltic fleets, merely said that his navy had found similar tracks on the sea bottom off the main naval base of Baltiysk in the Baltic and off Feodosiya in the Black Sea.[86] But then, who were the other potential intruders?

Unusual Suspects

New revelations, although still controversial, introduced new suspects. From September 22 to 26, 1982, the U.S. Navy had conducted a large-scale amphibious and naval exercise in the Baltic with the participation of three dock-landing ships: the LSD *Plymouth Rock* (LSD-29), the LPD *Raleigh* (LPD-1), and the LPD *Trenton* (LPD-14). All three had called in Kiel, Germany, the week before the incident. Eckernförde, near Kiel, was also a base for West German special underwater units (the Kampfschwimmers) which could have provided some logistic support for U.S. SEALs units operating in the Baltic. According to Tunander, a U.S. navy captain who was the operations officer of U.S. Navy SEAL Team 2 during the 1980s was based in Eckernförde with the German Kampschwimmers from 1980 to 1982.

Dock-landing ships had been previously used to transport the deep submersible NR-1 for some of its own classified missions. A coat of black paint was used to cover the usually bright-red sail. Any one of these dock-landing ships could have accommodated the small NR-1, which matched the description of the 45-meter-long submarine standing on the sea bottom and seen on the radar screen. Back in 1971, the NR-1 had secretly entered the Mediterranean inside the covered well dock of the USS *Portland* (LSD-37) to look for possible Soviet mines in several choke points.[87] Dock-landing ships could also carry any other kind of special delivery vehicles or midgets.

Even more puzzling, an American-flagged civilian tanker, the *Mormacsky*, was spotted by the Swedish navy outside of the Stockholm archipelago during the October 1982 submarine hunt. Tunander claims to have been told by retired intelligence officials that the *Mormacsky* was one of several rebuilt tankers used to carry midget submarines.[88]

During the Cold War, U.S. submarines participated in many daring operations inside the national waters of other countries. Several Sturgeon class nuclear attack submarines were earmarked in the 1980s for special operations and the handling of swimmer delivery vehicles.

In March 1984, John McWethy, the Pentagon correspondent for ABC TV, described the scope of clandestine U.S. submarine operations:

> [U.S.] submarines are repeatedly violating territorial waters of other nations while gathering intelligence. Most of the top-secret missions are into the waters of the Soviet Union, but according to both active duty and retired military sources, some missions have been run into the territorial waters of those nations considered friendly to the U.S. Even friendly countries, sources say, sometimes do things they don't want the U.S. to know about, things that could inadvertently threaten American security. The missions are conducted by specially equipped nuclear powered attack submarines and in some cases by a nuclear powered mini-sub called NR-1 (mini-sub). It has a seven-man crew, wheels on its underside for crawling along the bottom and is described by the Navy as a research vessel.[89]

Former commander of U.S. SEAL Team 3 Gary Stubblefield gave his explicit definition of covert operations:

> Covert operations involve operations where we don't look anything like U.S. SEALs (we hope) and there are no uniforms, no ID cards and no connection to the armed forces or government of the United States. Few SEALs, maybe one SEAL out of 200, will ever go on a real-world covert mission. . . . There are times where the government wants to take some kinds of direct action without publicizing it, and I have done such missions.[90]

Tunander also quoted Stubblefield, who added,

> [We conducted some covert operations in the early 1980s, some] really smart interesting training in the NATO and Atlantic theatres. . . . We set up and worked with support networks, E&E [escape and evasion] networks and we started getting smart about going into foreign areas.

All that involved looking like people who weren't in the U.S. Navy and doing things that people in the U.S. Navy weren't supposed to do.[91]

Tunander claimed that a CIA officer described the October 1982 incident as "something of an underwater U-2." Similarly, a senior U.S. Navy officer told the former director of the Norwegian Foreign Ministry's Political Division that the damaged submarine in 1982 was from the United States.[92] In 1993, former secretary of defense James Schlesinger confirmed to Tunander that a U.S. submarine had been damaged in the Stockholm archipelago in 1982.[93] According to Tunander, the operations were actually decided by a deception operation committee chaired by Bill Casey with Dick Allen and two representatives from the defense and the state departments. The operations were then run by the CIA-Navy National Underwater Reconnaissance Office (NURO).[94]

Since the United States had apparently decided to admit involvement, exactly how was it done and with what?

Built in Groton, Connecticut, in 1969 for the U.S. Navy, the NR-1 was the world's first nuclear deep-submergence submarine. Two retractable, rubber-tired, extendible bottoming wheels provide a fixed distance between its keel and the seabed, so that it can use its special instruments

The U.S. dock-landing ship USS *Pensacola* had secretly carried the 44-meter-long NR-1 in the Mediterranean. Three sister U.S. dock-landing ships were in the Baltic during the week preceding the October 1982 Harsfjarden incident.

The deep submergence vehicle NR-1 (1969), a U.S. Navy nuclear-powered submarine used for both civilian and military research. NR-1 is an unusual suspect in the October 1982 Harsfjarden incident.

to recover objects. Its dimensions closely match the Swedish description of a 35- to 40-meter-long submarine standing on the sea bottom and seen on the sonar screen.

The U.S. Navy had also earmarked larger attack submarines for special operations. The second-oldest U.S. nuclear submarine, USS *Seawolf* (SSN-575), was fitted in 1965 to 1967 with a special compartment to accommodate SEALs.[95] The Sturgeon class USS *Cavalla* (SSN-684) also deployed SEALs for clandestine missions. Both were based in the Pacific and both received distinctions in 1982 and 1983, the *Seawolf* being awarded the Navy Expeditionary Medal, a Battle Efficiency "E," an Engineering "E," a Supply "E," and a Damage Control "DC" before undergoing important repairs. Tunander suggested that the *Seawolf* might have been the larger submarine damaged in the Stockholm archipelago in October 1982.[96]

More Confessions

Weinberger's March 7, 2000, interview on Swedish television referred to bilateral contacts between the U.S. Navy and the Swedish navy to coordinate the intrusions. He also described the undertaking as a NATO operation.

By NATO, Weinberger clearly meant the participation of NATO members and not a planned NATO operation. A Danish lieutenant general, who formerly commanded BALTAP (the NATO wartime command for Denmark, northern West Germany, and the Baltic Sea) during 1993 to 1995, explained, "It was of interest to check if Sweden was first, capable, and second, willing to defend its territory. This was a legitimate NATO interest. The Norwegians and the Danes could tell the other NATO members: we trust the Swedes. They would certainly defend that flank. However, the great powers and the superpowers preferred to get their own information, to have it confirmed by their own means."[97]

More television interviews brought to light the apparently dominant British involvement. Sir Keith Speed, formerly a member of the Parliamentary Defence Committee (1983–1987), told the TV show *Strip Tease* that British submarines were testing Swedish coastal defenses. When asked whether the British had penetrated Stockholm harbor, he answered, "Not quite, but that sort of thing." A Swede commander, formerly in charge of the Swedish navy's acoustic intelligence, confirmed in November 2001 that he had provided the Ekéus Commission with the acoustic signature of a British Oberon class submarine.[98]

A former chief of staff of the Royal Navy Flag Officer Submarines told Tunander that he had made a couple of trips into the Baltic Sea onboard Oberon class submarines, passing the Danish Straits submerged and heading toward the Soviet Baltic coast before entering Swedish waters. He confirmed that SBS (Special Boat Service) frogmen were landed on the Soviet coast but refused to comment on SBS activities in Sweden.[99] A former Oberon skipper added that he had sailed into the Gulf of Bothnia, north of the Åland Islands. Ministerial approval was granted for every operation. One participant was HMS *Orpheus,* which had been refitted with a special compartment that allowed five Special Forces troops to exit the submarine while submerged.[100] West German submarines were also said to have used Swedish waters to practice for operations against the Soviet Latvian and Lithuanian coasts.[101]

Reagan and Thatcher's Psychological Victory

The submarine intrusions had a dramatic effect on the Swedish public's attitude toward the USSR. Tunander reported that in 1976, 6 percent of the Swedish population perceived the USSR to be a direct threat to Sweden.[102] In 1980, this percentage slightly rose to 8 percent after the Soviet invasion of Afghanistan. With the October 1981 "Whiskey on the rocks" incident, this perception changed dramatically, with 34 percent of the population seeing the Soviet Union as a direct threat and 71 percent as an unfriendly nation. After the October 1982 Harsfjarden incident and for the rest of the 1980s, 42 percent of the population looked at the Soviet Union as a direct threat and 83 percent as unfriendly. Submarine intrusions had totally altered the Swedes' perception of the Soviet Union, which was now seen as friendly by only 1 to 2 percent of the interviewees. Meanwhile, the popularity of the United States rose from 20 to 40 percent.[103]

One undeniable consequence of the intrusions, regardless of who conducted them, is that the Swedish public's attitude had in fact become much more sympathetic to the United States and NATO. Unless more participants of the actual intrusions come forth and confess their activities, the whole story may never be fully understood.

The topic is still hotly debated in Sweden, and Rear Admiral Emil Svensson maintains that the Soviet Union was behind most of the intrusions. During the course of several visits to Russia, he interviewed more than a hundred submariners and veterans. Some told him that they had indeed penetrated Swedish waters. Nevertheless, the comments made by former SEAL Gary Stubblefield and the interviews of former U.S. defense secretary Caspar Weinberger do support the theory that the Soviets were not alone. Still, Rear Admiral Svensson holds his reservations: "I can't understand why the Americans would go deep into the archipelagoes. If they want to 'show-up' a periscope it is easier and less risky to do that just outside the archipelago. There are lots of boats to get an 'audience' in that area."[104]

It remains an undisputable fact that both sides could not afford to ignore a potential zone of operation where the adversary could seek shelter or mount an attack coming from an unexpected direction. As Captain Lev Vtorygin remarked, however, these incidents affected Soviet-Swedish relations to such a degree that one wonders what interest Moscow could have had in continuing its alleged penetrations and risking another "Whiskey on the rocks" incident, which would have certainly been followed by a breaking of diplomatic relations and a closer association between Sweden and NATO.[105]

15

SPIES IN UNIFORM
1980s

The column of names of military attachés killed in the line of duty during the last twenty years is sobering. The list includes army lieutenant colonel Charles Ray, who was shot in Paris by a PLFP (Popular Front for the Liberation of Palestine) terrorist in 1983, and navy captain William Nordeen, the defense attaché in Athens, Greece, who was decapitated by a car bomb in 1988. Army warrant officer Kenneth Welsh and navy petty officer Mike Wagner died in the bombing of the American Embassy in Beirut, Lebanon, in 1984. A chief warrant officer, the Defense Attaché Office operations coordinator, burned to death in the embassy fire in Islamabad, Pakistan, in 1979. Four young women assigned to the Defense Attaché Office in Saigon died in a C-5 air crash on April 4, 1975, being among the last casualties of the long Vietnam war. The stereotypical portrait of the military attaché as a diplomat "cookie pusher" is refuted by this partial list of recent attaché casualties.

At the height of the Cold War, the diplomatic environment in the Warsaw Pact–Soviet bloc countries was openly hostile. Other countries with strict central control of information, such as India and Yugoslavia, which were nonaligned politically, also possessed forbidding environments. Even some Allied countries such as Israel, Pakistan, and Japan, although friendly and politically aligned with the West, have systems that control military and industrial information much more closely than most Western countries do.

The age of high technology has not lessened the role of the attaché but rather has altered the methods by which he operates. The paucity of accurate intelligence from reliable HUMINT (human intelligence) sources was glaring during the tragic events of the late 1970s and early 1980s, when diplomatic protection of U.S. embassies guaranteed by the Vienna Convention of 1949 was ignored and even violated by a number

Soviet attachés at work collecting brochures during a U.S. defense show. Former Soviet naval attaché Lev Vtorygin remembered, "We felt pity for you to give away so much."

of governments or political factions within those countries. The personal immunity guaranteed to diplomats in countries such as Iran, Lebanon, and Pakistan was grossly and tragically violated. Those turbulent times gave new emphasis to the job of HUMINT collection against international terrorists and fanatic religious and political groups.

A new emphasis on HUMINT followed the 1979 Teheran hostage experience. U.S. HUMINT was reorganized under the 1983 Department of Defense HUMINT Plan. The new approach focused the entire military HUMINT collection effort in the Defense Intelligence Agency's operations directorate, which also ran all service attaché operations. This change was first opposed by the individual service intelligence organizations, especially the army's, whose experience in the HUMINT arena far surpassed that of the other services. The opposition was understandable, but the army reluctantly joined the effort. The following accounts are based on Peter Huchthausen's recollections.

Balkan Attaché Adventures

The 1980s witnessed some remarkable U.S. military attaché successes, as well as major blunders. Most of the achievements are little known outside a small circle of those involved. In countries that maintain complete central control over information and where defense developments are tightly wrapped in secrecy, Western diplomats and especially military attachés are forced to travel a great deal. "Get out and get a feel of the people," said Ambassador David Funderfurk to his staff in the American Embassy in Bucharest, Romania. Travel enabled attachés to gain firsthand observation of what in more open countries could be culled from the pages of the open press. In the West, information concerning new trends in the economy, agriculture, maritime and naval construction, and weapons is openly available in the press. Attachés, like journalists, make their living from the flow of information.

During the Cold War, Romania, firmly in the grip of its state security organization, the Securitate, was among the Eastern countries that were most secretive about military information. Even the most trivial aspects of defense were hermetically sealed against foreigners finding out about them. Tourist maps of Romania were altered to hide military bases. Objects such as bridges, communications facilities, public transportation, power plants, and ports were designated strategic objects and placed on a list of locations that foreigners were forbidden to observe and photograph. Defense authorities censored information that pertained to military developments from appearing in the press.

For those reasons, in the 1980s, the U.S. naval attaché resident in Belgrade and dually accredited in Romania adjusted his frequent visits to Bucharest to coincide with the Romanian national and military holidays. That way he could gain maximum exposure to senior Romanian military officers and high-level government officials. He traveled to Bucharest often with other NATO attachés. Three major information objectives during that period were to uncover the flourishing Romanian navy ship-building program, disclose the true interface of the Romanian military machine with the Soviet defense forces, and debunk the publicly stated concept that Romania prohibited the transit of Soviet and other Warsaw Pact forces across its territory.

The Romanians were adept at concealing their military bases from the general public and foreign attachés. They used movable walls to mask the building projects from the open roads. The attaché was thus forced to insert himself into areas off the public roads, by car or by foot. This meant exiting the area without being observed or, worse, stopped and identified by nervous guards. Therefore, the value of the observation effort had to be quickly weighed against the risk of identification, exposure, and detention by the local security apparatus.

Among the accomplishments during the early 1980s was the timely action by the defense and army attaché in Bucharest, Colonel Frank Mastro. He personally repudiated Romanian dictator Nikolae Ceausescu's claim that Warsaw Pact troops never transited Romania. Mastro and one of his attaché staff NCOs intercepted two Soviet military convoys crossing Romania at night, one by road and the other by rail. After careful planning and acting on signal intelligence tips, Mastro and his assistants staged their vehicles at strategic road and rail crossings in the middle of the night to intercept the Soviet army convoys. Mastro and his troops actually joined one convoy and, passing truck after truck in a leap-frog maneuver, illuminated and photographed the Soviet unit markings. In the midst of this effort, local Romanian secret police (Securitate) vehicles attempted to run him off the road. Mastro's skill and his NCO's driving ability proved superior, and they ended their hair-raising midnight ride safely back at the U.S. Embassy. In retaliation, the next day all four tires on Mastro's vehicle were sabotaged by the Securitate.

During the same period, U.S. attachés carried out an extraordinarily daring operation in Romania in the naval port of Constanta. The strongly guarded naval base made it difficult for the Western attachés to observe the latest warships. The dictator Ceausescu was building a navy out of proportion to any threat that existed for his Black Sea country. New warships were rolling off the building ways in astonishing number, yet it was difficult for attachés to penetrate the heavy security to get a close look at the warships.

Nevertheless, the U.S. Embassy air force and naval attachés made a daring incursion into the guarded shipyard. After observing from a secure distance, the two officers noticed that one gate to the enclosed base was open to admit a large number of trucks carrying construction material. The trucks shuttled in a continuous chain, while the armed security guards waved them past with little notice. The Americans carefully smeared the diplomatic license plate of their native Romanian jeep, and, making sure that the surveillance vehicle that normally followed them was a safe distance away, they carefully drove into the midst of the endless chain of trucks. They passed through the base gates unnoticed and made several close passes to the navy piers, where lines of corvettes, mine-warfare craft, and support ships were moored. As they drove back to the gate, the naval attaché in an improvised observation nest in the rear of the jeep discovered that his camera's shutter speed had been accidentally set improperly.

"We have to go back for another pass," he exclaimed. The air force colonel who was driving made a discreet U-turn and returned for a second pass, even closer to the ships. They then returned to the column of trucks and slipped out of the gate. Their normal surveillance vehicle, which they had skillfully evaded, was waiting at the former observation point and caught sight of the Americans as they departed the base. The four-hour return drive to Bucharest and the relative safety of the U.S. Embassy turned into a wild ride during which the Romanian security vehicle tried to force the Americans into an accident. The American air force colonel, however, had been a B-52 pilot in one of his previous posts and proved to be the better driver. The attachés arrived unscathed with their take from the hunt.

Another intrepid envoy was the French military attaché in Bucharest, a flamboyant lieutenant colonel. When the colonel was apprehended in a forbidden zone taking photos of air operations at Craiova, a Romanian air force installation, and was in turn photographed by the Romanian Securitate surveillants in posed compromising photos with the base in the background, he thought it was the end of his career. Certainly, the inevitable persona non grata banishment would follow. After being released and on his return drive to Bucharest, while stewing about the uncomfortable predicament awaiting his return to his embassy, he stopped at a nearby military installation of high interest. Assuming that he was at the end of his attaché career, he decided to cast caution to the wind. With complete abandon, he climbed over a fence, walked inside the base, photographed the entire interior of the sensitive installation, climbed out over another wall, and drove away undetected. He was never dismissed—the earlier event was mysteriously forgotten.

Balkan Naval Ship Visits

During this period of the Cold War, U.S. Navy sailors, known as bluejackets, acquitted themselves well while on port visits for the U.S. fleet. The value of these visits was often doubted by many foreign-service officers. They were much more significant in Yugoslavia and Romania, however, countries that liked to stress their relative independence from Moscow, than they were in other countries. These visits were the only events during which the United States showed the flag, and the only times when thousands of Americans went ashore among the captive communist populations.

The navy visits thus played a major political role in U.S. foreign relations. In addition, for the American attaché whose job kept him somewhat isolated, the visits presented an opportunity to meet the officers and the men of the fleet, and to share his regional expertise as a diplomat. Arranging and carrying out a successful ship visit was a considerable challenge for an attaché serving in a totally closed society and where normal domestic products were scarce. In his hands rested the responsibility of setting the stage for the U.S. Navy to visit the host country. He had to guarantee navigational safety even when local charts were often considered secret by the hosts. The attaché arranged security, protocol schedules, and sporting and entertainment events for the ship's crew. Visits to normal Western countries were trying enough, but in closed societies, which sheltered their populations from the visitors, the job could be frustrating and sometimes infuriating. The attaché's greatest fear was that a well-meaning but careless act would precipitate a damaging international crisis.

Naval experience taught that when things go wrong in the fleet, they can go extraordinarily wrong. But during this Cold War period of port visits for the U.S. Navy in Yugoslavia, Romania, and finally in the Soviet Union, there were few cases of misconduct on the part of U.S. sailors. This is a tribute, not only to the leadership of the officers and the petty officers of today's fleet who prepared the men and the women sufficiently for these important visits, but to the quality of the modern U.S. sailor. The young American from a small town in the Midwest or from Brooklyn, who had seen the "night club strips" in Hamburg, Naples, Marseilles, or Norfolk, conducted himself with great credit while on independent liberty in these communist countries, where the average citizen enjoyed far fewer liberties than he.

The U.S. Navy during this period fulfilled an especially effective duty as a versatile instrument of American foreign policy: the role of the navy as a direct tool of diplomacy had not waned with the proliferation of diplomatic missions abroad or because the speed and the reliability of communications had improved. The image of a U.S. warship swinging at anchor, the commanding officer riding ashore in his gig in dress whites carrying a diplomatic dispatch to the chieftain of a remote republic, has changed only in the style of uniform and the lines of the ship and not substantially in importance.

The U.S. Sixth Fleet during the late 1970s and the 1980s was an essential instrument of the renowned but shaky U.S. foreign policy of "differentiation" during the darkest days of the Cold War. In Ceausescu's oppressed Romania, regular port visits by U.S. warships allowed scores of American bluejackets to come ashore in dismal Black Sea ports. The sailors became the only Western spectators who were permitted to peer into the windows of one of the most severely repressed outposts of the communist bloc. With Western tourism virtually nonexistent in Romania, its population had few other opportunities to interact with citizens of the free world. In a society starved of any knowledge of the outside world, this contact was crucial to keep the spark of hope alive. The juxtaposition of democracy with totalitarianism on the one-to-one level of sailor with citizen is a largely overlooked phenomenon that scored well for the West. It surely fostered faith in many citizens of a pathetic country where the aspiration for a better life depended on the slightest glimpse of evidence that a better life existed at all.

The U.S. Sixth Fleet's role in the equation began when regular annual visits were scheduled to the Black Sea port of Constanta in 1972. This was a period when, despite the periodic thaws in superpower relations, there were no regular Western naval visits to the rest of the Soviet bloc. For that reason, the U.S. Navy and consequently the Bucharest embassy attached great importance to the annual summer visits to Constanta, which usually included a cruiser with eight hundred American bluejackets.

The Romanian Black Sea coast had sprouted a string of holiday resorts constructed of immense and supremely ugly prefabricated concrete blocks. These large complexes resembled shabby prisons. They served as one of the few warm seashore meccas for the Warsaw Pact–privileged Party elite. The complexes, named after planets, accepted only group tours from the communist bloc countries. Only the International Hotel, which was run by the Romanian state tourist company, accepted the few Western vacationers who visited Constanta. The hotel was heavily guarded by the Securitate to keep Romanians and other Eastern visitors away from the Western guests. The brown dirt and sandy beaches along this stretch of coastline were lined with bathers from the otherwise sun-starved Soviet Union and other Eastern European countries. Into this milieu, at the height of the tourist season each year, came hundreds of American sailors from the U.S. Sixth Fleet.

During one of these navy visits, an entire orchestra, led by the first violinist, tried to defect to the U.S. attaché in full view of the entire

Romanian navy hierarchy. This unnerving event occurred during an intermission in a concert at the Mufatlar State Winery for visiting U.S. Navy officers. A half-dozen members of the orchestra gathered behind the violinist and confirmed their intention to seek political asylum with the U.S. Navy—not later, but before the concert ended. The American officer was somehow able to postpone the discussion until after the concert, when better reason might prevail. He later convinced them that their chances would be best if they delayed for several years. U.S. diplomats lived in constant fear that an attempted defection might suddenly end the strained goodwill shown by both sides during the annual port visits. While the ships were docked at the piers, Romanian security often intimidated local people to prevent their prolonged contact with Western officers and crew.

When the ships opened for public visiting, hundreds of curious young Romanians filed aboard. During this rare contact between Americans and Romanians, the crews served hot dogs, popcorn, ice cream, and cookies to create an atmosphere of hospitality. The Romanian navy liaison officers stood on the sidelines, dreading that something would go wrong during the public visits. To deter fraternization, hundreds of Romanian naval academy cadets were planted in two long ranks stretching from the ship's quarterdeck brow to the port gates. The police then forced all civilians to pass through this gauntlet after surrendering their personal identity papers before they could tour the ship. Every effort was made by the Romanian authorities to frighten and browbeat the crowd, without spoiling the event for the visiting Americans.

The Romanian liaison officers objected to the handouts the U.S. Navy gave to the local citizens. The Romanians claimed that the brochures had in the past contained vicious propaganda. One year, the defense ministry demanded a chance to review the pamphlets in advance for clearance. When U.S. diplomats first heard the charge while in Bucharest planning the visits, out of curiosity the embassy officers produced one of the traditional welcome-aboard pamphlets, which was to be translated into Romanian, and took it to the Ministry of Defense. The sinister Commander Nicku Padarariu, at the ministry's Defense Liaison Section, pointed out the offensive words. The pamphlets stated that the ships were a part of the U.S. Sixth Fleet, which was NATO's deterrent striking force in southern Europe. The Romanians preferred that visiting U.S. ships emphasize the bilateral relationship and not use the term *NATO*. The Americans ignored their protests and gave out the pamphlets unchanged. The Romanians, in turn, merely confiscated them from the crowds as they left the ships.

It was always a relief to see the same number of visiting Romanians leave the ships as had boarded for the tours. For the average U.S. sailor, who was usually seeing control and repression for the first time in his life, the display of open cruelty by the police and the military toward the public often provoked him to strike out subtly against what he saw. The American sailor sometimes did this in an ingenious way.

Once, in the tense atmosphere after a public visit, the identity cards of two young Romanian girls remained unclaimed with the police after the crowds had left the ship. Terror shone in the eyes of the senior Romanian liaison officer, Commander Niku Padarariu, as he confronted the U.S. attaché. "We must find those girls," he said, almost pleading. "We must, or there will never be another ship visit."

Certainly, the U.S. Navy would never permit the Romanians to search the ships, which were sovereign American soil. But the navy was also keen not to let the incident get out of control. After a thorough but quiet check by the ship's company, no girls were found, and the day's activities continued. The ingenuity of the American bluejacket triumphed, however. Shortly after the evening reception for the ship's company, hosted by the Romanian navy, two attractive Romanian girls mysteriously appeared on the arms of two U.S. sailors strolling in a Constanta park. The Romanians were visibly relieved, but no one ever solved the method of their clandestine departure from the ship.

A more vicious incident occurred in June 1984 during a navy port visit to Constanta, when a group of Securitate thugs attacked five Americans, including three U.S. Embassy personnel. The Securitate used rubber truncheons and brass knuckles in the bloody fracas. Sadly, one Romanian thrust a broken bottle into the face of a marine, who was a security guard at the U.S. Embassy It caused the loss of an eye. At three that morning in a Constanta hotel, the frantic survivor of the brawl woke up the attaché, who was the naval liaison with the Romanians. The attaché learned that the wounded marine had disappeared in a Securitate vehicle with one of the U.S. Embassy army personnel and was nowhere to be found.

The subsequent search through the dark and desolate streets of the sleeping Romanian town finally ended in a haunting scene. Embassy officers knew the young marine had been injured, so they began a systematic search of Constanta first-aid stations and hospitals. They were driven by the duty embassy chauffeur, who was known to have been compromised as a Securitate informer. On the third attempt, they found a hospital in an unbelievably deplorable state on an unlighted road. A sullen guard in a Securitate uniform summoned a frightened nurse, who acknowledged that an American had been brought in earlier. The nurse escorted the officers into a dimly lit office that smelled of sweet disinfectant and urine. After a short wait, an English-speaking female doctor appeared in the room wearing a smock that had probably once been white but was covered with old and darkened bloodstains, along with fresher ones. Ominously, the smock was still wet from recent soiling. The doctor spoke in a whisper and shook her head sadly, which led the diplomats to instantly fear the worst: that the poor marine had died. She led them into a room with ten beds. The walls were colored in two shades of what might have once been green, smeared with bloodstains.

The Americans were taken directly to the bedside of a figure covered in dark blankets. A young man lay with his head wrapped in gray bandages covering both eyes. A dark stain oozed from one eye. The poor marine was conscious and amazingly coherent. He answered to his name. "I'm fine sir, just can't see through these bandages." The ceiling was splattered with dark bloodstains, and several cockroaches scurried along the wall.

"I'm afraid I had to remove the left eye," the doctor whispered in good English. "It came out in five lacerated pieces. Hopefully, there was no damage to the optical nerve."

The diplomats assured the marine that he would be evacuated shortly. Leaving an embassy consular officer to stay with him, they quickly departed the dismal hospital, fighting the urge to vomit. They raced to the U.S. ships, from which they ordered the military medical evacuation "Nightingale" flight from Rhein Main Air Base in Frankfurt, Germany. The most unforgettable part of the entire episode was the Romanians' reaction when the U.S. Air Force medical evacuation aircraft arrived at the Constanta airport several hours later. Two U.S. Embassy officers stood next to two Romanian officers and watched the glistening C-9 jet, with the American flag emblazoned on its tail, taxi toward them. It stopped on the tarmac a few yards from an ambulance holding the injured marine. A large door opened in the side, and a platform slowly descended as if by magic. A Romanian standing with the Americans suddenly asked, "How many Americans are you evacuating?"

"One," they answered.

"Only one?" the officer asked, astounded. "You sent this huge airplane to pick up just one man?"

Three air force attendants, dressed in dazzling white, quickly wheeled the patient from the ambulance onto the platform, which ascended smoothly and disappeared into the fuselage. The aircraft began to taxi as the large door closed and sealed. It was in the air again after spending less than ten minutes on the ground. The embassy officers walked away with the astonished Romanians, feeling quietly thankful for being American.

Assignment in Belgrade

In Yugoslavia, following the May 1980 death of Marshal Josip Broz Tito, the dictatorship and the formula for ethnic unity he had established began to unravel. This allowed each republic to emerge as an independent state. Much of what follows was based on firsthand observations by Peter Huchthausen during his tour of duty as a naval attaché in Belgrade (and dually accredited in Romania from 1980 to 1984).

Although Tito sought refuge in the Soviet Union during the early World War II years, he soon disagreed with Soviet ideology. Stalin's support also began to cool toward the end of the war, as the Soviet dictator gazed longingly at Yugoslav ports on the Adriatic. The Soviet quest for warm-water ports dated back to Peter the Great. The West's worst fears were that Yugoslavia might fall into the warm bosom of the Warsaw Pact, by either force of arms or gentle seduction. Soviet use of Yugoslavia's Adriatic ports became a serious concern and catapulted the role of the U.S. Navy to the forefront of postwar Balkan geopolitics. These excellent ports included Rijeka near Trieste, Split on the central coast, and Tivat on the strategic Bay of Kotor in the south.

Soviet focus on the Adriatic seacoast developed into a key issue for the U.S. Sixth Fleet in the Mediterranean. After Tito's split with Stalin in 1948, Adriatic port usage remained of prime interest during the Cold War until the final collapse of the Soviet Union in 1991. The Soviet navy was stationed in the area on a relatively permanent basis, with the regular presence of a submarine tender and two diesel attack submarines in the port of Tivat in southern Montenegro. The Soviet navy's active participation in the warship visit program in Yugoslav Adriatic ports became a major diplomatic front-line contest with the U.S. Navy, and one that made the naval attaché job in Belgrade so rich and intriguing.

The West, delighted with Tito's 1948 break with the remaining communist bloc, hastened to Tito's side with economic and military assistance. Many Yugoslavs still fondly remember the American aid packages sent in the postwar years, the U.S. economic assistance, and the hundreds of U.S. tanks and other surplus military hardware that poured into the country to keep her on the right side of the new Iron Curtain.

To demonstrate true nonalignment, the Yugoslav government welcomed regular visits by both Warsaw Pact and NATO warships. Despite critical periods in relations with Yugoslavia, both the U.S. and Soviet navies continued to compete for her Adriatic ports. The main purposes of the fleet visits for the U.S. Sixth Fleet were to show the flag and to exploit the unique window on communism that nonaligned Yugoslavia offered. The United States ensured that the Sixth Fleet conducted the same number of visits as the Soviet navy so that Yugoslavia would remain truly nonaligned.

And so Adriatic ports became popular liberty calls for both the U.S. and the Soviet fleets. Soviet port visits in the Mediterranean during the period were limited to harbors in Syria, Libya, and Algeria. NATO had a corner on all of the classy ports on the French and the Italian Riviera and regularly visited Turkey and Greece. With Israel off limits and Egypt closed to the USSR since 1972, Soviet naval visits to Dubrovnik, Split, or Rijeka were a genuine treat for the Soviets. As a result, they became important intelligence-collection opportunities for NATO naval attachés in Belgrade. The attachés expended a great deal of effort observing the Soviet ships and their crews while in those ports. Given the secrecy shrouding all Soviet military operations worldwide, the attachés in Yugoslavia became the only Western military observers who could engage in close-range scrutiny of operational Soviet naval forces. The attachés boarded their ships often. It was a convenient accident of geography that both sides sent warships to visit the lush garden spots of the Adriatic and that duty demanded attachés' undivided attention and frequent physical presence on the coast. They were uprooted from the drudgery of the gray capital city of Belgrade, leaving army and air force colleagues behind to mind the store.

In the early 1980s, both the U.S. and Soviet navies conducted four visits per year to the Yugoslav Adriatic ports, most of them multiship detachment calls. During the port visits, the Belgrade attachés went to extremes to plot tactics to obtain answers for the naval analysts in Washington and London, who had endless questions about the Soviet fleet. The attaché was frequently tasked with identifying the senior officer in the visiting Soviet delegation by name, rank, and position in the fleet hierarchy. In those days, the slightest bit of information concerning Soviet forces was meaningful. Belgrade attachés made firsthand contact with visiting Soviet officers and often met their commanders and embarked flag officers. Since it was customary for warships visiting ports anywhere in the world to be open for general public visiting, the attachés were able to go aboard with the local citizens for a close look.

In 1984, at the end of four years in Yugoslavia, the U.S. attaché had visited most classes of Soviet surface ships that could be tied alongside a pier. When the Soviets' guided-missile helicopter carrier (CHG) *Moskva* first visited Dubrovnik, the Western attachés made every effort to get as close as possible. Having a deep draft, the carrier had to anchor about a mile from the Dubrovnik port of Gruz. On her first visit, the British and U.S. naval attachés obtained a dinghy with an outboard motor and approached the carrier for a close look. While concentrating on the mission, they neglected the change in the weather and were caught in a sudden *bura*, a classic Adriatic storm with severe winds blowing from the north. They pitched wildly in the blowing scud, hoping the shaky outboard wouldn't fail. In the distance, the attachés could see a motor launch from the *Moskva* heading toward them. The Soviet sailors in it obviously knew what the intruders were doing and escorted them back to port in case their dinghy swamped.

During another port call by a Soviet naval detachment, Western attachés set up a photo ambush for the Soviet officers. The detachment was led by a previously unknown admiral who was suspected of being the newly appointed commander of the Soviet Mediterranean Squadron. As the Soviet protocol party disappeared into the Dubrovnik town hall for the usual call on the town mayor, the NATO team deployed like hit men around the building. Their objective was to obtain a full frontal photo shot of the admiral as he emerged after the visit. The area was filled with summer tourists so the NATO team members were able to mingle with the crowd. The Western team noticed, however, that while poised for the attack, they were being filmed in action by plainclothes surveillants from the local security arm of the Yugoslav navy. It turned out that the admiral who was successfully photographed by the attachés was Vice Admiral Valentin I. Selivanov, the commander of the Soviet Mediterranean Squadron.

During another Soviet ship visit to Rijeka, Huchthausen was among the Canadian and U.S. officers who were arrested by a Yugoslav policeman while taking photos of visiting Soviet warships. The policeman demanded identification. When Huchthausen showed his diplomatic pass, the Yugoslav showed it to the Russian. "Look, I told you he was from your embassy."

"Comrade, he's from the American Embassy, not ours!" said the Russian indignantly. Huchthausen was then escorted from the pier area by a perplexed Yugoslav guard, who, as soon as the officers were behind a building and out of view of the Soviets, apologized and released Huchthausen.

During another Soviet visit, this time in Dubrovnik, the U.S. and Italian attachés were motoring together in a small boat and passed too close to a Soviet submarine moored in the port of Gruz. A voice speaking Croatian over a loudspeaker pierced the quiet harbor afternoon: "You, in the small boat, what is your business?" The two beat a hasty retreat when they noticed a group of Yugoslav surveillants in a police launch cast off from a nearby seawall and head their way. Thanks to a faulty engine, the police boat came to an abrupt halt and sat dead in the water in the central harbor, while the two Western officers disappeared quietly into a crowded fishing anchorage.

NATO attachés also observed and described the conduct of Soviet sailors enjoying liberty ashore in the Dalmatian towns. The sailors were required to remain together in groups of five or six with an officer or a senior petty officer in charge. They had very little money to spend, and what they had went toward the purchase of basic items such as cigarettes, postcards, Yugoslav magazines, and soap. The Soviet seamen usually showed a high degree of discipline, albeit most likely because they were threatened with severe punishment, often physical, for misconduct. The Soviets' behavior was a far cry from that of American sailors, who met local girls, bought the more expensive souvenir items, and mostly moved about in pairs. The Soviet groups were required to be back aboard ship by dark, while the Americans could generally stay out until midnight or longer, depending on their seniority. It was clear that the average Yugoslav citizens in those Dalmatian ports lived far better than did the Soviet sailors, who may have come from Siberia, Yakutia, central Asia, or Moscow. The difference in relative freedom of the Soviet and U.S. sailors was pronounced. Later, in Moscow, it became clear to the U.S. attaché that the Soviet warriors of the sea were the most envied of all the Soviet military because they had the luxury of seeing beyond the tightly sealed borders of the USSR.

There was a large Yugoslav ship-repair facility in the town of Tivat, near the entrance to the picturesque Bay of Kotor. This protected harbor already occupied a celebrated place in history as the home port for the once-powerful fleet of the Austrian-Hungarian Empire before its destruction in World War I.

In the port of Tivat, a Yugoslavia shipyard repaired Soviet, Libyan, and occasionally other foreign warships. During the 1970s and the 1980s, there were generally two Soviet naval units there, a submarine tender and a Foxtrot class diesel submarine, undergoing a routine refit. Libya also maintained two of its Soviet-built Foxtrot class submarines in Tivat for periodic overhaul. At any given time during the period 1980–1984, there would be three Soviet-made submarines in Tivat.

The U.S. Navy tried to cite this regular presence in Tivat as evidence that Yugoslavia was stretching her nonalignment. The Yugoslav navy skillfully pointed out, however, that such a presence was in concert with their law governing territorial waters and the port presence of foreign warships. This was the same law that defined the total tonnage, the number of vessels allowed, and the length of time foreign warships could visit their ports.

During that period, an increasing number of incidents flared up between the U.S. Navy and Libya, resulting from excessive territorial waters claims in the Gulf of Sirte and continued Libyan support for international terrorism. Libya was a key ally of Yugoslavia and employed thousand of Yugoslav engineers, while, in return, the Yugoslav armed forces trained many Libyans in their bases and facilities. Each time a crisis occurred in the Mediterranean between the U.S. Sixth Fleet and the Libyans, the navy alerted the Belgrade attaché office to conduct an immediate reconnaissance of Libyan diesel submarines in Tivat. Given the reckless way that Colonel Qaddafi used his Soviet weapons systems and the fact that he possessed six former Soviet Foxtrot long-range diesel attack submarines, in varying states of readiness, the U.S. Sixth Fleet was attentive.

The Libyan subs carried torpedoes and presented a threat, despite their usual poor state of material readiness. Since one-third of the Libyan submarine fleet was normally in the Balkan backyard of Tivat, and as the only U.S. naval officer who was in a position to report on their status, the Belgrade Western attaché was repeatedly tasked to report on short notice: How many Libyan submarines are in Tivat? Are they ready for sea? Did they have torpedoes aboard?

Tivat lies about 420 miles due south of Belgrade, across the mountains of Montenegro. In good weather, it was possible to reach Tivat by car in twelve to sixteen hours. On short notice, it could be done in several hours by air aboard a (JAT) Yugoslav airlines flight to Dubrovnik, then renting a car and driving to Tivat. But in the process of acquiring the tickets and procuring a rental car, a complex feat in Yugoslavia at best, the attaché's intentions would be tipped well in advance. And since attachés were required to file notice forty-eight hours beforehand for intended travel outside of Belgrade, the route to Tivat would be lined with a string of surveillants.

Nevertheless, when the navy called, the closely knit Allied attaché group acted. Using these travel procedures, they were able, within minutes of the completed reconnaissance, to put the report on the wire and out to the fleet. They certainly violated every security regulation governing the use of telephones in a communist country, but they accomplished the mission and never used the same routine more than once. The Sixth Fleet generally knew within several hours from the Belgrade-based naval attachés whether the Libyans were in port or not, and whether there were any major repairs under way or hull plates missing from the submarines, which could indicate the readiness of the units for sea.

During one particularly strained period, it appeared that the Libyan subs were scrambling to depart Tivat early. At the same time, one of the Sixth Fleet aircraft carriers was on a port visit in Venice, at the top of the Adriatic. To keep track of the Libyan activity, two NATO attachés spent an entire week drifting from one Dalmatian bed-and-breakfast lodging to another, while they waited to report the submarines' sortie. Each inn overlooked the very visible channel from Tivat to the open Adriatic.

Although the mission was demanding, the setting could not have been more beautiful. The Montenegrin and Bosnian coasts surrounding the towns of Herzegnovi and Tivat are breathtaking. The foreign officers often climbed high above the beautiful Bay of Kotor, surrounded by the smell of mimosa, and watched the islands far out in the azure Adriatic. It was a paradise setting for a job that seemed fanciful in the nuclear age. They were players in a high drama of Cold War intrigue, which had its moments of stark apprehension. During one alert, after sighting the Libyan submarines unexpectedly getting under way, two NATO officers raced for one of the town's few telephones where they could call the embassy in Belgrade to pass the coded report. They found the small post office with its single phone surrounded by local security personnel waiting to pounce. They retreated quietly and evaded the observers by slipping into a nearby tourist hotel. There, they managed to pass the information on the personal telephone of the hotel security office, while they feigned reporting the loss of a passport to the U.S. embassy. It took some imagination, but they succeeded at their task.

It was by chance that the U.S. attaché encountered a solution to this kind of assignment. Using a local informer saved time and the aggravation of complex travel and risk, yet still obtained the desired information. By good fortune, the attaché met a Yugoslav who worked in the Tivat shipyard as a foreman directing repair work on Libyan submarines. He was a young Jewish Bosnian from Sarajevo named Jacob, who hated Arabs with a passion and Libyans in particular. With careful planning and nurturing, the U.S. attaché set up a reporting network with Jacob. Thus, with very little cost, the attaché could get timely and accurate information directly from a source on the scene and swiftly forward it to the fleet. The scheme was practical but not without its tense moments.

Once, Jacob was passing along the weekly Tivat shipyard maintenance schedule to the attaché in a crowded restaurant in Belgrade's bohe-
mian Skadarlia district—a routine they established to always take place
at random locations. The young Bosnian was set upon by several rowdy
Serbs stoked with too much *slivovitz*, the local plum spirit. Fearing that
they had been caught in the act of espionage, the attaché and Jacob fled
in opposite directions, only to realize later that the assailants had attacked
merely at the sight of Jacob's Star of David necklace. No amount of coun-
sel could deter him from sporting his talisman, but at least in Belgrade he
began to wear it beneath his shirt. Nevertheless, both men felt jittery and
anxious during their meetings.

One night the attaché moved quietly along the wet cobblestones,
down the dark, winding Belgrade street. The smell of garlic, slivovitz,
and tobacco was strong in the night air. The attaché stepped behind an
empty kiosk, convinced that a single figure was following from behind.
He moved out and continued in the direction of the river, down toward
the foot of the crumbling walls of the old fortress Kalamegdan. He
slipped quietly into the shelter of a doorway and waited. Belgrade in a
late autumn rain was as cold as the dead. The shadow of the Kalamegdan
fortress loomed dark over the narrow street. The attaché stiffened as the
figure from behind paused. Surveillance had been heavy recently, and it
was becoming more difficult to make the trip all the way from the diplomatic quarter in Dedinje to the squalor and stench of the lower Belgrade
riverfront without picking up foot surveillance. He was used to vehicular surveillance; it was normal for suspicious cars to trail attachés. This
was more difficult. Despite moving carefully, he had picked up some
kind of surveillance: Serb police or the military counterintelligence,
he didn't know which. They were all equally ruthless, unprincipled, and
out of control. The American was tired after the long trip from Sarajevo
but would have to shake the follower at least briefly to make the contact.
He squeezed farther into the reeking, urine-soaked doorway and listened.
The sound of a river boat below on the Danube drifted above the silence,
as diesel engines chugged with a hollow *tonk tonk*.

The American officer came out of the doorway and quietly walked
back in the opposite direction. Young Jacob would be waiting by the side
of the old fortress, nervous and edgy, with only an hour before he was
due to deliver the submarine-repair status report to the Libyan Peoples'
Bureau, in their embassy in central Belgrade. For more than six months, the
attaché had received handwritten copies of the weekly readiness report
on Libyan subs in overhaul in the Adriatic port of Tivat. Finding the

willing source in the shipyard had been a major stroke of luck. The information proved valuable and was accurate in every case. It was a glorious arrangement: a happy Sixth Fleet intelligence officer learned the precise sailing dates of the Libyan submarines, and, in exchange for the data, the attaché parted with a great deal of American bourbon and lost many hours of sleep because of the harrowing late-night trips to the squalid Belgrade riverfront.

Jacob was reliable, exact, and intensely dedicated to his anti-Arab cause. His devotion, in fact, was extreme to the point of becoming a liability. Once when his comely sister substituted for him at a data rendezvous with the American attaché, she explained that her firebrand brother had attacked two Libyan sailors at Herzegnovi near Tivat after they had made a pass at her. In any case, the reporting arrangement lasted more than a year, then ended abruptly. Jacob again sent his sister to the attaché in Belgrade to pass his last report. Near tears, the dark-eyed beauty explained that her explosive brother had bashed another sailor, this time a Libyan officer and with a wrench instead of his fists. Jacob had fled and caught the ferry from nearby Bar across to Bari, Italy, from whence he planned to make his way to join the Israeli army. Although the U.S. Navy lost a good source, the Israelis certainly gained a determined scrapper. When the attaché opened Jacob's last envelope, which was curiously large, he found the blueprints of the entire Tivat shipyard, including the long-range schedule for the repair of Libyan and Soviet subs for the next six months.

In a similar stroke of good fortune, the U.S. attaché met another reliable source who worked at the Belgrade embassy of Libya as a telephone switchboard operator. This source's father, a former Yugoslav political figure, had suffered severely under Tito and had been jailed for many years. The grief and discomfort the affair brought to the family of this fine young woman provided the perfect motivation for her intelligence activities, which she focused against the communist Belgrade regime and the Libyan terrorist structure. Now long since emigrated from Yugoslavia, the woman not only provided timely reporting on Libyan government communications, but did so with open delight to be working against both the Libyan international terrorist network and the communist Yugoslav government—which she loathed with a passion. Her work made a significant contribution toward eventually dismembering Libyan terrorist training in Yugoslavia in the early 1990s.

A successful attaché lived with the constant fear that excess enthusiasm and energy might make him cross the delicate line and get caught in the middle of an incriminating act. That would cause the host government to pull out all the stops and send the officer home. The attitude within the U.S. Embassy was also a major restraint on initiative. The average foreign-service officer envisioned himself being assigned to a country to improve relations at whatever the cost. Thus, aggressive intelligence activities against the host country drew nearly as much hostility from the embassy officers as from those targeted. The U.S. ambassadors in Belgrade during the period, Larry Eagleburger and David Anderson, were both highly supportive in most cases. But it was always clear that if an officer were apprehended by the host country's counterintelligence services, the embassy would wash its hands of his activities and would declare that he had operated in violation of standing embassy policy. As such, it was truly independent duty. There were few accolades for an attaché who was sent home early, by either the host country or his ambassador. Yet the navy attachés were expected to facilitate the fleet, despite disapproval from the other service attachés and foreign-service officers within the embassy, all of whom suspected navy attachés of heading for the beautiful Dalmatian coast just to get away from the dreary and polluted capital.

More Successes

In the early 1980s, two other military attachés showed great skill in Poland during martial law and under threat of invasion by Soviet forces. U.S. Army colonel Charlie Williams and Canadian colonel Angus Paul-Duddy provided key intelligence by evaluating the day-to-day status of the advancing Soviet military. This vital information was not consistently available through satellite and SIGINT (signal intelligence) reconnaissance due to cloud cover and Soviet deception measures. The two men supplied the West with a ground-level analysis to help predict whether Soviet forces would invade Poland.

During the mid-1980s, the American defense and naval attaché serving in Madagascar achieved probably one of the most renowned intelligence-collection coups in recent history. The former navy P-3, long-range maritime aircraft pilot Tommie Thompson was an avid tennis player. In the course of his tour in the capital, he befriended the president of Madagascar, who was also a keen tennis player. The two scheduled regular games together, and a friendship soon blossomed. At the time, the Madagascar government had given the Soviets permission to erect a SIGINT site on a prominent height overlooking the Indian Ocean, in return for an undisclosed financial sum. Complete with complex

Wollenweber antennae, the project was designed to enhance Soviet intelligence coverage of the U.S. fleet in the Indian Ocean.

The president was a devout Muslim and was distrustful of the Soviet presence, and he confided to Captain Thompson that the deal for the SIGINT site was more a financial necessity for his country than a gesture of support for the infidel Soviets. Captain Thompson, being required to learn as much as possible about the site, elicited more and more information during and after his tennis games with the president. The president was quick to realize the importance of this Soviet site to the American. Surprisingly, one day after tennis he made a generous offer to Captain Thompson. On a certain date Thompson would meet the president's "man," who would escort Thompson to the site, which was now complete but, according to the agreement, locked and not yet manned by the Soviet intelligence monitors. Apparently, in stalling for more cash from the Soviets, the keen president was awaiting the additional funds before allowing the Soviets to man the site. In the meantime, he granted access to Captain Thompson and allowed him to bring a camera and inspect the entire empty site. Thompson did exactly what the president offered. He entered the huge site, which was loaded with electronic equipment but guarded only by local militia. He toured the entire installation and photographed everything, both the interior and the exterior. His long series of reports and impeccable photography earned him the award Collector of the Year from the U.S. director of central intelligence. And he was never confronted by the Soviet attaché in Madagascar. Truly, it was an intelligence collector's dream assignment.

Another attaché success was less spectacular but in retrospect produced valuable results. In 1984, the U.S. naval attaché in Belgrade lived in a comfortable residence in the plush Dedinje region near Tito's grandiose villa overlooking the Sava River. By chance the Soviet, Romanian, and East German attachés were neighbors just 100 meters away. Since the American was also accredited to Romania, he naturally befriended the Romanian colonel, who had been in Belgrade for an astounding thirteen years. The Romanian and his wife made annual trips to Bucharest, as he explained, to visit family and, yes, to be reindoctrinated by President for Life Nikolae Ceausescu's intelligence arm, the dreaded Securitate. Over time, it became apparent that the Romanian was fond of American bourbon. Prior to each return to Romania, the U.S. attaché discreetly gave his neighbor a case of Jim Beam bourbon. After several years, this gift became a ritual that finally paid off.

In the summer of 1984, the Warsaw Pact staged a large-scale military exercise in Bulgaria. Soviet, Romanian, and Bulgarian forces were to converge in Bulgaria and rehearse the seizure of the west side of the Turkish Straits. Times were tense, and naturally the Western attachés sought information about the pending maneuvers. Since the Romanian colonel was about to embark on his annual pilgrimage to Bucharest, the American upped the ante to two cases of bourbon. A month later, when the Romanian returned, he requested a meeting with his American neighbor, but necessarily, he insisted, outside in the garden. On a pleasant summer evening, the Romanian passed to the American the entire operation order for the forthcoming exercise. Later, at an attaché conference in London, the U.S. attaché was forced to defend his receipt of the invaluable intelligence.

"What secrets," he was asked, "did you pass to the Romanian for that information?"

"Lots of whiskey," was the response.

The Leningrad Shipyard

In the Soviet Union, NATO naval attaché teams obtained an extraordinary amount of valuable intelligence about naval construction in key locations under the most controlled and often hazardous conditions. Incidents frequently occurred, however, where the intelligence collector's fine line of discretion confronted the ham-handed surveillance techniques used by the local opposition. The era of the mid–Cold War was replete with stories of Leningrad, where the Canadian, British, U.S., and French naval coverage was often spiced with physical contact between the attachés and Soviet countersurveillance personnel, called "goons" by the attachés.

Soviet surveillance forces were often augmented when ship construction work necessitated exposing to public view some portion of a sensitive new unit, such as a submarine hull section or a whole new hull. The countersurveillance ranks then swelled with the addition of legions of off-duty shipyard workers and retired veterans employed as *druzhniki*. They donned red armbands and tried to confront Western attachés who lurked too close to sensitive military installations.

Leningrad was one of the few ports in the Soviet Union that was routinely open to foreigners. The vast port was the home of five major shipyards, where naval ships and submarines were constructed. Thus, with more than 90 percent of the rest of the USSR closed to foreign visitors, Leningrad naturally became a popular challenge for Western naval eyes to penetrate the massive screening by Soviet security organizations. The Soviets took big, often clumsy, steps to camouflage their new ships

A typical attaché photo taken in Leningrad during the late 1980s. In the distance is a Kirov class cruiser (center) and the most unique and largest intelligence-collection ship ever built: SSV-33 *Ural* (left). This photograph was not good enough to allow a detailed analysis.

and submarines as these took shape on the building ways or in gigantic submarine construction halls. Over the years, however, the NATO attachés worked hand in hand with U.S. and British analysts to achieve their objectives. These specialists in naval construction monitored all sources of intelligence, including satellite imagery, and could predict when certain ships and submarines would be launched. The launching generally was the most vulnerable period to attract prying eyes. Nevertheless, the Soviets devised ingenious and elaborate screens and false covers to thwart the Western attachés who tried to snap a few discreet photos of new units being launched.

Over the years, Schmidt's Embankment was the only road and walkway where the keen eye might catch a glimpse of a new submarine emerging from the yawning doors of the Sudomekh submarine halls across the Neva River. During the expected launch period, the Soviets resorted to brute force by flooding Schmidt's Embankment with pedestrian and vehicular surveillance, the ever-present "goon." Usually selected for their large size, rather than for their nimbleness of mind, the goons were the adversaries in an ongoing gentlemanly game of hide-and-seek. Sometimes, though, it was not so gentlemanly. For example, one year a U.S. Navy captain and a marine major from the Moscow Embassy were set upon by the goons and thrown to the ground, and their trousers were removed. The unfortunate officers were forced to walk in their underwear though the center of Leningrad in January toward haven at the U.S. Consulate. The next day the Leningrad newspaper carried a photo of the two officers on their frigid trek with the caption "U.S. Navy Spies Exposed Near Soviet Shipyard."

Regardless of caution, the daring challenge for an aggressive naval officer often overwhelmed common sense. Some naval attachés hazarded walking out on the frozen Neva River ice in Leningrad in mid-winter to obtain close-up photos and souvenirs from the hull of a newly constructed Kirov class nuclear missile cruiser, or they chanced entering the submarine base in Balaklava on the Black Sea to observe the hull of the experimental submarine *Beluga*, sitting high on a marine railway.

One notable intelligence-collection victory was achieved in Leningrad in 1986 by a determined assistant naval attaché, Lieutenant Commander Donn Northup. His timely photographs of a Soviet submarine propeller, first lying on the bed of a truck and then hanging by a crane over the Baltic Shipyard in Leningrad, provided crucial collateral evidence that was needed in a sensitive legal case against Toshiba. Northup's photos proved that the Japanese had sold vital milling technology to the USSR for its submarine propellers. This was in blatant violation of the Western restrictions of the Coordinating Committee for Multilateral Controls (COCOM), which regulated sales of high technology to the Warsaw Pact nations. The illegal Japanese sale of these computer-controlled machine tools to the Soviet Union cost the U.S. Navy an estimated $1 billion in necessary upgrades of detection equipment to match the new quieter Japanese-Soviet propellers.

The Soviet navy had learned of its woeful lag in sound quieting, regarding submarine propeller cavitation, most likely through information procured by the Walker spy family. The KGB was able to short-circuit the restrictions placed on acquiring the technology the Soviets needed to make better propellers. The KGB skillfully arranged for the Soviet Techmachimport company to approach the Moscow office of a small Japanese trade firm, Wako Koeki. This company then applied to Toshiba for export permission of the MBP-110 milling machine, which mills marine propellers on nine axes. Techmachimport ordered four of the milling machines in 1981. When Toshiba applied for the export license, it described the capability of the machine as limited to two axes milling, well within COCOM's three-axis maximum permitted for sale to Warsaw Pact countries. Toshiba listed the purpose of the export "to improve civil power utility in Leningrad." The Soviets ordered the NC-200 computer control systems to run the milling machines from a financially strapped Norwegian company, Kongsberg, which, like the Japanese, falsified the export documentation by stating that the equipment was limited to two-axes applications. Thus, at a cost of only $4 million, the Soviets purchased and installed top-of-the-line milling and control equipment in their Admiralty Sudomekh Shipyard in Leningrad. For that relatively minor outlay, they corrected a major flaw in their multibillion-ruble submarine fleet.[1]

Naval intelligence already possessed sufficient evidence of the Japanese violation in the trade restrictions, but it was of too sensitive a nature to be used in prosecuting the case in a court of law. Thus, when attaché Donn Northup's photos caught the end product, these could finally be used in a legal case against Toshiba.

Excess Bravado

Often, well-planned and discreet collection operations could backfire and blossom into serious acts of blatant spying. In 1988, two assistant naval attachés in the USSR were caught in exactly such a compromising situation while trying to pass as near as possible to the closed Soviet naval air base at Saki (the Soviet equivalent of the Patuxent River naval air test base). The two attachés had boarded a civilian bus marked with the destination of a small village near the sprawling airfield. The officers rode smugly along the airfield perimeter, looking through the windows to observe the activity over the fence. Then the bus suddenly turned into the base to cross the large field. The young attachés were identified and apprehended by gate guards. The widely publicized charge—attempted penetration of a high-security experimental naval air base by two U.S. military spies—resulted in two surprised and crestfallen naval attachés.

In another instance, a senior U.S. naval attaché and his wife were able to gain entry inside an ASW research facility on the Crimean coast near the outskirts of Yalta. It was not far from Soviet general secretary Gorbachev's summer retreat dacha at Foros. In this installation, naval intelligence suspected that experiments were being conducted with a new and unorthodox detection technique. The attaché had to leave his "partner in crime" outside. While the attaché was standing adjacent to the range in perfect lighting conditions, busily observing the range measurement equipment, he was accosted by a security guard wearing a red armband. In a risky gamble, he explained that his wife urgently needed the privacy of the nearby bushes. The embarrassed surveillant quickly withdrew, giving the couple enough time to retreat without being apprehended.

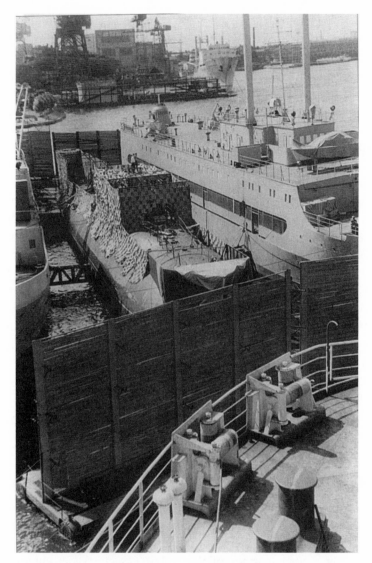

Maskirovka, the typical measures taken by the admiralty shipyard in Leningrad to hide a Victor I nuclear attack submarine from the prying eyes of naval attachés and satellites.

Latent fear, however, lurked in the back of every attaché's mind, sometimes in the form of images of the headless Commander Lionel Crabb. It was always possible that an unobserved sentry might react violently on discovering an intruder, despite the attaché's diplomatic immunity.

If a local sentry happened upon a foreign attaché in the act of penetrating a closed area with his camera hanging out, serious and even fatal consequences could result. The unexpected confrontation could provoke a guard who had little education and just enough intelligence to realize he had to do something. Such was the case in the unfortunate death of the U.S. Army major Arthur D. Nicholson Jr. in East Germany on March 24, 1985.

The U.S. Military Mission in Potsdam was one of the post–World War II quadripartite military missions that was permitted in the former occupied zones of conquered Germany.[2] Military personnel manned the liaison missions. Their duty was to observe the other side's military activity and report to their respective military headquarters. The missions were disbanded after Germany's 1990 reunification. In 1985, however, U.S. Army major Arthur Nicholson was on a routine observation mission near an area designated as permanently closed by the Group of Soviet Forces in East Germany, where the best of the Red Army's nineteen front-line divisions were deployed. Nicholson had successfully entered a tank storage structure and closely observed and took photographs of empty front-line vehicles. He had actually been inside them. He was surprised by a Soviet sentry, who panicked and shot him before he could leave the scene. The tragedy of the situation was that Soviet army personnel refused to allow Nicholson's travel partner to administer the first aid that might have saved his life. The major bled to death while the Soviets and the Americans watched, unable to assist.

Chasing Rainbows

In 1987, the defense attaché office in the U.S. Embassy in Moscow was adrift in scandal and uncertainty. A marine security guard sex scandal at the embassy had erupted the year before when authorities discovered that many of the young marines of the embassy security detachment were being regularly seduced by young, attractive Soviet employees working in the chancellery.

The yellow building on Moscow's Garden Ring Road had been constructed by German prisoners of war in the 1950s. It housed most of the senior attachés and the Foreign Service officers in ornate apartments in a gloomy, ten-story edifice. At that time, the embassy employed roughly 244 Soviet citizens as administrative help, travel assistants, cleaners, and nannies for the younger families. As a matter of course, these employees were working for the KGB and enjoyed nearly free access to the dwelling areas of the chancellery, although they were not allowed in the classified offices, where many secrets lay hidden.

The marine security detachment, which was made up of young, specially selected noncommissioned officers, was subordinate to the embassy administrative counselor. This chain of command was jealously guarded by the State Department, and a strict chasm separated the marines from the officers and the men of the defense attaché office. In 1986, this office consisted of thirteen senior army, navy, and air force officers and a handful of enlisted personnel. While the marines were busily bedding the young Soviet nannies and secretaries, the KGB was industriously worming its way into more areas of the embassy. Then, in 1986, President Ronald Reagan summarily dismissed fifty-five Soviet diplomats from their missions in Washington, D.C., and New York for undiplomatic behavior— namely, spying. The Soviets retaliated by declaring five U.S. diplomats in Moscow persona non grata and sending them packing. Among them was the senior naval attaché, a former submarine captain.

As a result of the suspected Soviet penetration of the chancellery, and in retaliation for the expulsion of the diplomats—but mostly to save face—the State Department fired all 244 of the Soviet employees and forbade visits to the embassy by any Soviet citizen. The battle lines were drawn. But during the ensuing Moscow winter, the coldest in fifty years, it was the height of comedy to see senior U.S. diplomats, now bereft of Soviet help, shoveling snow, making their own travel arrangements, and driving their chaufferless cars around the frigid capital. Amid the chaos and uncertainty in this period, which was the lowest point in diplomatic relations between the United States and the USSR, a most bizarre intelligence-collection program reached a pinnacle of folly.

For years, U.S. intelligence had received snippets of information about an extensive underground railway system being constructed beneath the cavernous and ornate Moscow metro lines. This was reportedly a mini-rail tunnel 9 meters in diameter that linked the Kremlin and the ministry of defense buildings with the KGB and other sinister headquarters below the capital. Construction had begun in Stalin's regime but was interrupted by the war. The underground rail tie was designed to spirit senior Soviet leaders to various command bunkers, safe havens, and evacuation centers in the event of nuclear war. Highly classified satellite photos provided the attachés with evidence that the construction was still in progress in the 1980s, and the U.S. attachés were ordered to find the missing pieces.

Upon close scrutiny, one can observe that the Moscow skyline is spotted with strange-looking sheds perched 50 to 180 feet high atop steel structures. These were apparently erected to house earth-removal conveyer belts that extracted soil from deep excavations. U.S. attachés were regularly dispatched at night to observe and time the cycles of buckets dumping soil from the tunnel excavations. This intelligence-collection effort, conducted in absolute secrecy, was given the highest priority and urgency. On the way to or from diplomatic receptions or dinners, many attachés ducked away to stake out the excavation sites. Often clothed in full dress uniform covered by a black raincoat, an attaché might spend long hours on a cold, wet Moscow night lurking near the conveyer belts to count and time the swinging buckets of mud. This process extended well into the Gorbachev "thaw" period and seemed to indicate a sinister plan by the Soviets to prepare for a surprise nuclear exchange. That the underground VIP-evacuation system had been conceived during Stalin's time and continued at a haphazard pace since then completely befuddled the U.S. intelligence analysts.

Finally, after the fall of the Berlin Wall and the implosion of the Soviet Union, the detailed plans for this extravagant hidden rail system—resembling the subway linking the U.S. Senate and the Capitol Building in Washington, D.C.—were nonchalantly given to U.S. businessmen working with the World Bank during the post-Gorbachev period. Not knowing the significance of the underground plans to the U.S. military, the World Bank employees filed the engineering plans as one of many anachronistic shards of the communist period. This is but one example of the amazing fragments of intelligence that were lost in the scramble for a piece of postcommunist Russia. Massive sums were invested on both sides: the Soviets spent a fortune constructing the rail system, and the United States wasted time, money, and energy trying to learn about the project that was never completed.

16

SOVIET, U.S., . . . OR OTHERS?
1940s–1980s

In 1990, the Soviet press published a series of articles on UFOs and air defense. They expressed concern that the U.S. Strategic Defense Initiative (SDI) systems might confuse unidentified flying objects (UFOs) with incoming missiles and might launch an unprovoked counterattack against the Soviet Union. These articles referred to the "tens of thousands of UFOs" seen over the USSR and for which the Soviet Academy of Sciences could offer no explanation.[1] The articles echoed a similar concern expressed thirty years earlier in the columns of the *New York Times* by a former director of the Central Intelligence Agency, Rear Admiral Roscoe Hillenkoetter. An experienced naval intelligence officer, a secret courier for Franklin D. Roosevelt in Europe, and an attaché in Paris and Vichy after the German invasion, Hillenkoetter had been appointed as the first director of the newly created CIA (1947–1950). He made his famous statement on UFOs in February 1960:

> Behind the scenes, high ranking Air Force officers are soberly concerned about the UFOs. But through official secrecy and ridicule, many citizens are led to believe the unknown flying objects are nonsense. . . . I urge immediate Congressional action to reduce the dangers from secrecy about Unidentified Flying Objects. . . . Two dangers are steadily increasing: 1) the risk of accidental war from mistaking UFO formations for a Soviet surprise attack; 2) the danger that the Soviet government may in a critical moment falsely claim the UFOs as secret Russian weapons against which our defenses are helpless. I urge immediate Congressional action to reduce the dangers from secrecy about Unidentified Flying Objects.[2]

After this statement, Hillenkoetter resigned from the National Investigations Committee on Aerial Phenomena, which he had cofounded in 1956. Observers hinted that the admiral might have been asked to do so.

Both articles suggested that the sighting of UFOs might trigger a nuclear war out of confusion. The 1970s had been characterized by so many sightings that defense officials had voiced their concern publicly. In France, the defense minister Robert Galley (1973–1974) and in the United Kingdom, the British joint chief of staff (1971–1973) Admiral Norton Hill both made straightforward statements on the reality of the mysterious objects.[3] In a foreword to a book by ufologist Timothy Good, the chief of Defense Staff of the Ministry of Defense in Great Britain Lord Hill-Norton wrote the following:

> The evidence that there are objects which have been seen in our atmosphere, and even on terra firma, that cannot be accounted for either as man-made objects or as any physical force or effect known to our scientists, seems to me to be overwhelming. . . . A very large number of sightings have been vouched for by persons whose credentials seem to me unimpeachable. It is striking that so many have been trained observers, such as police officers and airline or military pilots. Their observations have in many instances . . . been supported either by technical means such as radar or, even more convincingly, by . . . interference with electrical apparatus of one sort or another.[4]

Less than a decade later, on September 17, 1987, President Ronald Reagan again raised the issue before the United Nations. While opening the forty-second session of the UN general assembly, the U.S. president stated that the United States and the Soviet Union would join forces "if the Earth faced an invasion by extraterrestrials."[5]

Soviet Rockets Violating Sweden's Airspace?

Following the February 1942 mysterious "air raid" above Los Angeles (see the introduction), the first air-defense confusion of the Cold War may have happened during the summer of 1946. Witnesses in Sweden and Finland reported seeing rocket-shaped objects crossing the skies at incredible speeds. The initial explanation was that they were captured and reconstructed German V-1 and V-2 missiles fired from the Soviet occupation zone of Germany. According to witnesses, however—including a Swedish air force pilot on August 14, 1946—the cigar-shaped rockets had no wings or rudders and left no flames in their trails. Speeds in excess of 2,000 miles per hour were observed.[6]

Seeking a rationale for the event, the U.S. Embassy in Stockholm blamed the Soviet Union, whose political motive would have been "to intimidate Swedes in connection with loan negotiations or to offset a supposed increase [in U.S. military pressure] on Sweden resulting from the U.S. Navy visit."[7] This explanation was repeated in a subsequent telegram dated August 29, addressed to the State Department: "While over 800 reports have been received and new reports come daily, Swedes still have no tangible evidence. Full details of reports this far have been sent to Washington by our military and naval attachés. My own source is personally convinced some foreign power is actually experimenting over Sweden. He guesses it is Russia. He has promised to notify me before anyone else if anything tangible should be discovered."[8]

Jones Questions Their Soviet Origin

The famed and brilliant head of British technical intelligence Professor Reginald Victor Jones dismissed the reports and faulted the witnesses. In his memoirs, he recalled:

> We had one diversion of an intelligence nature, which no doubt arose from the general atmosphere of apprehension that existed in 1945 regarding the motives of the Russians and which anticipated the flying saucer; . . . our bomber crews had reported single engine night fighters with yellow lights in their noses over Germany at times when we knew that no single-engine night fighters were flying. So we were not unduly surprised when incidents started to be reported in Sweden, which regarded itself as in the front-line should the Russians attempt to move westward."[9]

Jones disagreed with the general interpretation that these were long-range flying bombs being launched by the Soviets to intimidate Sweden. He could not see any purpose in alerting the West to the fact that the USSR had a controllable flying bomb with a longer range and superior reliability to the V-1 or V-2. Despite the hundreds of sighting reports, no rockets had been confirmed to have crashed, although some pieces were said to have fallen off the "ghost rockets" and recovered. According to Jones, they were irregularly shaped pieces or lumps unlikely to have been associated with a mechanical device. Sweden lent some of these pieces to Jones, who sent them to the Chemical Analysis Section at Farnborough. British air commodore Vintras, the director of air intelligence, was astounded by the results and overexcited when he called Jones: "There, what did I tell you! Farnborough has analyzed the stuff that you sent, and one of the lumps consists of more than ninety-eight percent of an unknown element!"

Jones was amazed: "Not only had the Russians a flying bomb of fantastic performance, but they were driving it with a fuel made from an element that was new to the world of chemistry!" Jones's skepticism would not accept that the rockets were Soviet. He dismissed the lump as an ordinary piece of coke and the ghost rockets as meteors bright enough to have been visible by day.[10]

In October 1946, Swedish military authorities admitted that they had been unable to ascertain the origin and the nature of the ghost rockets: "Radar had detected some [200] objects, which cannot be the phenomena of nature or products of the imagination, nor can be referred to as Swedish airplanes."[11]

Soviet Disks . . . over New Mexico?

Other mysterious objects would soon invade the U.S. airspace and fly above military facilities, including over the proving ground of White Sands, in Las Cruces, New Mexico, where the V-2 was being tested. During the months of June and July 1947, the U.S. Army Air Force and the U.S. Navy began to collect witness reports on strange sightings. These were compiled in the now declassified secret document S-03537 issued by the Air Force Base of Wright Field Dayton in 1949 and available at the National Archives, College Park, Maryland.[12] Many sightings were reported by military personnel. Some of those reports described shrewdly controlled flying disks that were observed in clear weather and that could not match any known aircraft.

During one instance, two small UFOs were seen whirling around the rocket. On two other occasions, navy personnel were able to track the objects with theodolite stations.[13] Commander Robert B. McLaughlin, an expert in naval ordnance and guided missiles, directed the test. Since August 1946, McLaughlin had been in charge of the navy unit that assisted in classified projects at the White Sands Proving Ground. He had begun his research in the guided missile field in 1939, four years after his graduation from the U.S. Naval Academy. By 1941, he had outlined the technique for a "beam rider" guided missile, which, by the ability to "change its mind" after being fired, heightened antiaircraft accuracy against high-altitude evasive aircraft. During the war, he was a gunnery officer aboard the aircraft carrier USS *Intrepid* and saw action when she underwent an attack by thirty-three Japanese kamikaze suicide planes in three and a half minutes. McLaughlin gave the following account of the incident:

This day we were firing a Navy upper atmosphere missile. Shortly after its take-off, two small circular objects, guessed to be approximately 20 inches in diameter, appeared from no place and joined the Navy missile on its upward flight. (Similar small disks have also been previously reported as well as the larger types mentioned earlier.) At about the time the Navy missile was doing well over 2,000 feet per second, the object on the west side passed through the exhaust gases and joined its friend on the east. They then apparently decided the missile was not going fast enough for them. They accelerated, passed the Navy missile and sailed off upward and eastward. Some eight minutes after the Navy missile had fallen back into the range, I received a radio report from a very powerful optical observation post located on a mountain top. The navy missile, it said, had just passed over the mountain and was going out of the range to the west. This could have been one of the two objects that we had seen and which had changed direction, or it could have been a third one. The odd thing is that before long I had reports from eleven men in five separate OP's, none of which could communicate with each other and which were located at different points of the compass. All had seen the two objects perform as I have described.[14]

The three UFO incidents, which had occurred above White Sands, drove MacLaughin to risk a conclusion: "Putting together all the data observed in three appearances, one of which I had seen for myself, and all which I believe beyond doubt . . . I think . . . that there is too much evidence from too many reliable sources for us to be content with inconclusive explanations, and we must press on to find an answer. I think that the saucers are piloted space ships . . . because of their flight performance."[15]

MacLaughlin's "conclusive" explanation, which he made public in the magazine *True*, was not very different from the one reached by the Air Material Command at the "Secret" level and declassified in January 1969. In a well-known September 23, 1947, letter, General Nathan Twining, the commander of the Air Material Command, stated:

a. The phenomenon is something real and not visionary or fictitious.
b. There are objects probably approximating the shape of a disc, of such appreciable size as to appear to be as large as man-made aircraft. . . .
d. The reported operating characteristics such as extreme rates of climb, maneuverability (particularly in roll), and motion which must be considered evasive when sighted or contacted by friendly aircraft and radar, lend belief to the possibility that some of the objects are controlled either manually, automatically or remotely. e. The apparent common description

is as follows: (1) Metallic or light reflecting surface. (2) Absence of trail, except in a few instances where the object apparently was operating under high performance conditions. (3) Circular or elliptical in shape, flat on bottom and domed on top. (4) Several reports of well kept formation flights varying from three to nine objects. (5) Normally no associated sound, except in three instances a substantial rumbling roar was noted. (6) Level flight speeds normally above 300 knots are estimated.[16]

It is a fact that in 1947 high-ranking army air force and navy officials took the appearance of the "saucer phenomenon" very seriously. There was no doubt in their minds as to the physical reality of the crafts that had been observed flying above ultra-sensitive continental U.S. military facilities and demonstrating outstanding performances.

Study Groups and Naval Intelligence Connections

With the creations of Projects Sign and Grudge in 1947 and 1948 followed by Project Blue Book in 1952, the newly established U.S. Air Force was officially tasked to study the "flying saucers" phenomenon and collect the information from the public. But a document officially released to Canadian researcher Scott Foster by the National Archives of Canada in 1979 confirmed the existence of a parallel secret study group under Dr. Vannevar Bush, a chief U.S. scientist. In this November 21, 1950, "Top Secret" Memorandum, Wilbert B. Smith, a Canadian government scientist, gave the following account of his visit to the United States: "I made discreet inquiries of the Canadian Embassy in Washington and obtained the following information about flying saucers. (1) the subject is the most classified in the United States, even more than the H Bomb; (2) flying saucers exist; (3) their modus operandi is as yet unknown, but there is a small group working under Doctor Vannevar Bush; and (4) the entire subject is considered of enormous significance by the United States authorities."[17]

Bush had been recruited by President Franklin D. Roosevelt after the U.S. naval attaché in London, Captain Alan G. Kirk, had explained to the president how the British were using one of their best scientists— Professor Reginald Jones—to study and counter secret German weapons.[18] Bush previously chaired the National Advisory Committee on Aeronautics and the Office of Scientific Research and Development before being appointed as chairman of the Joint Research and Development Board of the War and Navy Departments in July 1946.

Other facts shed light on these circumstances, revealing the deep involvement in UFO studies of men connected to naval intelligence during the 1940s. Rear Admiral Roscoe Hillenkoetter and Thomas Townsend Brown were two founders of the National Investigations Committee on Aerial Phenomena (NICAP) in 1956, a private investigation organization that opposed the government's UFO secrecy.

Thomas Townsend Brown was a navy scientist and a pioneer of magneto-hydrodynamics studies. Researcher Paul Schatzkin has retraced Brown's connections with both the navy and the intelligence community.[19] In 1930, Brown joined the Naval Research Laboratory and participated as a staff physicist in the U.S. Navy International Gravity Expedition to the West Indies in 1932. He then served on the yacht *Caroline* as a radio and sonar operator, socializing with the owners of RCA Victor and with William Stephenson, a Canadian businessman working for the British intelligence service and a future intelligence liaison officer between England and the United States during World War II.[20]

In 1939, Brown joined the Glenn L. Martin aircraft company before returning to the navy. Promoted to lieutenant, he became the officer in charge of magnetic and acoustic minesweeping research and development under the bureau of ships. He then resigned from the navy under mysterious circumstances and joined the Lockeed-Vega aircraft corporation. Schatzkin believes that Brown had been reassigned to a secret project until 1950, when he started to advertise his self-designed air purifier. Brown became publicly known as the inventor of the ionic breeze. His apparatus looked like a window but had no electric motor or moving parts. Electricity would pump a fluid, air, or water across the frame. The system could be applied to a water craft, replacing a propeller, by allowing the water to move alongside the hull. In 1953, Brown managed to set in motion a small disk tethered to a high-voltage wire by creating an interaction between electrical and gravitational fields.[21]

Brown's research was related to the earlier works of the enigmatic and flamboyant Serb genius Nicolas Tesla. Tesla, after having invented the radio, devised alternating current, rotating magnetic field, and remote-controlled boats and torpedoes for coastal defense. He also theorized radar—fifteen years before its realization—and a so-called death ray. Tesla's biographer Margareth Cheney has discovered that when the Serb scientist died in New York in 1943, Dr. John Trump, a subordinate of Dr. Vannevar Bush in the Office of Scientific Research and Development, and Willis George from the Office of Naval Intelligence, reviewed and copied Tesla's papers. The two officials concluded that "they did not include new sound workable principles or methods." Yet the Army Air Technical Service Command, Wright Field, Ohio, obtained and kept some of Tesla's papers "for evaluation." This is the same place where the controversial object recovered in Roswell was sent, according to the initial U.S. Army Air Force press communiqué.[22] The remaining papers were given back to Yugoslavia, where Tito later allowed the GRU to study them.[23]

Like Tesla, Brown conducted fundamental research into electromagnetism, gravity, antigravity, radiation, field physics, and spectroscopy. Brown worked for the National Defense Research Committee (NDRC) and for the Office of Scientific Research and Development under Dr. Bush before joining private industry. In this context, Brown's statement that he could explain the operation of unidentified flying objects is significant, although his 1953 experiments apparently failed to convince his U.S. colleagues.[24]

Generalissimo Stalin and Lord Mountbatten of Burma Comment on the New Mexico Events

Through its spy ring embedded in the Manhattan Project, Moscow was well informed of the events taking place in New Mexico. On August 13, 1991, the *Rabochaya Tribuna* published an interview with Professor V. Burdakov from the Russian Academy of Sciences. Remembering a conversation with Sergei P. Korolev, Burdakov revealed that in July 1947, the rocket scientist had been invited to the Ministry of State Security (MGB) in the Lubyanka Prison, where an apartment had been readied for him. He was given foreign publications and documents related to flying saucers and was told to study them, with the help of a team of translators. After a few days, Korolev met with Joseph Stalin, who asked for his opinion. Korolev answered that the phenomenon was real but did not represent a threat. Stalin then told Korolev that his conclusion was similar to that of other physicists who had been consulted, namely, Mstislav Keldysh and Alexandre Topchiyev, the future president and vice president of the Soviet Academy of Sciences.[25]

Philip Ziegler, the official biographer of Admiral Lord Louis Mountbatten, quoted a conversation that the admiral—then the commander in chief of the Royal Navy Mediterranean Fleet—had with a friend in 1950. Contrary to Korolev, Mountbatten saw the mysterious objects as a potential threat. He specifically referred to a statement by Dr. Vannevar Bush on October 6, 1949, concerning the incident at White

Sands that was described by Commander McLaughlin: "The fact that they can hover and accelerate away from the earth's gravity again and even revolve round a V-2 in America (as reported by their head scientist) shows that they are far ahead of us. If they really come over in a big way that may settle the capitalist-communist war. If the human race wishes to survive they may have to band together."[26]

Flying Saucers Greet the Navy Secretary

On February 10, 1951, the four crew members of a U.S. Navy transport plane on its way from Keflavik, Iceland, to Argentia, Newfoundland, reported a near miss with a glowing sphere. Lieutenant Kingdon, the second in command, gave the following account:

> The object was very large and was circular with a glowing yellow-orange right around its outer edge. This object appeared to be climbing and moving at a tremendous speed, and it appeared to be on a more or less collision course with our aircraft. When it appeared that there was a possibility of collision the object appeared to make a 180-degree turn and disappeared over the horizon at a terrific speed. During the course of events LTJG A. L. Jones [the commanding officer] had come to the cockpit and he made a turn in the direction of the object but it went out of sight in a short period of time.[27]

UFO observations were not reported only by junior officers. Navy secretary Kimball and Admiral Arthur Radford, the chief of naval operations, were the witnesses of a UFO sighting while they were flying on board two separate planes to Hawaii. According to the now unclassified U.S. Air Force Blue Book files, the incident took place on March 14, 1952, near Hawaii and involved two disk-shaped craft. In a popular book, published in 1973, retired U.S. Marine Corps major Donald Keyhoe gave the following description of the encounter:

> "Their speed was amazing," [Secretary of the Navy Kimball] told me later, in Washington. "My pilots estimated it between fifteen hundred and two thousand miles an hour. The objects circled us twice and then took off, heading east. There was another Navy plane behind us, with Admiral Arthur Radford on board. The distance was about fifty miles. I had my senior pilot radio a report on the sighting. In almost no time Radford's chief pilot called back, really excited. The UFOs were now circling their plane—they'd covered the fifty miles in less than two

minutes. In a few seconds the pilot told us they'd left the plane and raced up out of sight." After landing, Secretary Kimball had a report radioed to the Air Force, since it was officially in charge of the UFO investigation.[28]

Back in Washington, Secretary Kimball was denied an explanation by the U.S. Air Force. He then directed Rear Admiral Calvin Bolster, the chief of naval research, to make a full investigation of all navy and marine reports from then on and also to try to get duplicate reports from witnesses in unexplained earlier cases. This navy investigation was to be kept separate from the Air Force Project Blue Book.[29]

Five months later, three naval aviators flying a Privateer over the North Pole saw three disks just below a Skyhook balloon that was loaded with a package of sensitive photographic plates to capture cosmic rays. In 2004, one of the witnesses, Commander Edward P. Stafford, U.S. Navy (Retired), described in the U.S. Naval Institute magazine *Naval History* how the three objects then detached themselves from the tail of the balloon and formed up into a compact vee. They climbed from an altitude of 90,000 feet and disappeared with blinding speed in just about three seconds.[30]

The Uninvited Take Part in Operation "Mainbrace"

In September 1952, NATO's first major exercise, nicknamed Mainbrace, saw the unexpected participation of a series of flying disks and spheres. They were witnessed and photographed by naval and air force personnel, as well as by international press correspondents who had been invited for the event. This occurred just one month after the Privateer's sighting. The maneuvers commenced on September 13 and lasted for twelve days. According to the U.S. Navy, "units of eight NATO governments and New Zealand participated, including 80,000 men, 1,000 planes, and 200 ships . . . in the vicinity of Denmark and Norway." Directed by British admiral Sir Patrick Brind, it was the largest NATO maneuver ever.

On September 13, the Danish destroyer *Willemoes* was north of Bornholm Island. During the night, Lieutenant Commander Schmidt Jensen and several crew members saw an unidentified object, triangular in shape, emitting a bluish glow while moving at a high speed toward the southeast. Commander Jensen estimated its speed at more than 900 miles per hour. His report was forwarded to the U.S. Air Force Project Blue Book.

Just before 11 A.M. on September 19, a British Meteor jet aircraft returning to Topcliffe airfield in Yorkshire, England, was followed by a silvery disk that swayed back and forth like a pendulum. Lieutenant John W. Kilburn and other observers on the ground said that when the Meteor began circling, the UFO stopped. The disk-shaped craft rotated on its axis while hovering. The disk suddenly took off westward at high speed, changed course, and disappeared to the southeast.

On September 20, personnel of one of the largest participants in the exercise, the aircraft carrier USS *Franklin D. Roosevelt*, observed a silvery spherical object. Reporter Wallace Litwin took a series of color photographs, which were examined by naval and air force intelligence. The U.S. Air Force Project Blue Book manager, Captain Ruppelt, stated, "[The pictures] turned out to be excellent. . . . Judging by the size of the object in each successive photo, one could see that it was moving rapidly." The possibility that a balloon had come from one of the ships was immediately checked out and dismissed. But that was not all. At about 7:30 P.M. on that same day, three Danish air force officers sighted a shiny disk above Karup Air Force Base in Denmark. The seemingly metallic object passed overhead and disappeared to the east. Again on the next day, six RAF (Royal Air Force) pilots flying a formation of jets above the North Sea observed a shiny sphere approaching from the direction of the fleet. The UFO eluded their pursuit and vanished. When returning to base, one of the pilots looked back and saw the UFO following him. He turned to chase it, but the UFO also turned and sped away.

On September 27 and 28, more flying objects were spotted throughout Western Germany, Denmark, and southern Sweden. A brightly luminous object with a cometlike tail was visible for a long period of time, moving irregularly near Hamburg and Kiel.[31] After Operation Mainbrace, Captain Ed Ruppelt of the U.S. Air Force UFO investigation group was told by an RAF exchange intelligence officer in the Pentagon that these sightings had "caused the RAF to officially recognize the UFO."[32]

"An Unexplained Phenomenon of the Sea"

UFO phenomena was apparently not restricted to the skies. The following account was published in the January 1952 edition of the Naval Institute magazine *Proceedings*. It described an odd occurrence in the Arabian Sea, witnessed in 1949 by Commander J. R. Bodler, U.S. Naval Reserve, and master of the Liberty ship *Charles Tufts:*

> The vessel had passed through the Strait of Hormuz, bound for India. Little Quoin Ils Light was still in sight on the starboard quarter, bearing

305° T, distance 20 miles. The night was bright and clear, with very good visibility, no moon. The Third Mate called me to the bridge, saying that he had observed something he thought I should see. About four points on the port bow, toward the coast of Iran, there was a luminous band—which seemed to pulsate. Its appearance suggested the aurora borealis, but much lower; in fact on or below the horizon. Examination with binoculars showed that the luminous area was definitely below the horizon, in the water, and drawing nearer to the vessel. With the approach of this phenomenon it became apparent that the pulsations seemed to start in the center of the band and flow outward towards its extremities. At a distance of about a mile from the ship, it was apparent that the disturbance was roughly circular in shape, about 1000 to 1500 feet in diameter. The pulsations could now be seen to be caused by a revolving motion of the entire pattern about a rather ill-defined center; with streaks of light like the beams of search-lights, radiating outward from the center and revolving (in a counterclockwise direction) like the spokes of a gigantic wheel.

For several minutes, the vessel occupied the approximate center of the phenomenon. Slightly curved bands of light crossed the bow, passed rapidly down the port side from bow to stern, and up to the starboard side from aft, forward. The luminosity was sufficient to make portions of the vessels upper works quite visible. . . . The shape of the "pinwheel," the well-defined "spokes" the revolutions about a center, and the speed with which each band of light traversed the water, all preclude the possibility of this phenomenon being caused by schools of fish, porpoises, or similar cause.[33]

It was Commander J. R. Bodler's conviction that he had been "privileged to witness one of the rare instances of a most curious and impressive natural phenomenon." But was this a natural phenomenon?

Inexplicable as these events were, both East and West suspected the other of using these phenomena as subterfuge.

Very Earthly Disks

While the Canadian and French firms Avro and Couzinet attempted to develop their own operating flying disks, the U.S. intelligence services had been reviewing the possibilities of the USSR producing similar aircraft.

On January 21, 1948, the army G2 had distributed to all U.S. attachés abroad a memorandum on unconventional aircraft. The army had a report indicating that the Soviets were planning to build a fleet of eighteen hundred jet-propelled Horten VIII–type (six-engine pushers) flying wings in the Gotha factory that was now under Soviet control. Among a long list of requirements, army intelligence wanted to know the whereabouts of the Horten brothers, Walter and Riemar, and their sister, and about their eventual association with the Soviets. The army also harbored specific questions on disks:

> Are any efforts being made to develop the Horten "Parabola" or modify this configuration to approximate an oval or disc?
>
> For any aircraft whose shape approximates that of an oval, disc, or saucer, information regarding the following items is requested:
>
> a. Boundary layer control method by suction, blowing, or a combination of both. b. Special controls for effective manoeuvrability at very slow speeds or extremely high altitudes.

None of these Soviet-sponsored Horten flying wings or disks materialized.

Even though it had never been proved that the Germans had met with any success in their flying disks experiments, the fear that the Soviet Union could embark on a similar path was still alive in 1954. An article from the December 1954 issue of the U.S. Air Force *Air Intelligence Digest* (AID), titled "The Flying Disc," addressed the idea that the Soviets might be developing disk-shaped aircraft, while suggesting that the United States might soon have its own flying disks:

> A new aircraft configuration with a circular platform is taking shape on the drawing boards of Western aircraft designers that may well be the beginning of a new era of flight. . . . New type of jet aircraft, powered by a turbine larger than any now in use, is expected to take off, land vertically, and be able to hover. . . . One of the big questions now facing the United States is this: What are the Soviets doing in the disc-aircraft field? If the United States accepts the possibility of success of circular-shaped aircraft, then it must also conclude that the Soviet Union is capable of developing such aircraft.[34]

This article expressed concerns about the Soviet Union's future capabilities to launch disks from submarines against North America. A disk's ovoid design would allow it to take off from a relatively small platform, which could be mounted on a deck with a circular shelter. Accordingly, this December 1954 issue of the *Air Intelligence Digest* was referred to in an informal January 13, 1955, Office of Naval Intelligence letter and list of requirements, addressed to the U.S. naval attaché in Moscow with the accompanying comment:

> "The disk aircraft is a vertical riser which means that it could be housed aboard most naval vessels, including submarines." The article further states: "Obviously a disk aircraft and a sub would make a deadly combination in an offensive operation. It is estimated that in the space of thirty minutes, a disk could take off from its nesting place on the deck of a sub, climb to 65,000 feet or higher, make a run on a target area 200 to 300 miles inland, drop its bomb load and zip back home to the sub." While it is obvious that the Naval Attaché is alert to reporting data on circular type aircraft should he ever be in a position to gain such information, the above may be of interest on a subject which may assume greater importance in the perhaps not too distant future.[35]

Man-made flying disks failed to meet expectations. But it is interesting to note that Thomas Townsend Brown was recruited in 1955 by a French firm—La Societe Nationale de Constructions Aeronautique Sud Ouest (SNCASO)—to conduct electro-gravity research work until the company was absorbed by a competitor.[36]

Scientists attempt to imitate the "flying disks." As with the Canadian Avro, the French-built Couzinet saucer proved unsuccessful. The United States feared that the USSR would mass-produce flying disks and launch them from submarines.

Continental Defense and UFO Reporting

The fear of a surprise Soviet air attack led to the establishment in 1949 of a barrier of radar picket ships guarding the seaward approaches to the northeastern United States. Since the inception in 1948 of JANAP-146 "instruction for the reporting of vital Intelligence sightings," the navy provided reports on "unidentified targets" that might be Soviet secret weapons attacking the United States. After the well-publicized Mainbrace 1952 UFO sightings, a more specific OPNAV instruction was issued directing the navy to send UFO reports to the U.S. Air Force Project Blue Book and to the Pentagon.[37] Then in 1954, a revised JANAP-146 instruction commanded all navy and government ships, including those of the coast guard, to report "unknown airborne targets" to the Continental Air Defense (CONAD).[38] In 1955, merchant ships and fishing vessels were also enlisted to assist in this intelligence-collection effort, which was designated merchant intelligence (MERINT). "Aircraft formations flying toward the United States and sightings of military vessels, submarines, unidentified flying objects (UFOs), and guided missiles" were to be made to the nearest military or civilian radio station for relay to the sea frontier commander.[39]

Reporting of UFOs to the U.S. Air Force Project Blue Book was also becoming standard procedure for all ships, naval or civilian, and some of the observations were sent twice to comply with both sets of instructions. Among the 258 Blue Book cases collected during 1952 to 1969 that involved ships, 43 percent were reported by the U.S. Navy, 37 percent by civilian ships, and 19 percent by the U.S. Coast Guard.[40] The U.S. Air Force, however, did not release the totality of the reports produced under these instructions. This supports the view that Project Blue Book was a front window hiding further secret research on outlandish cases that were not being given publicity because they could not be easily explained away.

The Cigar and the Carrier

One such remarkable case was publicized by former crew member Chet Grusinski of Clinton, Michigan. He wrote several articles in which he described what he saw above the USS *Franklin D. Roosevelt*'s flight deck:

> I was myself a member of her crew from 1958 to 1960, and I was one of the many eyewitnesses in one of the UFO sightings, in September or October of 1958. The ship was on a shake-down cruise off the U.S.

Naval Base of Guantanamo in Cuba, at between 8.00 and 10.00 pm, when a small light appeared and started following us and then headed straight for us. The sky was clear, the air warm, and, as I was not on duty I had gone up on to the flight-deck (where I should not rightly have been). Suddenly the thing was overhead; it was cigar-shaped, with a row of windows, through which we could see the figures of people watching us. I could feel the heat from the craft on my face. It was quite silent. Altogether I estimate that up to 25 other men who were there on the flight deck also saw it. Through the ship's intercom someone from the bridge watch was screaming for the officer of the deck or bridge to get up there at the double. Then suddenly the object turned to a red-orange color. (This was when, against the dark night sky, I was able to see its "cigar" shape.) And it vanished rapidly. Meanwhile the ship's own movement had stopped suddenly. I estimate that the sighting may have lasted five minutes or more. Special investigators . . . subsequently came aboard the carrier (on the pretext that they were allegedly investigating "gambling below-deck"!). But they questioned many sailors about this sighting. My own Petty Officer First Class questioned me about what I had seen and I replied that I did not know what it was. After that, I heard nothing more about the matter.[41]

The Phenomenon Takes on the Soviet Navy

As the Soviet navy expended its radius of operations, its vessels were also confronted with a series of inexplicable phenomena and objects.

In 1962, Captain Third Rank Kviatkovskiy, the future chief of Soviet naval intelligence (1987–1992), saw through his periscope a moving bright red-orange light near the Faeroe Islands. He knew that it could not be the aurora borealis.[42]

In 1965, Captain Second Rank Voronov reported that his submarine had surfaced in the Atlantic to rendezvous with a support ship. Some of the crew members were on deck when they saw a cigar-shaped object about 100 to 150 meters long, 1 mile from the submarine. This object did not show up on the submarine's air- or surface-search radar display. Before the eyes of the crew, three rays of light appeared from the bottom of the object and illuminated the ocean's surface while the cigar dropped into the sea and submerged.[43]

In August 1965, the crew of the Soviet merchant ship *Raduga* sighted a flying and burning disk at a distance of 2 miles. It emerged from the sea and hovered at a height of 100 to 150 meters above the surface. The object was completely illuminated. While ascending, the disk lifted a considerable amount of water with it, which then fell back into the sea afterward.[44]

Retired Soviet rear admiral Chefonov remembered two very strange encounters with invisible objects during the late 1960s:

> At that time I was the commanding officer of a nuclear submarine. While returning to our base north of Nakodka, sailing on the surface, we were surrounded by a thick fog. No visibility. Relying on our radio locator we were proceeding towards base when suddenly the radar operators detected an object following a perpendicular course at a tremendous speed. Our main command post tried to estimate its speed and trace its route. First I augmented our speed but the object remained on a collision course. And then I decided that I could not take such a risk and I stopped. We blew our horn, fired signals and directed our flash lights to illuminate our submarine. I listened to the distances given by our radar operator. 5 yards, 4, 3, 2, 1. . . . All those standing on deck stared through the mist to no avail. The radar operator screamed that the object had now entered the dead zone: 0.5 yards. Nothing. Nothing happened. The object had vaporized. We waited for another 15 minutes. Then I ordered the submarine to augment its speed. The target never reappeared. We returned to base. The second case took place the following year. We were heading back to port from the polygon where we had exercised. Visibility was excellent. Again a contact showed up on the radar screen. There were six of us on the bridge. We looked intensively but we could not see a clue. Again the main command post had determined that we were on a collision course: nothing, absolutely nothing. Again I followed the regulations by the book and ordered to blow our siren and fire signals. The invisible target was again in our dead zone. The radio operator counted the remaining yards: nothing. When I reported back to the flotilla commander he just waved his hand with skepticism. "We have enough trouble. If we report this, controllers from the main naval staff will come and we don't want that."[45]

Frogs from the Deep

In the early 1960s, the newest classes of Soviet nuclear submarines were able to operate at depths far greater than their predecessors. Many of the new Soviet submarines detected persistent phenomena that still defy rational explanation. Their passive sensors recorded strange signals that sounded like a frog, or *kviakier* in Russian. This clearly distinct sound was unlike any other underwater sound known or recorded earlier. Kviakier became the official designation for this mysterious underwater phenomenon. On various occasions, the kviakier reacted to the submarine's active sonar and seemingly "communicated" with the submarine by sending back strange noises. The kviaker seemed to behave intelligently, following Soviet submarines and playing with them, swirling around, crossing their wake, and tracking them whenever they took evasive actions. Soviet captains had the unpleasant sensation that they had lost the element of surprise and that they could not evade their followers, except when they were returning to port.[46]

Soviet Naval Intelligence's Special Study Group

In 1969, as the U.S. Air Force was preparing to discontinue its well-publicized UFO study under Project Blue Book, Admiral of the Fleet Sergei S. Gorshkov, the Soviet navy commander in chief, initiated a secret investigation into the phenomena of unidentified flying and submerged objects. The Soviets suspected that some of the inexplicable encounters represented new reconnaissance and antisubmarine devices developed by the U.S. Navy. Gorshkov launched a full and systematic investigation of all unknown phenomena observed at sea to date. The Soviet defense minister Marshal Gretchko created a special study group under the fleet intelligence and reconnaissance bureau at the Soviet Main Naval Staff in Moscow. Gorshkov seemed determined to solve the mystery: "The problem is so new and so difficult for us that we will spare no efforts and no resources to give answers to our sailors and our ships," he said "The result is important."

The task set up by Gorshkov was not easy: provide a systematic survey and analysis of all unexplained phenomena, including UFOs, that had been or were being reported around the world and that could put Soviet vessels at risk. Secret orders were sent to the fleet to the effect that the special group would be given full support. Procedures were established in the fleet headquarters of each military action theater (TVD). An officer from the naval intelligence and reconnaissance bureau was tasked to gather these materials and have them ready for the special group. The latter repeatedly visited the fleet, interviewing commanding officers, taking notes from the logs, and collecting photographs.

The Main Naval Staff also launched a series of seagoing expeditions. During the month of April 1970, the Soviet intelligence trawler *Xaripton Latviev* was dispatched to the North Atlantic to gather acoustic signatures from the oceanic depths in an effort to better understand the kviakiers

The Soviet Black Sea Fleet tender *Volga* was overflown by nine flying disks for eighteen minutes on October 7, 1977, off the Kola Peninsula in the Barents Sea. As a consequence, the Soviet navy issued a special set of instructions to study the objects.

mystery. The mission was aborted however, when the *Xaripton Latviev* was diverted to rescue survivors from the nuclear attack submarine K-8. It had surfaced after suffering a major fire and eventually sank off the Bay of Biscay. (The crew of the K-8 had abandoned the submarine but was ordered by Moscow to reboard the stricken submarine to try to tow it to safety. Upon reboarding, however, the submarine suddenly sank, taking fifty-two crewmen men to their watery graves.) Other missions were conducted, and Gorshkov was briefed of their results on a regular basis.[47]

A key observation, however, happened during the fall of 1977. Admiral Smirnov, the first deputy commander in chief of the Soviet fleet, stated that on October 7 the submarine tender *Volga* had been overflown by nine flying disks during an eighteen-minute period while in the Barents Sea, some 200 miles from the coast. As a consequence of this event, Admiral Smirnov and Rear Admiral Peter Navoytsev, the first deputy for operations, ordered the promulgation of "Methodological instructions to organize the observation of physical and abnormal phenomena in the fleet and monitor their influence on environment, living organisms and technical equipments."[48]

The Soviet Navy's Set of Instructions

Signed by Rear Admiral M. Iskanderov, the first deputy to the chief of staff of the Northern Fleet, and Captain First Rank E. Gachkenitsch from the Main general staff, this special set of instructions gave an exhaustive definition of the "Abnormal physical phenomena and objects":

Abnormal physical phenomena and objects (which are different from known phenomena and objects) have been observed in our country and in foreign countries by fortuitous observers as well as by professional specialists: aircrews, ships crews and various observation services.

We can divide the reported abnormal phenomena into two groups:

- local phenomena which display different geometrical shapes with sharp contours
- extended phenomena observed at the same time over a large territory and of rather large dimensions

The origin and the appearance of these phenomenon [are] characterized by changing geometrical shapes, changing courses, maneuvers, light effects, and the presence of electromagnetic fields and other emissions [that] have an influence on environmental conditions, living organisms, and electro-technical and radio technical installations.

The results of the observations enable us to report distinct features [that] characterized abnormal phenomena:

- Shape: spherical, cylindrical, discoid with one or two sides convex, discoid with a cupola; other external details such as windows, portholes; and the ability to split into parts flying independently
- Light: light on a dark background, stable or glowing light, pulsating light, multicolor light simultaneously or successively, rainbow like and others
- Smell: unusual
- Sound: distinct
- Maneuvering specifics: very high speed, unusual course, some hovering as if suspended in the air, brisk maneuvers, vibrations, rotations, capacity to move back and forth from the air into the water [49]

Explaining the rationale for the investigation, the instructions related the data that was collected to future space exploration: "Information on abnormal phenomena allow us to believe that this problem is worth serious investigation in order to better understand the Earth atmosphere and the Cosmic space for future use."

The methodological requirements aimed at presenting "undisputable information on abnormal physical phenomena and their influence on the environment, the living organisms and the technical equipments." It explained that the navy and its warships and sensors were ideally suited for this mission: "These phenomena can be observed visually and with the technical means available onboard naval vessels and among the forces and the establishments of the fleet."

The instructions further detailed the procedures to train the investigators and to react to an event that by definition could not be predicted: "It is impossible to anticipate the occurrence of abnormal phenomena (or objects). In order to obtain a more complete information on their apparition and behavior it is imperative to use all available technical means: radar stations, cameras and other filming equipments; radio stations for initial contact; acoustic ranges; dosimeters; compasses; binoculars, telescopes and other optical apparatuses."

The instructions also required monitoring the effects on equipment and humans: "The apparition and the behavior of abnormal physical phenomena require that we pay a special attention to the following: operating conditions of the weapons and sensors, presence of radiological interferences, suspension of radio communications, intense vibration of needles in compasses and other electromagnetic equipments, failure of internal combustion engines, interferences with the fleet data transmission, malfunction of electronic and mechanical clocks, and other effects on the environment, living organisms and human beings."

The next paragraphs of the instructions dealt with the collection of imagery and the use of color filters, spectral emulsions, and stereoscopic cameras, as well as the precise recording of the hydro-meteorological, aero-logical, and geophysical conditions. The last part specified the content of the reports, their classification "Secret," and their communication to headquarters (in three copies) and to the hydro-meteorological centers within five days of the ship's return to port.[50]

The GRU Takes Over

Meanwhile, a succession of disconcerting cases continued to affect the Soviet navy. For about six months during the winter of 1978–1979, a UFO was regularly seen hovering above the Zapadnaya Litsa submarine base in the Kola Peninsula. A smaller disk flew above the ammunitions depots on at least one occasion. In the Barents Sea, the captain, the officer of the watch, and the signalman of a surfaced nuclear strategic submarine saw through the periscope a disk hovering about 15 meters over the conning tower and projecting a telescopic ray of light onto the deck.[51] In May 1979, in the Black Sea, a few yards from a naval ship, a huge disk took off from the sea. At first, it lifted up a certain amount of water, which fell back into the sea. The Soviet navy reportedly got a good photograph of the event. In December 1980, in the Mediterranean 200 miles west of Gibraltar, the crew members of the survey vessel

Victor Bougaiev saw a cigar-shaped object hovering over the poop deck. The cigar was white with a black stain. Then the cigar split in two: one section flew to the northwest and the other to the northeast. The observation lasted for four minutes and was recorded in the ship's log.[52]

In the early 1980s, however, the Soviet navy's investigation into "abnormal phenomena and objects" was suddenly halted. After a celebrated UFO sighting in Petrozavodsk on September 23, 1977, that could not be ignored, the Soviet Academy of Sciences had initiated a special study on a parallel course with that of the Soviet Defense Ministry. At first, both programs were called Setka. For this reason, it had been decided to centralize the military data and the analysis within the GRU. The officers who were previously involved with the navy's special group did not want to be transferred to the GRU, and they took new assignments in naval intelligence. But the data were handed over to the GRU. They were classified top secret and not releasable. Studies on kviakiers, which had involved at least one institute in Leningrad, were suspended at the same time. It is likely that the study of kviakiers was also handed over to the GRU, alongside the material related to UFOs.[53] The GRU study group was later renamed "Gorizont" and "Galactica." One hundred thirty scientific institutes were involved in the study of the physical effects noted during the observations. The existence of this GRU UFO study group remained classified until 1989, when glasnost removed censorship on the UFO topic. According to Captain Evgueni Litvinov, who is in charge of the "anomaly" commission at the geographical section of the Russian Federation's Academy of Sciences, Galactica was terminated after the collapse of the USSR.[54]

Electromagnetic Interference

Electromagnetic interferences and the loss of radio communications and electrical power were the most common physical effects noted onboard the ships visited by the mysterious objects. These effects were reported both in the United States and in the Soviet Union. A case file from NICAP provides the testimony of Jim Kopf, who described how the communications of the USS *John F. Kennedy* (CV-67) were suddenly cut off by a hovering UFO on July 2, 1971:

> I was assigned to the communications department of the *Kennedy* and had been in this section about a year. The ship was returning to Norfolk, VA[,] after completing a two week operational readiness exercise (ORE) in the Caribbean. We were to stand down for 30 days, after arriving in

Norfolk, Virginia, to allow the crew to take leave and visit family before deploying to the Mediterranean for six months. I was on duty in the communications center. My task was to monitor eight teletypes printing the "Fleet Broadcasts." . . . It was in the evening, about 20:30 (8:30 p.m.) and the ship had just completed an eighteen hour "Flight Ops." I had just taken a message off one of the broadcasts and turned around to file it on a clip board. When I turned back to the teletypes the primaries were typing garbage. I looked down to the alternates which were doing the same. I walked a few feet to the intercom between us and the Facilities Control. I called them and informed them of the broadcasts being out. A voice replied that all communications were out. I then turned and looked in the direction of the NAVCOMMOPNET and saw that the operator was having a problem. I then heard the Task Group operator tell the watch officer that his circuit was out also. In the far corner of the compartment was the pneumatic tube going to the Signal bridge (where the flashing light and signal flag messages are sent/received). There is an intercom there to communicate with the Signal Bridge and over this intercom we heard someone yelling "There is something hovering over the ship!" A moment later we heard another voice yelling, "It is God, it is the end of the world!" We all looked at each other, there were six of us in the Comm Center, and someone said, "Let's go have a look!" The Comm Center is amidships, just under the flight deck, almost in the center of the ship. We went out the door, through Facilities Control and out that door, down the passageway (corridor) about 55 feet to the hatch that goes out to the catwalk on the edge of the flight deck (opposite from the "Island" or that part of the ship where the bridge is). . . .

As we looked up, we saw a large, glowing sphere. Well it seemed large, however, there was no point of reference. That is to say, if the sphere were low; say 100 feet above the ship, then it would have been about two to three hundred feet in diameter. If it were say, 500 feet about the ship then it would have been larger. It made no sound that I could hear. The light coming from it wasn't too bright, about half of what the sun would be. It sort of pulsated a little and was yellow to orange. We didn't get to look at it for more than about 20 seconds because General Quarters (Battle stations) was sounding and the Communication Officer was in the passageway telling us to get back into the Comm Center. We returned and stayed there (that was our battle station). We didn't have much to do because all the communication was still out. After about 20 minutes, the teletypes started printing correctly again. We stayed at General Quarters for about another hour, then secured. I didn't see or hear of any messages going out about the incident. Over the next few hours, I talked to a good friend that was in CIC (combat information center) who was a radar operator. He told me that all the radar screens were just glowing during the time of the incident. I also talked to a guy I knew that worked on the Navigational Bridge. He told me that none of the compasses were working and that the medics had to sedate a boatswain's mate that was a lookout on the signal bridge. I figured this was the one yelling it was God. It was ironic that of the 5,000 men on a carrier, that only a handful actually saw this phenomenon. This was due to the fact that flight Ops had just be (*sic*—been) completed a short time before this all started and all the flight deck personnel were below resting. It should be noted that there are very few places where you can go to be out in the open air aboard a carrier. From what I could learn, virtually all electronic components stopped functioning during the 20 minutes or so that what ever it was hovered over the ship. The two Ready CAPs (Combat Air Patrol), which were two F-4 Phantoms that are always ready to be launched, would not start.[55]

Similar interference was being reported by the Soviet navy. In the 1977 flying saucer sighting by the Northern Fleet, the submarine tender *Volga* lost radio communications and was unable to make contact with her base for the duration of the observation. Once the flying disks had departed, all systems functioned normally. In 1979, the crew of a Tu-142 long-range reconnaissance aircraft, having launched from Olenogorsk, in the Murmansk region, observed a lenticular object with a diameter of about 30 meters and a row of portholes. The instruments onboard the Tu-142 suddenly ceased to function, and the plane had to drop 100 tons of fuel into the sea before making an emergency landing. The object then disappeared vertically.[56]

In September 1982, during the Pacific Fleet exercises off the coast of Kamchatka, the ship-to-shore high-frequency radio communications were mysteriously cut off for thirty minutes while unidentified flying objects were observed above the ships. In October 1988, 3 miles south of Sutkotka Island, the crew members of the aircraft-carrying cruiser *Novorossiysk* saw a gigantic object with thirty-six lights arranged symmetrically but with undefined contours flying above their vessel. All of the electronic systems onboard the *Novorossiysk* went dead, while the diesel generators and the radio accumulators also ceased to operate. The object then departed southward at high speed. Forty seconds later, the radios functioned again, followed in three minutes by the radar and fire control systems. In total, this phenomenon and its aftereffects lasted for fifteen minutes.[57]

In a repeat of the case observed by an American ship's master in 1949, a retired Colonel V. recalled that two landing ships repatriating troops from Luanda, Angola, were confronted with three underwater lights positioned forward, under, and aft of one of the two ships. The three lights followed the landing ship for about 15 miles before moving away. During this event, the two ships lost radio contact, and the second landing ship could no longer see the first landing ship on its radar screen. These disturbances disappeared with the lights.[58]

Making Sense of the Incredible

With the closure of the U.S. Air Force Project Blue Book in 1969 and the official end of air force investigations into UFOs in 1974, the U.S. military seemed to be no longer interested in the phenomena. Even during the life of Project Blue Book, the U.S. military generally adopted an ironic tone in its public comments and its intelligence bulletins. UFOs have remained, however, on the JANAP and NORAD reporting instructions for the defense of the United States and Canada. A declassified DIA (Defense Intelligence Agency) attaché report on a spectacular UFO sighting above Teheran in September 1976 showed a distribution list that included the chief of Naval Operations, which proved that the interest had certainly not faded away.

The DIA analyst made the following comment: "An outstanding report. This case is a classic which meets all the criteria necessary for a valid study of the UFO phenomenon." The analysis called the UFO performance "awesome," noting that the objects displayed "an inordinate amount of manoeuvrability." This DIA report, released in 1977 under the Freedom of Information Act, proved, perhaps inadvertently, that the U.S. military was still monitoring and studying the UFO phenomenon despite its denials.[59]

In a late and indirect tribute to Thomas Townsend Brown and his research on electric thrusters, in November 2001 NASA posted on its Web site the astounding but largely unnoticed news that it had patented a new electric thruster. Brown was convinced that electric thrusters could explain UFO propulsion and performances. NASA commented "that this technology will provide improved reliability due to the use of no moving parts to create thrust in a gaseous environment. Thrust is expected in aqueous environments as well." The potential applications listed by NASA include a linear accelerator to launch payloads, shaft-driven applications, stepping actuator, and near-earth orbital manoeuvring. But the implication is that this mode of propulsion could supplement and eventually be substituted for rocket propulsion for space travel.[60] As Napoleon once said, "Impossible is a word only to be found in the dictionary of fools."[61]

More UFO cases affected the Soviet navy during the 1980s, apparently with a peak in 1989. In 1982, during a naval exercise off Balaklava, Crimea, a flying object was chased by an interceptor before it submerged into the Black Sea. For an hour and a half, in December 1983, the commanding officer and crew members from the submarine rescue ship *Sprut* observed an oval object flying slowly above the Gulf of Kolskiy in the Barents Sea, at an altitude of 500 to 1,000 meters. Before the crews' eyes, this object split into three spheres, which sped westward to the Norwegian border.[62] These observations left little doubt in the minds of Soviet navy leaders as to the non-American and unearthly origin of these objects, which were capable of flying at incredible speeds and submerging at will. It was the judgment of Admiral of the Fleet Chernavin, the former commander in chief of the Soviet navy, that "according to the Marxist Leninist theory, the universe has no borders and is endless: that's why we cannot deny the existence of life on other planets."[63]

17

THE LEGACY OF THE SOVIET NAVY
1989 AND AFTER

The Loss of *Komsomolets*

On April 7, 1989, the U.S. Atlantic Fleet Ocean Surveillance and Intelligence Center intercepted a burst signal transmission indicating that a Soviet submarine was in distress. A closer look by ELINT (electronic intelligence) satellite located the submarine in the northern Norwegian Sea. The signal was received simultaneously at the Royal Navy Intelligence Center DI-3 Navy, in Whitehall, which quickly commanded a Nimrod ASW long-range patrol aircraft to fly to the position for a look.

Shortly thereafter, a signal was sent to the NATO intelligence center in Bodo, Norway, alerting Norwegian patrol aircraft that a Soviet submarine appeared to be in trouble roughly 200 nautical miles west of North Cape. The ocean surveillance locating system was finely honed and soon picked up a number of surface contacts departing Soviet Northern Fleet waters and heading at high speed for the source of the intercepted transmission. Several Soviet merchant ships were also detected altering course and heading for the same position.

Within two and a half hours, the British patrol aircraft was on the scene. Its pilot stated via encrypted message that indeed a Soviet nuclear-powered attack submarine, known by its NATO name as a Mike class, was on the surface apparently fighting a fire, and there were sailors in the frigid water. The Norwegian air force sent a clear signal to the Soviet government offering helicopter rescue assistance from Bodo Airfield. The Soviet navy refused the offer of assistance. Although the ocean surveillance and intelligence system was working well, U.S. and NATO intelligence centers could only piece together enough data to know that the submarine had surfaced on fire and, after several hours, apparently had suffered an explosion, and that many crewmen were struggling in the icy-cold waters. Soviet long-range aircraft had arrived on the scene and dropped additional rescue equipment to the survivors.

It was the second year of Soviet president Mikhail Gorbachev's campaign of openness, and despite the efforts of the Soviet navy to suppress details of the accident, bits and pieces found their way into the Moscow newspapers and national television news. The true facts of what transpired were finally revealed, and the Soviet navy hierarchy emerged with a badly marred reputation. Four days after the sinking, a spokesman for the Northern Fleet appeared on Moscow national television and blandly said that there were no nuclear weapons aboard and the reactor had been safely shut down. Within two days, the navy admitted that there were indeed two nuclear-tipped torpedoes in the forward torpedo room. Soon after this rather poor performance, the Political Directorate of the Soviet navy was disestablished, a casualty of reform, when Gorbachev dismantled the parallel political hierarchy within the entire Soviet armed forces. The political commissar, the *zampolit* office, was finally dead. At the time, Peter Huchthausen was the U.S. naval attaché in Moscow (through 1997), and the following is an account of what occurred on that cold day in April 1989.

The *Komsomolets* K-287 was a revolutionary titanium-hulled nuclear submarine. This boat was capable of diving to more than 3,000 feet and was designed with the latest in Soviet nuclear-submarine technology. She caught fire in the northern Norwegian Sea while on an operational deployment. The fire, which was started by hydraulic fluid in the rudder-steering mechanism, was never completely contained. When electrical surges ignited scores of servo-driven instruments, the fire spread quickly to five additional compartments. The crew was unable to isolate the fire, particularly after a high-pressure air line ruptured and fanned the fires into blow-torch intensity. The extreme temperatures caused equipment paint to flash and protective rubber masks to melt on the crewmen's faces. Many of the crew members were poisoned when extreme pressure in one burning space forced carbon monoxide into the shipwide emergency breathing system, into which many of the fire fighters were plugged with flexible breathing hoses.

A few minutes after the fire started, the stricken submarine drove herself to the surface from a depth of 500 feet. But the sub wallowed helplessly on the surface while the crew fought the fires. Her oxygen and lubricating oil tanks suddenly exploded, which ruptured the pressure hull and caused progressive flooding of the boat. The submarine settled slowly at first and then sank rapidly, stern first, before rescuers arrived on

The Project 705 Alfa-class attack submarine, perhaps the best submarine hull design ever manufactured.

the scene. Most of the crew of *Komsomolets* abandoned ship into the arctic waters without survival equipment while she was still on the surface.

The commanding officer, Captain Second Rank Yevgney Vanin, remained in the submarine in a valiant effort to save those still aboard as she slowly sank. Most of the forty-two crewmen lost in the disaster succumbed to exposure in the near-freezing seawater.

Six months after the *Komsomolets*'s sinking, Captain Peter Huchthausen was visiting aboard the *Kirov* nuclear-powered cruiser in Severomorsk. I learned in a conversation with the ship's medical staff that the cruiser had been the first navy ship to arrive on the scene when the *Komsomolets* sank and had taken aboard some of the survivors who were pulled out of the frigid waters. Two of those who were rescued died suddenly, after receiving clean bills of health from the cruiser's medical officer. The two survivors had subsequently eaten a large meal and succumbed immediately after lighting and inhaling deeply on cigarettes. According to the doctor, they had suffered from extreme smoke inhalation, which had not been initially diagnosed. Other survivors had told of inflatable life rafts sinking right after being tossed from rescue aircraft and survival clothing that leaked, exposing the sailors to the icy waters, where most of them died of exposure. Even more horrible was the story that the submarine's integral emergency rescue chamber had malfunctioned. According to the cruiser's medical officer, who treated the sole survivor of the five who had surfaced from the depths in that steel cell, it had been a terrifying experience.

The *Kirov* doctor, who appeared to be a remarkably intelligent young officer, was visibly moved as he told me the story. I wasn't sure whether it was the terrible memory of the event or the fact that he was recounting the details to a U.S. naval officer that made him so tense. He was a thin man in his mid-thirties, with hair rumpled as if he had just pulled a sweater over his head. He wore glasses so thick his eyes were magnified far too large for his face. Looking at him, I could make out the pores in the doctor's eyelids through the thick lenses, which were so smudged it was a wonder that he could see properly. At first, the doctor merely stared wide-eyed, apparently adjusting to the idea that he was talking to a U.S. naval officer in uniform aboard a Soviet cruiser and in his medical office. After confirming that we were alone in the sick bay, the doctor pushed me farther inside by the arm. "You're the first American I've ever met," he said. "But you don't look like a capitalist."

The doctor ended his words by humming several notes—all in the same key, as if singing to himself. Before I could decide whether I should be glad I didn't look like a capitalist, the doctor guided me farther inside the medical infirmary and motioned for me to sit next to a bunk. "I want to tell you something, umm, umm. It's important you hear this. I assume you'll report what I say; attachés are spies aren't they?" He used the Russian term *razvyedchik*, which loosely includes anything from a scout to an informer. The doctor was a bundle of nerves and glanced repeatedly at the main door to the passageway. Before I could answer, the doctor gave me a cup of tea. Then the doctor poured himself one and sat on the edge of a small gray desk to begin his story.

"You know we arrived on the scene just after the *Komsomolets* went down and retrieved some survivors out of the sea and from aboard a merchant ship in the area? Well," he continued, humming his monotone notes after each phrase, "the captain of *Komsomolets*, Yevgney Vanin, remained aboard his sub when it suddenly began to sink. Vanin was a brave man. When all others were trying to save themselves, he tried to save those still inside the sub by putting them into the rescue chamber— even as the sub went down like a rock. Warrant Officer Slyusarenko was the only one of the five to come out of the chamber alive."

As the doctor related the account of the sole survivor of the chamber, he spun his tea glass in its metal holder.

"While lying in that bunk," he jabbed his finger toward a small bed beside his chair, "Slyusarenko told me how Captain Vanin found four others still aboard and helped them, on their hands and knees, to enter the rescue chamber; mind you, the submarine wasn't sinking on an even keel. They had to move uphill, crawling hand over hand up the passageways, over hatch combings, opening hatches above their heads. Can you imagine?

"Another officer, Yudin, and Warrant Officer Chernikov closed the bottom access hatch after climbing inside the sphere. The captain told Yudin to prepare to detach the capsule from the sub's hull. But just before attempting release, they heard a frantic clanging on the bottom hatch. One more crewman, apparently the submarine's engineer, was trying to enter the sealed rescue sphere. Those inside the chamber tried to reopen the lower hatch but couldn't. The sphere already had several meters of water covering the bottom hatch, which made it virtually unmovable."

The doctor continued as if he were there inside the sphere, and he twirled the tea glass faster and faster in its metal holder. "To the horror of the men inside the sphere, it kept plunging deeper. They couldn't detach from the submarine hull. Yudin tried the release lever again, but nothing happened. 'How does it work?' Yudin yelled.

"'Read the instructions,' the captain answered. Can you imagine?" the doctor asked, staring at me wild-eyed. "They never once practiced with the rescue chamber. They had to read the instructions as they sank! Someone, by sheer unusual forethought, had posted written instructions next to the release mechanism. Can you imagine," the doctor repeated, "sinking in a burning submarine, and poor Yudin had to stop and read the instructions?" The doctor stammered to the end of the sentence. "And still worse, the knocking continued as the submarine's compartments popped one by one with a sickening *crunch*, as they plunged past their crush depth of three thousand feet—the depth meter in the sphere broke at just twelve hundred feet."

The doctor glared, his head at a crazy tilt. "Well, according to Slyusarenko, Yudin was calmly reading the posted list of steps to take when they crashed on the sea bottom and the rescue chamber detached itself and began to float to the surface." The tea glass spinning in the doctor's grip suddenly shattered with a snap, accenting his description of the imploding submarine. He continued, "After finally separating from the submarine hull, the rescue chamber soared to the surface. But that didn't save the five inside. No, even though the sphere was designed to compensate for the pressure gradients, it shot to the surface too quickly, killing four of the five men inside with barotrauma—same as the bends. Their bodies were ejected like rag dolls through the small hatch in the chamber, which popped open on the surface, venting the pressurized chamber like a deflating balloon. It filled with water and sank immediately.

"All died except for Slyusarenko; he was thrown through the hatch when it popped open. He was nearly frozen and had broken limbs and a smashed head when I got him in here. It was a crime, the lack of training."

A commotion started near the door, and a senior officer suddenly pushed into the cruiser sick bay. He berated the doctor, who now stood as if in shock. "You shouldn't detain foreign visitors here," the senior officer mumbled and ushered me quickly out of the sick bay, leaving the doctor completely overcome with emotion and unable to speak. He stared at his broken glass as blood dripped from a fresh cut on his hand.

The loss of the *Komsomolets* became a celebrated indictment against faulty Soviet submarine design and safety measures. It was disclosed during the investigation following the disaster that the submarine's crew had never drilled with the modern, sophisticated, automatically releasing life rafts or the escape chamber or even in basic damage and casualty control. Automatically releasable signal transmitters had been designed to float free from a sinking submarine, rise to the surface, and transmit burst distress signals via satellite, but the crew had spot-welded them to the hull so that the transmitters wouldn't get lost. The deplorable details of the failure to train the *Komsomolets* crew are contained in the book *Loss of the Komsomolets: Arguments of the Constructor*, written by one of the people who had originally constructed the project, Dmitri Romanov.

As more accounts of accidents and near-disasters emerged from the former USSR, the chilling facts exposed the narrow gap that existed between safety and sheer cataclysm in the Soviet armed forces. Other versions of the naval tragedies have appeared since then. Most of them are in fragments that give more details of the nuclear submarine accidents and document the appalling history of Soviet nuclear waste dumping.

More than a mile under the northern Norwegian Sea, Russians installed nine seals to the bow of *Komsomolets* to contain the spread of plutonium leaking from two torpedo warheads. According to the expedition leader, Dr. Sagalevitch, nine holes in the bow were covered by seals made of rubber and titanium. Six of the holes were torpedo tubes, and three were fractures caused by the sinking.[1]

The Russians abandoned as too risky a plan to raise the entire hull. They were so concerned about the condition of the *Komsomolets* wreck in the Norwegian Sea that they turned to the West for assistance in solving the nuclear-effluent threat. As a result of a joint Russian, U.S., and European study of the site, the Russians erected a concrete sarcophagus around the shattered hull to contain the twenty-six pounds of plutonium that are still present in the two nuclear-tipped torpedoes, which were found to be corroding at an alarming rate. The area where the submarine rests is among the most biologically productive in the world's oceans. Minimal transport of radionuclides along the food chain from seawater to plankton to fish could result in grave environmental, economic, and political consequences.[2]

More than thirty years had passed since the loss of the Black Sea fleet battleship *Novorossiysk*, and apparently nothing has been learned. As Dmitri Romanov wrote in his book *Tragedy of Submarines*:

> On 29 October 1955 battleship *Novorossiysk* sank at her moorings at her Black Sea base due to an external explosion, with the loss of more than six hundred men. The governmental commission for investigating the causes for the loss of the battleship determined that the central structure of the navy was ultimately responsible because of insufficient attention to detail, and that responsible cadres were guilty of not sufficiently training the ship's company for combat readiness and casualty control, and the lack of documentary instruction in damage control. Thirty-three years later tragedy occurred aboard submarine *Komsomolets*, for exactly the same reasons. A ship sinks, people are lost, and the central structure of the navy continues as if nothing happened.[3]

A Recipe for Disaster

The growing revelations about the sorry state of safety in the Soviet navy during the Cold War inspired closer investigation. During research in Russia in September and October 1995, Huchthausen interviewed survivors of several submarine losses, including *Komsomolets* K-287, Navaga K-219, and officers who commanded the squadron and the flotilla from which Golf II class K-129 sailed and never returned. He also listened to the arguments of senior constructors of various submarine projects from the Rubin Central Design Bureau for Marine Engineering and senior officers from Soviet navy logistics and engineering. He interviewed senior officers from the operational navy, both retired and of the current Russian fleet, including the chief of the Main Navy Staff Admiral Valentin Selivanov and Admiral Valentin Ponikarovsky, a former nuclear submarine commander and chief of the Soviet Naval War College.

After the discussions, it became clear that despite a sharp polarization between the constructors and the operators regarding the causes of the fires, reactor failures, and other accidents, both sides agreed that the extreme pressure generated by the Cold War arms race—the motivation to out-produce the Western navies in top-of-the-line ships and submarines—played a significant role in the frequency of accidents. In addition, the operational forces were under extreme pressure to "make the deployment" at any cost during periods of an increasingly high tempo of operations dictated by the state of relations during the Cold war. Operators consequently cut corners in the safety of operations and maintenance.

One major flaw in the navy's material readiness equation resulted from the system that governed Soviet military industry. While Western defense budgets were determined by economics, the Soviet military-industrial machine was dictated by a bureaucratic system of central planning. From arguments presented by both the operators and the shipbuilders, it is clear that the technical information feedback flow from the operators to the builders and the designers failed miserably in the Soviet system. Five- and ten-year-plans, drafted by Kremlin political leaders with little input from engineers or military officers, drove each design bureau and shipyard. This resulted in a tendency for managers to cut corners to keep to a scheduled production plan at the expense of seeking quality control and insisting on safety margins that are so central to equivalent Western industries.

An additional peculiarity of the Soviet submarine design and construction process was the fiercely driven personal involvement of one or two senior individuals with each project. In an ironic inconsistency with what might be expected, given the communist ideology of collective effort, three individuals were solely responsible for the design and the construction of the *Komsomolets* from its first planning stages in the late 1960s until its completion and launching in 1983.

Project 685 Plavnik is the Russian name for the Mike class program that created the deepest-diving nuclear attack submarine in the world and possibly one of the fastest. It was the personal project of chief designer and constructor Vladimir Kormilitsin, assistant constructor Dimitri Romanov, and assistant Anatoli Chuvakin, who died of a heart attack when he heard that their prototype, *Komsomolets*, burned and sank in April 1989.[4]

The constructors' detailed investigations into the causes of the disaster placed the blame fully at the feet of the operational navy for inadequate training in fire-fighting and damage-control and for failing to provide operational feedback about the problems of the submarine's material condition to the design bureau and the shipyards. Romanov blamed the fleet entirely for losing control of the fire aboard the *Komsomolets*.

This occurred, according to Romanov, because the crew did not activate the installed fire-suppressant systems. During an interview in September 1995, Romanov claimed that systems aboard the submarine *Komsomolets* had been disabled by crew members to avoid frequent accidental activation. The crew members did this because they feared being punished if damage resulted from their unintentionally triggering the sensitive devices. The systems were similar to the automatic magazine sprinkler systems installed aboard the ill-fated Kashin class destroyer *Otvazhniy*, which exploded and sank in the Black Sea in 1974.

In the submarine forces, the emphasis was to steam on, operate at sea, deploy at all costs, *pakhat'*, in Russian—to "plow on mindlessly." While in port, crewmen were used for useless menial tasks such as painting rocks, loading trucks, and other political work, instead of undergoing much-needed training in simulators. Therefore, when submarines deployed or went on maneuvers, the crew had to learn key functions on the job and under real and dangerous conditions. Practicing safety procedures was considered a negative endeavor that undermined morale and conveyed distrust of the system and of the central control authorities.

Conversely, Soviet navy operating forces and most accident survivors blamed the shipyards and the designers for the failure of the fire-fighting systems and the installed safety equipment, as well for as the lack of adequate rescue equipment. Painful investigations into the *Komsomolets* accident also brought flaws to the surface that were separate from both the design and the operation of the submarine. These implicated the organization of search-and-rescue forces, command and control, and outmoded operational secrecy practices that prevented timely assistance from being requested from nearby Norwegian rescue assets.

In summary, the failures that came to light during the *Komsomolets's* sinking were the same as those in the long-suppressed but recently publicized investigations into the sinking of the battleship *Novorossiysk* in 1955, the destroyer *Otvazhniy's* fire and sinking in 1974, and numerous earlier submarine accidents. All point to the operational forces' inability to keep abreast of the rapidly growing fleet technology and being too hasty to deploy new ships and submarines. The armed forces threw caution to the wind and sailed away into the sunset, practicing only to shoot their weapons.[5]

During this long period of Soviet naval expansion, which was fraught with calamity, the relationship between the Soviet ship and submarine designers and constructors, on one hand, and the fleet operators, on the other, began to publicly diverge. Feedback concerning shoddy construction and design weaknesses seldom was communicated from the fleet to the design and construction organizations. Romanov, from the Rubin Central Design Bureau for Marine Engineering in Leningrad, grew so distraught following repeated submarine accidents that he recorded his views in a sensational 1995 book titled *Tragedy of the Submarine* Komsomolets: *Arguments of the Constructors*, giving the designers' and constructors' arguments for the many failures.

Early reactor accidents at sea aboard first-generation nuclear submarines K-3, K-19, K-11, and K-27 in the 1960s and the 1970s set the precedent for the widely diverging opinions of the constructors and the fleet. The causes behind the dramatic sinking of the Kit K-8 attack submarine off the coast of Spain in 1970, with the loss of fifty-two crewmen, and the Navaga K-219 off Bermuda in 1986 were revealed in full detail only after the collapse of the Soviet government. In 1989, well into the Gorbachev reform years, the sensational *Komsomolets* sinking was reported piecemeal in the Soviet media as it happened but not in full gory detail until six years later.

Romanov's charges of lack of adequate training for fleet personnel were answered angrily by the Soviet navy's top two operational commanders. Fleet Admiral Vladimir Chernavin and his first deputy, Ivan Kapitanets, printed their views in the widely read *Soviet Naval Digest*, the Russian equivalent of the U.S. Naval Institute *Proceedings*. The two senior commanders cited lack of proper political motivation and shoddy submarine construction as prime causes for the mishaps. Admiral Kapitanets wrote that faulty material, poor quality control, and indifferent shipyard workers contributed to the navy's poor safety record. He pointed out that the diesel fuel used in the Soviet navy ignited at temperatures 20 degrees lower than did American diesel fuel. He also claimed that the rescue equipment had failed during the *Komsomolets*'s sinking, not when the men deployed the gear.

Since Gorbachev's 1985 dismissal of Fleet Admiral of the Soviet Union Sergei Gorshkov, the navy commander in chief, and his subsequent death in 1988, the true story of the deplorable losses suffered during the twenty-seven years of dynamic naval expansion on his watch has come to the surface. During the first several years following the 1991 collapse of the USSR, Russian disclosures received only fragmentary attention in the West. But, ironically, these disclosures forced the exposure of submarine operations that were long held secret on both sides of the former Iron Curtain. Russian accounts now point openly to numerous incidents with foreign attack subs, many of which caused damage on both sides, and any of which could have resulted in lost missiles, damaged reactors, and crew falling deep into watery graves.

The change that followed the accelerating process of openness of the Gorbachev era, beginning in 1985, had a profound impact on the Soviet navy. For years, the West had suspected that the Soviet Union was concealing its new fleet's marginal material condition, and finally it was being proved. In Moscow, precise information about the fleet's appalling material state and safety, especially the nuclear-powered ships, began to surface. This confirmed what the West had surmised, based on earlier fragmentary intelligence reports and accounts leaked during increasing contacts between Western diplomats and Soviet citizens.

By 1989, the customary fear of repression had deteriorated significantly, and precise accounts of naval accidents were made public. It is now documented that during the first ten years of the stampede into the nuclear power arena, from 1958 to 1968, the Soviet navy lost eight submarines and two hundred men, with more than four hundred crewmen being gravely irradiated.[6] During those dynamic years of Soviet naval expansion, serious radiation casualties were spirited away to isolated hospital wards, where their symptoms were disguised as nervous disorders. Many victims were classified as suffering from "trauma caused by stress." Scores died of radiation poisoning, their remains hidden in unmarked graves, with others still unaccounted for. The Supreme Soviet, the Soviet legislature, which normally rubber-stamped Party Central Committee decrees into law, made it unlawful for medical authorities to enter "radiation poisoning" or "irradiation" as the cause of death on official death certificates.

The USSR's nuclear power misfortunes began aboard its prototype atomic submarines and surface ships. The world's first nuclear-powered icebreaker, the much-advertised *Lenin*, had been built in the late 1950s. She suffered a fuel element meltdown in one of her three reactors that killed at least thirty crewmen late in 1966. Her nuclear engineering plant, a test plant for future submarines, was gutted and discarded in the Kara Gulf near Novaya Zemlya, portions of which are still highly contaminated.

Nuclear submarine pioneer Captain Leonid Ossipenko's celebrated *Leninsky Komsomol,* the first Soviet atomic-powered submarine, was launched in 1957. The prototype submarine, K-3, was a Project 627 Kit (NATO name November class) and on July 17, 1962, was the first Soviet submarine to navigate under the ice and surface at the north pole. For that achievement, Ossipenko and his second-in-command, the well-known nuclear-power innovator Lev Zhiltsev, were awarded the order Hero of the Soviet Union by Nikita Khrushchev. The fact that during the epic cruise to the pole, a leak in a reactor steam generator mortally radiated a dozen K-3 crewman was quietly suppressed. Four years later, the same sub lost thirty-nine crewmen in a fire, limped back to port, was decommissioned, and was similarly dumped in the Barents Sea near Novaya Zemlya, in a growing graveyard of failed Soviet atomic projects.

The name of the chief engineer of that first sub, Nikolai Mormul, who was mentioned in chapter 10, became a legend in the long road of Soviet submarine misfortunes. After the disasters aboard the icebreaker *Lenin* and the first sub K-3 and a series of major submarine accidents resulted in the loss of more than 250 navy men, nuclear engineer Mormul became an open critic of Soviet submarine safety. He began his futile attempts to correct nuclear power by submitting detailed procedures for operating and improving submarine power plants to the Main Navy Engineering Directorate in Moscow. He immediately fell into disfavor with the naval high command for steadfastly refusing to stop his attempts to rewrite nuclear engineering procedure. Soviet commander in chief Admiral of the Fleet Sergei Gorshkov ordered Mormul jailed for lack of patriotic zeal and for attempting to air the sorry plight of the early Soviet submarine force. Mormul's real misdemeanor was daring to advocate improving engineering safety. In 1986, shortly after the Chernobyl accident, while Mormul was still incarcerated, he sent Gorbachev a written procedure to correct deficiencies in the woefully flawed nuclear power plants throughout the USSR. But again his expertise was ignored, and he was labeled a nuisance.

Mormul was released by Gorbachev in 1987, during a mass pardon of political prisoners. He was promoted to rear admiral and returned to the Northern Fleet as senior engineer. But it was too late; his skills had been ignored mostly for political reasons. Mormul later wrote a book exposing the hidden past of the Soviet submarine force and the deplorable record of nuclear dumping at sea. The book was finally published in Russia in 1996, after being suppressed by security censors since its completion in 1992.[7]

More ominous than the gruesome accounts of the individual naval disasters suffered at sea are Mormul's exposures and the official revelations of the profusion of nuclear reactors and atomic weapons scattered in various states of decay on the sea bottom. The alarming accounts, which began to surface after 1991, confirm the number and the locations of wrecked nuclear reactors and lost atomic warheads lying on the ocean floor, the remains of forty years of hidden Soviet submarine disasters.

In the Kara Sea dumping area alone, near Novaya Zemlya in the Barents Sea—the largest Soviet nuclear graveyard—more than 3.5 million curies of nuclear waste were disclosed to already be on the seabed in 1992. This is the equivalent of one-tenth of the radiological contamination leaked to the atmosphere during the Chernobyl incident. The residue exists in the form of 8 scuttled submarine hulls, 16 discarded reactors—6 with fuel still inside—and 9,000 additional tons of discarded fuel assemblies and liquid nuclear waste, all in water no deeper than 150 feet.

Between the late 1950s and 1993, normal procedure for the Soviet and then Russian navy was to dump liquid and solid radioactive waste and spent nuclear fuel at sea in designated sites in the Barents Sea and the Pacific Ocean. Testimony by Mormul and other witnesses in the disposal business, however, claim that vast quantities of nuclear waste disposed of during the Soviet years were done without records, at night, and in areas not authorized for dumping, both in the Barents Sea in the north and in the Sea of Japan off Pacific Fleet ports. Thus, a completely accurate accounting of early nuclear waste disposal is impossible.

Unauthorized secret dumping of radioactive waste at sea was a result of the prevailing attitudes during the height of the so-called stagnation period in Soviet history, when senior naval and shipyard personnel avoided accepting responsibility for their actions for fear of repercussions. The pressure to produce according to unrealistic centrally controlled planning took precedence over everything else. Party officials and naval construction leadership falsified records to show compliance with centrally planned production. Nuclear engineering work was completed with little or no quality control. Shipyard safety during construction did not exist. Personal advancement was based on meeting production quotas, with little regard for quality or safety.

Following the disclosures of this disastrous Soviet naval record and increasing international pressure to end nuclear dumping at sea, Russian president Boris Yeltsin commissioned a team in 1992 to study and report on the nuclear-dumping situation in the areas around the former Soviet naval bases. The study, known by its short name the *Yablokov Report*, was released a year later and revealed details of the Soviets' acknowledged nuclear dumping in the North Atlantic, the Barents Sea, and the Pacific. But according to nuclear engineer Nikolai Mormul, the *Yablokov Report* documents merely a fraction of the total materials dumped without records.[8]

The *Yablokov Report* also expands on sketchy data known earlier from intelligence holdings and leaked by Admiral Mormul about several major submarine reactor accidents, among them the Chazma Bay incident.

The proud legacy of Admiral Gorshkov: the former battle cruisers *Kirov* (right) and *Kalinin* (left), side by side in the Northern Fleet base of Severomorsk in 1994. Western intelligence was very concerned with the safety condition of the ex-*Kirov* after a reactor incident during her 1990 deployment to the Mediterranean.

This frightening accident occurred on August 10, 1985, aboard an atomic submarine that sustained an uncontrolled spontaneous uranium fission chain reaction in the Pacific port of Shkotovo-22, which is 30 miles north of Vladivostok. A Project 675M (NATO name Echo-II class) cruise missile submarine suffered an accident during nuclear core refueling. During the preparations to fuel, a leak was discovered under the reactor's cover, a defect that normally required dry-docking to fix. Nevertheless, to save time, the crew and the yard workers attempted repairs while the submarine was still in the water.

During the process of raising the reactor cover to remove the spent fuel, the workers accidentally extracted two fuel rods, triggering an uncontrolled chain fission reaction. It blew the one-ton reactor cover more than 100 yards high, which then fell back onto the submarine. The explosion scattered radioactive fuel and the bodies of ten crew and yard workers throughout the bay. The blast ejected the nuclear fuel and the latticelike neutron-absorbing quench baffle in a huge plume of contamination that contained an estimated 4 cubic meters of highly radioactive material over an area as large as the Chernobyl tragedy acreage, stretching 6 kilometers toward the city of Vladivostok. Shortly after the explosion, dock workers and navy men were ordered to mop up and conceal the accident.

According to witnesses, after the blaze was extinguished, the gates of the shipyard were closed and workers who had been involved in the cleanup were directed to shave off their hair, surrender their clothes, and take repeated showers. The fragmentary remains of the ten men, along with other contaminated debris, were buried in specially prepared pits. The submarine was towed to a remote pier and abandoned, stripped of its reactors. It was still there in 1994, awaiting removal. Censors forcibly excised a detailed report of that accident from Mormul's book before it was published.

According to the official *Yablokov Report*, eight whole submarines lie scuttled in the Atlantic, the Barents, and the Pacific, along with eighteen reactors in various states of damage—and six still contain partially spent nuclear fuel. In addition, more than thirty-eight nuclear warheads are strewn about the world's seabed. Thirty of them are in 18,000 feet of water off the coast of Bermuda, where they were lost inside the hull of the world's first nuclear-powered ballistic-missile submarine to sink at sea, the Project 667 (NATO Yankee Class) K-219.[9]

Unstable liquid missile fuel caused numerous accidents in the Soviet fleet. In a similar case, the crew on another Yankee, with improved-range missiles but still liquid fueled, intentionally opened one missile silo door and ejected a ballistic missile, complete with atomic warhead, into the Pacific Ocean to preclude an onboard explosion. Frantic efforts by the navy successfully retrieved the errant warhead a few days later.[10]

For years, the official U.S. and Soviet navy response to criticism of dumping nuclear debris has been that the best place to dispose of old reactors and warheads is in deep mud on the sea bed—the deeper, the better. This, according to experts, is valid only if the reactor container or missile casings are well sealed and intact. But how long can such containers remain unchanged on the sea bed? The U.S. Navy carefully monitors the bottom effluent near the sunken USS *Thresher* (SSN-593) and USS *Scorpion* (SSN-589) wrecks, the only U.S atomic submarines lost in accidents, and publishes the results every few years.[11]

Russia inherited the world's largest submarine nuclear fleet—245 boats—from the USSR in 1991 and faced the immense task of decommissioning and then dismantling the aging nuclear boats. A report released in 1996 by the Bellona Foundation, a Norwegian-based environmental group with offices in Russia, expressed the international concern that Russian nuclear naval facilities posed a serious environmental threat to the surrounding regions and that Russia is unable and unwilling to make the necessary expenditures to remedy the situation.[12] The report stirred up considerable acrimony within Russian political circles, to the extent that the Russian researcher participating in the study, retired navy captain first rank Alexander Nikitin, was arrested and charged with theft of secret materials concerning nuclear submarine dispositions and information on naval reactors deemed vital to Russian national security.

The sudden crackdown by the new Russian Federal Security Service—the successor to the KGB—began in October 1995 with the arrest of Captain Nikitin. The target of its offensive appears to be primarily foreign researchers who had roamed with relative impunity throughout Russia's formerly closed military areas, collecting data for an analysis of nuclear safety and environmental damage caused by the outrageous nuclear waste–disposal practices of the Soviet years.

According to the *Bellona Report*, the disposal of solid and liquid radioactive waste is the main problem plaguing the decommissioning program. One of the more serious threats to safety is the failure to properly maintain the storage facilities for spent nuclear fuel and radioactive waste.[13]

In 1995, former commander of the Northern Fleet Admiral Oleg Yerofeev stated that "the problems of storing spent nuclear reactor fuel, radioactive waste, inactive submarines and the lack of servicing for the submarines in active service are a problem not only for the Northern Fleet, but also for the Russian state. Therefore, it would be natural not only for the Fleet to take necessary action, but also for the Ministry of Emergency Situations, Emercom, also to act. If measures are not taken to address the situation today, over a period of time the situation could become critical and lead to an ecological disaster."[14] Who would know better?

In an effort to ease the problem of dismantling the reactors, Russia requested and obtained Western assistance. Over the past fifteen years, U.S., Norwegian, Japanese, British, French, and Italian programs have significantly helped reduce the problem.

According to *Bellona*, the radiation monitoring and emergency response to nuclear accidents in the Northern Fleet area are complete to the point that in 2008 potentially dangerous installations are under round-the-clock supervision.[15] Andrei Ponomarenko, the Bellona-Murmansk's nuclear and radiation safety coordinator, assures that "the 'light projects,' in which we include breaking up submarines are fundamentally finished." Almost nothing, however, has been done to decommission surface vessels with nuclear power installations, while the situation at Andreyeva Bay near Murmansk remains critical, with radioactive fuel in storage continuing to degrade.[16]

St. Andrew's Flag Flying Anew

The dramatic shift in Soviet behavior from normal sullen secrecy to creeping kindness after the 1988 Gorbachev-Reagan Moscow summit came as a shock, albeit a welcome one to Westerners who were then living in the Soviet Union. Most were apprehensive about whether the warm feelings would continue into the following year. But the dramatic unraveling of Soviet Communism kept snowballing and would culminate in the autumn 1989 destruction of the Berlin Wall. As the end to the Cold War became inevitable, incredible scenes unfolded in Moscow.

Since the summer of 1988, the Americans had witnessed the Immediate Nuclear Force (INF) Treaty breakthrough, the Marshal Sergei Akhromeyev–Admiral William Crowe visit exchanges, the consummation of the Agreement for Joint U.S./USSR Military Cooperation, and the resumption of warship exchange visits. The Soviet army was out of Afghanistan and Defense Minister Dmitry Yazov had apologized for the death of army major David Nicholson, who was shot by a Soviet sentry while he was working with the U.S. Potsdam Military Mission. In 1989, it was hard to surprise anyone in the U.S. Embassy in Moscow.

The naval attaché had just returned from another precedent-setting event, escorting the chief of naval operations Admiral Carlisle Trost and his delegation on a tour of key naval sites in the Soviet Union. He was the first of the U.S. military service chiefs to visit under the new agreement. In Severomorsk, the headquarters of the Northern Fleet, Trost and his small staff, plus the attaché, were taken aboard a Victor III nuclear-powered attack submarine, the Soviets' most advanced attack submarine. The Americans were treated to a small buffet in the crews' dining compartment and were surprised when they were served wine with the food. Admiral Trost said to his Russian host that the U. S. navy served no alcohol aboard ship. The Soviet host responded that red wine had inherent qualities of absorbing low-level nuclides, a true but damning statement as to the radiological tightness of their reactors.

For the U.S. attachés, it was an especially heady time. After years of spooking around by peeping over fences, watching shipyards, and spending millions of dollars traveling to see what little was to be seen in the closed system, they were suddenly faced with unexpected charm and camaraderie on the part of their Soviet hosts.

One of the most dramatic events was the visit to Moscow by the U.S. Marine Band, the president's own White House band, which is arguably one of the best military bands in the world. It made a grand, six-city concert tour of major cities of the Soviet Union, beginning with a joint concert in Moscow with the renowned Red Army band. The naval attaché drew escort duty for the cities of Kiev, Minsk, and Leningrad. In each city, the marines performed first in the public concert halls and second in a nearby military base. It was an extraordinary experience to be suddenly ushered in grand style into the areas closed to foreigners ever since World War II. Russians can be wonderful hosts when they are permitted to be.

The news of the growing unrest in Eastern Europe and its citizens' flaunting of the central authorities was broadcast daily on the Moscow evening television program *Vremya*. With Mikhail Gorbachev's glasnost ("openness") in its fourth year, more real life was being shown to the public. The people's blatant confrontation with communist authorities that remarkable October in Hungary, Czechoslovakia, and East Germany was reported in surprising detail on the Moscow news. Those events and the government's release of control over the Russian Orthodox Church built into a whirling crescendo. On Christmas Day 1989, *Vremya* showed the bullet-riddled bodies of Nicolae and Elena Ceausescu lying in the Romanian courtyard at Tirgoviste. That scene on television in every home in the USSR was an omen that loomed over Moscow's communist leaders.

The dramatic destruction of the Berlin Wall was echoed in the Moscow diplomatic community by the reactions of the military and naval attachés from Eastern European capitals. During the previous two and a half years, the attaché had developed a close personal friendship with the East German naval attaché Captain Rolf Franke. Rolf was the senior naval officer in the Embassy of the German Democratic Republic, GDR. He had served as an intelligence officer his entire career and probably hadn't seen a ship or the sea since he became a naval cadet in Berlin. He had served as an attaché and an adviser in Cuba, Peru, and Vietnam and in several other overseas assignments that he preferred not to mention. As with most Warsaw Pact attachés stationed outside their native countries, he was permitted to take only his wife. For purported reasons of schooling, the Eastern attachés were required to leave their children or one other family member behind, especially when they served in the West. Some called it ransom, others called it insurance to make sure the Eastern attachés returned home.

Rolf had grandchildren in Berlin who visited on occasion. His wife grew more somber and morose the longer she lived in Moscow and the more the situation in her native East Germany deteriorated. Rolf frequently invited the American to his embassy, located nearby on Leninskiy Prospect, and he became more outspoken as the situation in Berlin changed.

Rolf began to talk openly of his retirement plans, his salary, and his situation at home. He professed to be a devout communist but repeated that their system had been making grave errors for years. He was frankly critical of East German president Eric Hoenecker when he was alone with the American attaché. Had the times been normal, the U.S. attaché would have categorized him as a perfect case of an official preparing to defect. But if that was his intention, it was completely overtaken by the rapidly unfolding events in Berlin. As it turned out, Rolf's entire homeland defected to the West before he could. Rolf confided that he feared for his safety and that of his family. When the news later broke about the sacking of the East German Security Police (*stasi*) headquarters in Berlin, Rolf was in the American's apartment visiting. Rolf outlined his plans to bring his entire family to Moscow until the situation clarified. He said that his career record would certainly preclude a normal retirement and might even result in his being jailed if he went back. It was truly a sorrowful experience for him.

The leaders of the Communist East deserve no pity for their fate and are rightly held responsible for the policies that resulted in many tragic deaths of their citizens who tried to escape and for the broken homes and physical and mental suffering of their populace. Nevertheless, many Western diplomats serving in the East truly felt pity for the other Warsaw Pact military officers, like Rolf, whose lives came unraveled during that period.

The crescendo of the East-West military relationship occurred during the 1989 annual East German Armed Forces Day celebration. That year was the fortieth anniversary of the German Democratic Republic and the founding of its armed forces. A grand diplomatic event was scheduled just one month before the Berlin Wall crumbled, and the future of Germany rested on the whims of Soviet general secretary Gorbachev and the Politburo. The GDR anniversary became more than just their national day; it was the last great fling in Moscow staged by any Warsaw Pact military organization.

Early Christmas morning in 1991, the phone woke the former U.S. naval attaché in his Moscow River apartment. He had retired from the navy in September the year before and had returned to Moscow with a new business firm called Consult America. Captain Valentin Serkov, a retired Soviet naval officer, a former Incidents at Sea agreement founder, and a law of the sea expert, was on the line. In a voice charged with excitement, he insisted that they meet to observe a monumental event. Mikhail Gorbachev had resigned the day before—and Serkov was expecting something unusual. He wanted to meet at first light in Red Square near the Spassky Gate.

He donned his fur hat and overcoat and walked the 3 kilometers to Red Square in the frozen morning from his apartment near Taganskaya. Snow flurries blew across the Moscow River and stung his face as he leaned into the cold wind and trudged along the deserted embankment. He wondered what was up. His breath froze in short clouds of mist as he walked in the early light. The sky turned a brilliant pink as the sun rose through the broken puffy clouds. Moscow was at its best in the early morning hours when the streets were empty.

He arrived at a nearly deserted Red Square. Only the Kremlin guards in their gray overcoats and fur *shapkas* were visible moving slowly on the square. He could make out the broad figure of Serkov standing with his back against the black wrought-iron fence on the west side of St. Basil's. He was looking high above the top of the Lenin mausoleum. He waved and began to walk toward the American.

"Look up there," he said. The American noticed frosted tears smeared across the Russian's ruddy face as Serkov walked toward him. "It's gone. Look!"

The American turned to follow his gesture. High atop the Kremlin dome fluttered an oversized white, blue, and red Russian tri-color! The solid-red flag of the USSR, with its hammer and sickle, was gone! The two men stood in the Red Square in the early sunrise and the blowing flurries, two former naval intelligence officers, one Russian, the other American. The Kremlin guards watched in silence and stomped their feet in the light snow. The Soviet Union was no more. Sometime later, during the navy days of July 1992, the ancient Russian naval blue-and-white cross of St. Andrew replaced the Soviet naval ensign. Until 1997, the red jack with hammer and sickle remained the flag of the Black Sea Fleet, which was being disputed between Ukraine and Russia. It is still displayed in parades and on navy days with the pride attached to a flag that was never surrendered in battle and was raised on the Reichstag in Berlin in 1945.

The U.S. naval attaché post in the USSR had a long history of high adventure and intrigue ever since navy commander Hugo Koehler won recognition traveling in disguise throughout Russia while reporting on Bolshevik activities during the civil war of the 1920s. His assignment

Relaxed faces on a Soviet scientific vessel photographed from a Western helicopter in 1991 indicate that the Cold War is over.

was to "find out . . . whether Bolshevism is a real force, a workable idea, one that will endure, or whether it is the false doctrine it appears to be."[17] On that frosty morning in the Red Square, seventy-one years and many attachés later, the U.S. attaché sensed that Koehler's assigned question had finally been answered. The red banners of the Bolshevik fleet, unfurled in 1917, were finally taken down.

The nearly seventy-year confrontation with Communism was 1945. Yet are the seeds of antagonism completely gone?

After the Wild West days of Russian president Boris Yeltsin in the early 1990s, there followed an apparent retrenchment of the former disgruntled intelligence services under President Vladimir Putin, who restored the authority of the state. Secret naval bases that were opened to Westerners during the slack security period of Yeltsin have been closed again. In 1999, a retired U.S. naval intelligence captain visiting Russia, apparently on business, was found in possession of the blueprints of an earlier version of an astonishing weapon: a 200-knot underwater missile known as the *Skvall* and based on a World War II Italo-German concept. The officer was arrested, tried, convicted of espionage, and sentenced to twenty years in prison.[18] Although the American's poor health resulted in his being released by Putin, it seemed as if the former antagonists were back at their old games. This incident did not help to calm down lasting suspicions of Western intentions. The tragic loss of the Russian Oscar II class missile submarine *Kursk* near the Kola Gulf in 2000 was initially blamed on a collision with a U.S. submarine. Although later investigations attributed the sinking and the loss of life to a defect on a torpedo, the 65-76, now withdrawn from service, many in the Russian navy still suspect that it was the fault of the U.S. Navy. (The *Skvall* was apparently not involved in the sinking.) Old habits die hard, especially when dealing with the Russian conspiracy mentality that has permeated Russian culture since the times of the czars.

The differences that have emerged in 1999 and in 2008 between the Atlantic alliance and Russia over the Balkans and the Caucasus leave Moscow with the impression that the West is practicing a double standard. While NATO used military force to redraw the borders of Serbia and grant its independence to Kosovo, Russia was shunned for encroaching on Georgia's sovereignty to liberate the people of Abkhazia and Ossetia. Moscow has not displayed, though, the same magnitude toward its Chechen separatists who were bloodily subdued to safeguard access to the strategic resources of the Caspian Sea. Accordingly, the official Russian naval doctrine issued in 2001 stressed the necessity to deter and oppose foreign enterprises that would disregard Russian national interests. The navy still focuses on Cold War anti-aircraft-carrier tactics to protect the approaches of Continental Russia, but the real balance is struck by the modernization of the aging strategic submarine fleet to ensure minimal deterrence. In this context, the installation of a U.S. anti-ballistic missile shield in Eastern Europe, aimed at Iran, is seen as an alteration to that balance. Despite this flagrant opposition, it is unlikely that Russia and the West will embark in a mutually disadvantageous new Cold War.

NOTES

Introduction

1. Carter P. Hydrick, *Critical Mass: The Real Story of the Atom Bomb and the Birth of the Nuclear Age* (New York: Whitehurst, 2004), pp. 127–138.

2. Richard Billings, *Battleground Atlantic: How the Sinking of a Single Japanese Submarine Assured the Outcome of World War II* (New York: NAL Caliber, 2006).

3. From a letter from Japanese ambassador Yoya Kawamura to Peter Huchthausen dated November 17, 2007.

4. Joseph Mark Scalia gives a detailed account of U-234's adventures, her full cargo, and the exploitation of her captured material and German experts in his book *Germany's Last Mission to Japan: The Failed Voyage of U-234* (Annapolis, MD: U.S. Naval Institute Press, 2000).

5. Deborah Shapely, "Nuclear Weapons History: Japan's Wartime Bomb Projects Revealed," *Science* 199 (January 1978): 155; and Robert Wilcox, *Japan's Secret War* (New York: Marlowe, 1995), p. 104.

6. Shapely, "Nuclear Weapons History," p. 155.

7. Reporter Bill Henry of the *Los Angeles Times* published the following account: "I was far enough away to see an object without being able to identify it. . . . I would be willing to bet what shekels I have that there were a number of hits scored on the object." Editor Peter Jenkins of the *Los Angeles Herald Examiner* claimed that he had seen "the V formation of twenty-five silvery planes moving slowly across the skies towards Long Beach." Five people died on the ground from heart attacks or being hit by unexploded shells. A Japanese American truck driver was arrested for failing to black out his lights. Already criticized for the unchallenged submarine attack on Santa Barbara, Secretary of the Navy Frank Knox blamed the air alarm on "jittery nerves," while General Marshall suggested to the president that "If unidentified airplanes were involved they may have been from commercial sources, operated by enemy agents for purpose of spreading alarm, [and] disclosing locations of anti-aircraft positions."

8. The type XB mine-laying submarine displaced 2,177 tons submerged, had an overall length of 89.80 meters, had a 9.20-meter beam, and could dive to a depth of 120 meters. The boat had thirty vertical mine shafts 134.6 centimeters in diameter and had been modified for a maximum cargo capacity, plus it had a fuel capacity for a range of 18,450 nautical miles.

9. This information was confirmed to the authors in October 2006 by retired Japanese ambassador and naval historian Yoya Kawamura. When Kawamura first tried to include the data on the Japanese nuclear program and the three German submarines after interviewing several U-234 crewmembers for an article, the editors of a popular Japanese periodical tried in 2006 to suppress the article. Most Japanese refuse to believe

365

that their military was making serious efforts to produce nuclear weapons of its own. In the fall of 1969, former Japanese lead pilot and Pearl Harbor planner Lieutenant Colonel Minoru Genda spoke at the U.S. Naval Institute in Annapolis at a conference that Peter Huchthausen attended. When asked whether Japan would have used an atomic bomb in 1945 had it possessed one, Genda responded, "Yes, of course we would have used an atomic bomb." Upon his return to Japan, Genda was summarily fired from his position in the Japanese Self-Defense Forces, and he retired.

10. Scalia, *Germany's Last Mission*, p. 57.

11. According to some records, in March 1945 the U.S. stock of fissionable material was short of that needed to attain a critical mass for a weapon. With the arrival of the material from U-234, the U.S stock suddenly acquired the amount that was necessary for a bomb. Furthermore, Dr. Schlicke's arrival enabled the United States to incorporate his infrared proximity fuse into the Hiroshima and Nagasaki bombs. Hydrick, *Critical Mass*, p. 23.

12. Ibid., p. 99.

13. PRO, ADM, First Sea Lord's papers, 1958, visit of Lord Mountbatten of Burma to Canada and the United States.

1. Victor's Plunder, 1941–1945

1. Minutes prior to the U-234 surrender to the U.S. Navy, the scientist and passenger Dr. Heinz Schlicke was observed by a German crewman jettisoning several small tubes of microfilm and saying, "There goes the rocket that could fly the Atlantic." Geoffrey Brooks and Wolfgang Hieschfeld, *The Story of a U-Boat NCO, 1940–1946* (Annapolis, MD: Naval Institute Press, 1996), pp. 212–213.

2. Public Records Office, Kew Gardens, ADM1/18254 report by Royal Navy Commander Curtis of 30 Advance Unit upon entering the German Walterwerke submarine development center in Kiel on May 7, 1945.

3. Quoted in Townsend Hoopes and Douglas Brinkley, *Driven Patriot: The Life and Times of James Forrestal* (New York: Vintage Books, 1993), p. 198.

4. Bradley F. Smith, *Sharing Secrets with Stalin: How the Allies Traded Intelligence, 1941–45* (Lawrence: University Press of Kansas, 1996), pp. 16–17.

5. Ibid., p. 38.

6. Ibid., p. 147.

7. Ibid., p. 160.

8. Ibid., pp. 60–61.

9. Ibid., p. 139.

10. Ibid., p. 130.

11. Ibid., pp. 130, 131, 175, 220.

12. Ibid., pp. 197, 217, 219.

13. "Ultra" was the name used by the British for intelligence resulting from decryption of coded German radio communications in World War II. The term became the standard designation in both Britain and the United States for all intelligence from high-level cryptanalytic sources. "Magic" was the United States codename for intelligence derived from the cryptanalysis of PURPLE, a Japanese foreign office cipher.

14. Smith, *Sharing Secrets with Stalin,* pp. 238, 239.

15. Ibid., p. 221.

16. Ibid., p. 220.

17. Yu L. Korshunov and A. Strokov, *Torpyedy VMF SSSR* [Soviet naval torpedoes] (St. Petersburg, Russia: Gangut, 1994), pp. 25–27.

18. Andrew Lycett, *Ian Fleming* (London: Phoenix, 1995), pp. 138–158.

19. Ibid., p. 145.

20. Tom Bower, *The Paperclip Conspiracy: The Hunt for the Nazi Scientists* (Boston: Little, Brown, 1987), pp. 79–80; plus, the Navy Yard orders establishing the Naval Technical Mission.

21. Bower, *The Paperclip Conspiracy*, p. 78.

22. Carter P. Hydrick, *Critical Mass: The Real Story of the Atom Bomb and the Birth of the Nuclear Age* (New York: Whitehurst, 2004), p. 52.

23. Later, in testimony before investigators while the esteemed scientists von Braun and Wagner were applying for U.S. citizenship, they stated that although they knew workers had died, they had nothing to do with the forced labor. Dr. von Braun offered the fact that his one-time arrest by the Gestapo and subsequent release mitigated his functions as an SS officer. Most likely, his arrest was a result of jealousy by SS leader Heinrich Himmler, who resented the special treatment that Hitler accorded to all German scientists and had been fighting to wrest full control of the secret wonder weapons production from von Braun. It is surprising that almost none of the German scientists captured at the end of the war, despite their brilliant minds, admitted knowledge of the death camps that were attached to the rocket construction sites.

24. PRO ADM 1/18254 Admiralty Report, *The Final Stages of the War in North West Europe*, p. 22.

25. Bower, *The Paperclip Conspiracy*, pp. 97–98.

26. Ibid., p. 99

27. Ibid., pp. 99–100.

28. L. N. Mikhailov, *Admiral Flota Sovietstkogo Soyuza N.G. Kuznetsov* (St. Petersburg: Sudostroynye, 2004), pp. 241–279.

29. Jurgen Rowher and Mikhail Monakov, *Stalin's Ocean Going Fleet* (London: Cass, 2001), pp. 180–184.

30. Ibid.

31. Ibid., p. 205.

32. Franz Kurowski, *Allieierte Jagt auf deutsche Wissenschaftler, Das Unternehmen Paperclip* (München, Germany: Kristall bei Langen Müller, 1982), pp. 120–122.

33. Boris Chertok, *Rockets and People* (Washington, DC: NASA History Division, Office of External Relations, January 2005), NASA SP-2005–4110.

34. Botho Stuewe, *Peenemuende West, Die Erprobungsstelle der Luftwaffe fuer geheime Fernlenkwaffen und deren Enwicklungsgeschichte* (Augusgburg: Bechtermuenz, 1998), pp. 330–413.

35. Peter Huchthausen's interview with Dr. Hans Muehlbacher and Mrs. Brigitte Wagner, the widow of Herbert Wagner, Vienna, Austria, October 20–23, 2006.

36. Stuewe, *Peenemuende West*, pp. 330–413.

37. Norman Polmar and K. J. Moore, *Cold War Submarines: The Design and Construction of U.S. and Soviet Submarines* (Washington, DC: Brassey's, 2004), p. 103.

38. Chertok, *Rockets and People*, pp. 289–290, 291.

39. P. Kennedy, *Germany's V-2 Rocket* (Atgen, PA: Schiffer, 2006), pp. 94–95.

40. Ibid., p. 96.

41. PRO, ADM 178/404, Intelligence Report from HMS *Royal Rupert*, Wilhemshaven, to the DNI, February 1, 1946.

2. Penetrations, 1945–1952

1. Between August 7 and August 10, 1945, a group of eleven Soviet experts was allowed by Naval Intelligence to tour the New York Shipbuilding Company; the Bethlehem Shipbuilding Company in Staten Island; Westinghouse Electric Company in Pittsburgh; and the Aluminum Company of America in New Kensington, Pennsylvania. And between December 1944 and November 1945, Soviet personnel were also admitted into twenty-four naval facilities, including the Naval Gun Factory at the Washington Navy Yard; the Naval Research Laboratory; the David Taylor Model Basin, the Hydrographic Office in the Washington area; the Naval Proving Ground at Dahlgren, Virginia; the Engineering Experiment Station in Annapolis, Maryland; the Fleet Sonar School in Key West, Florida; and various other, less sensitive, sites.

2. NARA 1945 907003, Box 16, Memorandum to the Naval Attaché, American Embassy, Moscow: American Scientific and Technical Personnel now in the USSR, January 16, 1946.

3. CNO to naval attachés, Washington, D.C., January 29, 1946.

4. Amy Knight, *How the Cold War Began: The Igor Gouzenko Affair and the Hunt for Soviet Spies* (New York: Carroll and Graf, 2005), pp. 199–200.

5. PRO, *Monthly Intelligence Report*, April 1946, pp. 47–49.

6. PRO, *Monthly Intelligence Report*, August 1946, pp. 76–77.

7. Richard J. Aldrich, *The Hidden Hand: Britain, America, and Cold War Secret Intelligence* (London, John Murray, 2001), p. 109.

8. Robert Louis Benson and Michael Warner, *VENONA Soviet Espionage and the American Response, 1939–1957*, NSA, 1996 booklet, www.nsa.gov/museum/index.html, VENONA Historical Monograph #4.

9. NARA 1945 907003, Box 16, Chief of Naval Intelligence to Distribution List, Secret, August 9, 1946.

10. Ibid.

11. Wyman H. Packard, *Century of U.S. Naval Intelligence* (Washington, DC: Naval Historical Center, 1996), pp. 256–257.

12. *ONI Review* (July 1951), pp. 290–292.

13. Steven T. Usdin, *Engineering Communism* (New Haven, CT: Yale University Press, 2005), pp. 111–113.

14. Ibid., pp. 114–143.

15. NARA 907003, Box 16, Chief of Naval Intelligence to All U.S. Naval Attachés and Naval Liaison Officers, February 14, 1947.

16. See "Part I: The American Response to Soviet Espionage," the Central Intelligence Agency Web site, https://www.cia.gov/library/center-for-the-study-of-intelligence/csi-publications/books-and-monographs/venona-soviet-espionage-and-the-american-response-1939–1957/part1.htm.

17. Packard, *Century of U.S. Naval Intelligence*, p. 258.

18. NARA 907003, Box 16, Chief of Naval Intelligence to Distribution List, February 13, 1948.

19. NARA 907003, Box 16, Chief of Naval Intelligence to U.S. Naval Attaché, Mexico, Greece, Moscow, Paris/France, Rome/Italy; Assistant U.S. Naval Attaché, Odessa; Intelligence Officer, Cinclant, U.S. Naval Forces Mediterranean; Officer in Charge, Atlantic Fleet, Intelligence Center, Norfolk; District Intelligence Officer, Third Naval District, September 16, 1947.

20. NARA 907003, Box 16, Assistant Naval Attaché, Odessa, to U.S. Naval Attaché, Moscow, March 7, 1947.

21. NARA, NND 957307 RG38 POW Desk, Box 4, ComNavGer to ONI, *Information Report on Poland—Communications in Merchant Marine*, October 19, 1953.

22. NARA, NND 957307 RG38 POW Desk, Box 3, *Air Intelligence Information Report*, 7050 AIS to ONI, September 11, 1952.

23. *ONI Review* (February 1951): p. 84.

24. *ONI Review* (November 1951): pp. 473–474.

25. PRO, ADM 223, *Quarterly Intelligence Report*, April to June 1955, pp. 47–48.

26. OENO party representatives and operation centers were in Antwerp, Cardiff, Liverpool, London, Dublin, Rotterdam, Hamburg, Copenhagen, Trieste, Genoa, Marseilles, Rouen, Le Havre, Alexandria, Port Said, Buenos Aires, Rosario, Rio de Janeiro, New York, San Francisco, and Calcutta. More were also active in Canada and Australia.

27. NARA, NND 957307 RG38 POW Desk, Secret, Security Information, Box 4.

28. *ONI Review* (December 1951): p. 490.

29. Aldrich, *The Hidden Hand*, p. 120.

30. Christopher Andrew and Vasili Mitrokhin, *The Mitrokhin Archive and the Secret History of the KGB* (London: Basic Books, 1999), p. 600.

31. Renaud Muselier, *L'amiral Muselier* (Paris: Perrin, 2000), pp. 194–195; also Georges Fleury, *De Gaulle, De l'enfance à l'appel du 18 juin* (Paris: Flammarion, 2007), pp. 13–17.

32. Interview with Captain Claude Huan, French navy retired, Vincennes, April 15, 2008.

33. Peter Wright with Paul Greengrass, *Spycatcher: The Candid Autobiography of a Senior Intelligence Officer* (New York: Dell, 1987), p. 239; Thierry Wolton, *Le grand recrutement* (Paris: Grasset, 1993), pp. 231–235.

34. Jean-Paul Eyrard, *L'Eminence rouge, Un opposant au général de Gaulle au sein de la France libre: le capitaine de vaisseau Raymond Moullec, dit Moret*, unpublished manuscript, 2004, p. 47.

35. Wolton, *Le grand recrutement*, pp. 231–235; also Wolton, *Le KGB en France* (Paris: Grasset, 1986), pp. 168, 202.

36. Eyrard, *L'Eminence rouge*, p. 138.

37. Wolton, *Le grand recrutement*, p. 241.

38. Eyrard, *L'Eminence rouge*, p. 138.

39. Ibid.

40. Aldrich, *The Hidden Hand*, p. 147.

41. Ibid.

42. Ibid., p. 148.

43. Ibid., p. 173.

44. *ONI Review* (April 1951): p. 169.

45. Ibid., p. 176.

46. Ibid., p. 177.

47. NARA 907003, Box 16, Memorandum for Mr. Reinhardt, Moscow, USSR, January 17, 1948.

48. *ONI Review* (April 1951): pp. 166–167, 169.

49. NARA, NND 957307 RG38 POW Desk, Box 4, CIA Seattle to ONI, June 23, 1953.

50. NARA, NND 957307 RG38 POW Desk, Box 3, *Air Intelligence Information Report*, 7050 AIS to ONI, September 11, 1952.

51. NARA, NND 957307 RG38 POW Desk, Secret, Security Information, Box 4, Foreign Service Dispatch, Am. Embassy, Tel Aviv to the Department of State, August 12, 1955.

52. NARA, NND 957307 RG38 POW Desk, Box 3, DIO-4ND to ONI, *Information Pertinent to the Port of Gdynia*, June 5, 1951.

53. NARA, NND 957307 RG38 POW Desk, Box 3, COMNAVGER to ONI, *Background Information on the Most Popular Brothel of Wismar*, November 23, 1951.

54. NARA, NND 957307 RG38 POW Desk, Box 4, various reports.

55. Packard, *Century of U.S. Naval Intelligence*, p. 195.

56. Ibid., p. 196.

57. Mario de Arcangelis, *Electronic Warfare: From the Battle of Tsushima to the Falklands and Lebanon Conflicts* (Dorset, UK: Blandford Press, 1985), p. 122.

58. Packard, *Century of U.S. Naval Intelligence*, p. 196.

59. Ibid.

60. Radar and electromagnetic emissions associated with the testing of foreign civilian and military systems such as telemetry, beaconry, electronic interrogators, tracking, arming, fusing, and command signals, video data links in Gerald K. Haines and Robert E. Leggett eds., *Watching the Bear: Essays on CIA's Analysis of the Soviet Union* (Langley, VA: Center for the Study of Intelligence, Central Intelligence Agency, 2003), p. 111.

61. De Arcangelis, *Electronic Warfare*, p. 109.

62. Packard, *Century of U.S. Naval Intelligence*, pp. 113–114.

63. Sherry Sontag and Christopher Drew, *Blind Man's Bluff: The Untold Story of American Submarine Espionage* (New York: Public Affairs, 1998), pp. 23–24.

64. "New Cold War Revelations: British Used Spy Trawler," *Sea Power*, March 1998. See also "The Gaul Mysteries," www.offmsg.connectfree.co.uk/OffSHELF/offRAMES.htm.

65. Ibid.

66. *Junge Welt*, DDR, October 29–30, 1966; *Junge Welt*, DDR, November 19–20, 1966.

67. Indrek Juro, *Operations of Western Intelligence Services and Estonian Refugees in Post-War Estonia and the Tactics of KGB Counterintelligence: The Anti-Soviet Resistance in the Baltic States* (Vilnius: Genocide and Research Center of Lithuania, 1999).

68. Stephen Dorril, *MI-6: Inside the Covert World of Her Majesty's Secret Intelligence Service* (New York: Free Press, 2000), p. 200.

69. Tom Bower, *The Red Web, MI6 and the KGB Master Group* (London: Aurum Press, 1989), pp. 101ff; Zanis Vasilevski, *Armed National Resistance Units—the Medium of the Games of the SSSR/Latvian SSR and Great Britain Special Operative Services in 1945–1956* (Genocidas ir Reziztancia, Nr. 2, 1997).

3. Allies to Antagonists, 1945–1952

1. The January 26, 1934, pact of nonaggression that was concluded between Poland and Germany, followed by the November 25, 1936, Anti-Komintern Pact signed by Berlin and Tokyo and their respective revocation of the naval treaties, exposed the impotence of the Soviet fleet in its ability to support the Spanish Republic.

2. Jurgen Rowher and Mikhail Monakov, *Stalin's Ocean Going Fleet* (London: Frank Cass, 2001), pp. 58, 67.

3. Natalia I. Yegorova, "Stalin's Conception of Maritime Power: Revelations from the Russian Archives," *Journal of Strategic Studies* (April 2005): 157–186.

4. Ibid., p. 159.

5. V. P. Kuzin and V. I. Nikol'skii, *Voenno-Morskoi Flot SSSR 1945–1991: Istoriia sozdaniia poslevoennogo Voenno-Morskogo Flota SSSR i vozmozhnyi oblik flota Rossii* (Saint-Petersburg: Istoricheskoe morskoe obshchestvo, 1996), p. 18.

6. May 19, 1946, decree of the Soviet Council of Ministers No. 1017–419, on "Questions of Rocket Weapons," in Kuzin and Nikol'skii, ibid., p. 18.

7. Ibid., p. 62.

8. Rowher and Monakov, *Stalin's Ocean Going Fleet*, p. 190.

9. Ibid., p. 191.

10. Yegorova, "Stalin's Conception of Maritime Power," pp. 159, 165.

11. Kuzin and Nikol'skii, *Voenno-Morskoi Flot SSSR 1945–1991*, p. 20.

12. Rowher and Monakov, *Stalin's Ocean Going Fleet*, pp. 221–224.

13. Ibid., p. 191.

14. *ONI Review* (January 1946): p. 13.

15. NARA 907003, Box 16, Office of the U.S. Naval Attaché, American Embassy, Moscow, Office Memorandum: Organization for Preparing Reports, November 20, 1947.

16. NARA 907003, Box 16, Chief of Naval Intelligence to Distribution List, including U.S. Naval Attaché, Moscow, USSR, September 30, 1948.

17. Within the U.S. naval attaché office, the work was shared as follows: Stevens would read the main newspapers—*Pravda*, *Literaturnaya Gazyeta*, *Ogonyok*, and *Krokodil*—as well as the State Department and military attaché dispatches. His deputy, Lieutenant-Colonel McMillan, USMC, exploited *Komsomolskaya Pravda*, *Isvestiya*, *Vechernyaya Moskva*, and the news translations; he collected data on landing forces and amphibious warfare, the military geography of coastal areas, hydrography and oceanography, resources and trade, industry and science in general, coastal cities and towns, climatology and meteorology, the arctic area and the far north, and port facilities. Lieutenant Levy's area of responsibility included inland waterways, and cable and radio communications, merchant marine, shipyards, and shipbuilding; accordingly, his readings were *Trud*, *River Transport*, and *Morskoi Flot*. Ensign Golloway reported on the Soviet navy and air force, atomic energy, guided missiles, aircraft construction, jet propulsion, industry, and science as applied to the navy. In order to fulfill this agenda, he had to process the pertinent data in *Krasniy Flot*, *Krasnaia Zvezda*, *Nauka I Zhizn'*, and *Moskovski Bolshevik*.

18. NARA 907003, Box 16, Office of the U.S. Naval Attaché, American Embassy, Moscow, Office Memorandum: Organization for Preparing Reports, November 20, 1947.

19. These included twenty first-class science awards of 200,000 rubles each; twenty second-class science awards of 100,000 rubles each; and thirty "first class invention and basic improvements in productive work awards" of 150,000 rubles each.

20. A. Petrzhak Kiril and G. N. Flerov, another senior scientific research worker, at the Academy of Sciences; NARA 1945 907003 Box, AGWAR/USFET to Bissel WDGBI, January 30,1946.

21. NARA 907003, Box 16, Chief of Naval Intelligence to U.S. Naval Attaché, Moscow, USSR, May 12, 1947.

22. NARA 907003, Box 16, Maples to the Chief of Naval Intelligence, Exchange of Courtesies and Information with USSR, Moscow, June 14, 1947.

23. NARA 907003, Box 16, Chief of Naval Intelligence to U.S. Naval Attaché, Moscow, USSR, September 11, 1947.

24. Marius Peltier, *Attaché naval à Moscou* (Paris: Éditions France Empire, 1954), p. 54.

25. Ibid., pp. 69–70.

26. NARA 907003, Box 16, Assistant U.S. Naval Attaché, Odessa to Intelligence Officer, Moscow, USSR, December 5, 1946.

27. NARA 907003, Box 16, Airgram, Embassy Moscow to State Department, January 18, 1948.

28. NARA 907003, Box 16, U.S. Naval Attaché, Moscow, USSR, to Chief of Naval Intelligence April 10, 1947.

29. NARA 907003, Box 16, undated document on the naval schools.

30. NARA 907003, Box 16, Intelligence Report 232–8–48, Intelligence Division, Office of the Chief of Naval Operations, Navy Department, July 13, 1948.

31. NARA 907003, Box 16, Assistant U.S. Naval Attaché, Odessa to Intelligence Officer, Moscow, USSR, December 5, 1946.

32. NARA 907003, Box 16, Assistant Naval Attaché, Odessa to Chief of Naval Operations, September 30, 1946.

33. NARA 907003, Box 16, U.S. Consulate, Vladivostok to USNA Moscow, July 20, 1948.

34. NARA 1945 907003 Box, Harshaw to Maples, Odessa, April 17, 1946.

35. NARA 1945 907003 Box, Poullard to Maples, Vladivostock, January 22, 1946.

36. NARA 907003, Box 16, Assistant Naval Attaché Odessa to U.S. Naval Attaché, Moscow, March 5, 1947.

37. NARA 907003, Box 16, Assistant U.S. Naval Attaché Odessa to U.S. Naval Attaché, Moscow, May 29, 1947.

38. PRO, Naval Intelligence Department, Monthly Intelligence Report, August 1946, p. 89.

39. NARA 1945 907003 Box, Poullard to Maples, Vladivostock, January 22, 1946.

40. NARA 907003, Box 16, Navy Department, Office of the Chief of Naval Operations, Attachés in USSR and USA, comparison of treatment of. Instructions for Naval and Air Attachés and Assistant Attachés Accredited to the Navy Department, March 15, 1947.

41. *ONI Review* (April 1951): pp. 166–169.

42. NARA 907003, Box 16, Report on Visit to Riga during March 1947 by Captain S. B. Frankel, USN.

43. NARA 907003, Box 16, Report on Air and Rail journey from Moscow via Tiflis, Erivan, and Leninakan to the Turkish Border, April 19 to 23, 1947, submitted by Captain S. B. Frankel, USN.

4. The Korean War, 1950–1953

1. Joint Intelligence Committee, March 1, 1946, in P. Hennessy, *The Secret State: Whitehall and the Cold War* (London: Allen Lane, 2002), p. 26, n. 99.

2. Richard J. Aldrich, *The Hidden Hand: Britain, America, and Cold War Secret Intelligence* (London: John Murray, 2001), p. 275.

3. Curtis A. Utz, *Assault from the Sea: The Amphibious Landing at Inchon,* No. 2 in the series *The U.S. Navy in the Modern World* (Washington, DC: Naval Historical Center, 1994).

4. Aldrich, *The Hidden Hand*, p. 286.

5. Edward J. Marolda, "Cold War to Violent Peace," in W. J. Holland Jr., ed., *The Navy* (Washington, DC: Naval Historical Foundation, 2000).

6. *ONI Review* (March 1951): p. 118.

7. NARA, NND 957307 RG38 POW Desk, Secret, Security Information, Box 4, U.S. Naval Attaché, Seoul, Korea, to ONI, December 25, 1950.

8. Edward J. Marolda, "The Hungnam and Chinnampo Evacuations," in Spencer C. Tucker, ed., *Encyclopedia of the Korean War: Political, Social, and Military History* (Santa Barbara, CA: ABC-CLIO, 2000). Air force and marine aircraft airlifted out another 3,600 troops, 1,300 tons of cargo, and 196 vehicles.

9. Ibid.

10. *ONI Review* (August 1951): p. 317.

11. *ONI Review* (March 1951): p. 119–120.

12. Wyman H. Packard, *Century of U.S. Naval Intelligence* (Washington, DC: Naval Historical Center, 1996), p. 182.

13. Aldrich, *The Hidden Hand*, p. 279.

14. NARA, NND 957307 RG38 POW Desk, Box 4, General Headquarters Far East Command, Military Intelligence Service Group, Enemy Documents, Korean Operations, General Staff, August 10, 1951.

15. *ONI Review* (January 1951): p. 10–11.

16. NARA, NND 957307 RG38 POW Box 3, Information Report, Commander Naval Forces Far East to the Office of Naval Intelligence, October 8, 1951.

17. NARA, NND 957307 RG38 POW Desk, Box 3, Information Report, Commander Naval Forces Far East to the Office of Naval Intelligence, February 22, 1951.

18. *ONI Review* (January 1951): p. 10–11.

19. NARA, NND 957307 RG38 POW Desk, Secret, Security Information, Box 3, Commander Naval Forces Far East to ONI, Prisoner of War Interrogation, Soviet Magnetic Homing Torpedo and North Korean Naval Activity, PW Interrogation, January 10, 1953.

20. Ibid.

21. Packard, *Century of U.S. Naval Intelligence*, p. 123.

22. U.S. Navy Historical Center, Naval Operational Archives, *Sea-Based Airborne Antisubmarine Warfare 1940–1977*, vol. 1, 1940–1960, February 17, 1978, prepared for Op-095 under ONR Contract N00014–77-C-0338, 2nd ed., R. F. Cross Associates, Declassified, p. 101.

23. NARA, NND 957307 RG38 POW, Box 3, Allied Translator and Interpreter Section, Research Supplement, Interrogation Report no. 105, Soviet Submarine Facilities, General Headquarters Far East Command, Military Intelligence Section, General Staff, August 10, 1951.

24. Ibid.

25. U.S. Navy Historical Center, Naval Operational Archives, *Sea-Based Airborne Antisubmarine Warfare 1940–1977*, p. 106.

26. Ibid., p. 108.

27. Ibid., p. 116–117.

28. Sergo Beria and Francoise Thom, *Beria, My Father* (London: Duckworth , 2001), pp. 222–232.

29. Aldrich, *The Hidden Hand*, p. 292.

30. Ibid.

5. Requiem for a Battleship, 1955

1. *ONI Review* (1958): p. 116.

2. Junio Valerio Borghese gives his account of his World War II swimmer operations in *Decima Flottiglia Mas* (Milano: Garzanti, 1950).

3. O.P. Bar-Biryukov, *Chas X dlya linkora* Novorossiysk [X hour for battleship *Novorossiysk*] (Moscow: Centerpoligraph, 2006), pp. 277–279.

4. Ibid., pp. 277–279.

5. Ibid., p. 279.

6. Boris Karzhavin, *The Loss of the Otvazhny* (St. Petersburg: Korvet, 1994), pp. 3–20.

7. Ibid., p. 227.

8. Norman Friedman, *Guide to the World's Naval Weapons* (Annapolis, MD: Naval Institute, 2005), p. 24.

9. Alexander I. Kolpakidi and Dmitri Prokorov, *Imperia GRU, Otcherkiy historii rosiskoi voennoi ravedkii* (Izdatelstvo: Olma Press, 1999), pp. 142–146.

10. Ibid., pp. 142–146. The four major naval *spetsnatz* units are 4 INDEP SPETZNAZ PT Parusnoe (Baltyysk) (Baltic Fleet), formerly Viljandi, Estonia (transferred from Army GRU to Navy GRU); 431 INT SPETZNAZ PT Tuapse (Black Sea Fleet), formerly Kronstadt (Baltic Fleet); 42 SPETZNAZ PT Russkyy island (Pacific Fleet); and 420 INT SPETZNAZ PT Polyarnyy (Northern Fleet).

11. V. Vorobieva, "Morskie mlekopitayuschie v vooroujennoi bor'be na more," *Morskoi Sbornik*, March 2005.

6. Khrushchev and Crabb, 1953–1960

1. PRO, ADM 223, *Quarterly Intelligence Report*, January to March 1956, pp. 24–25.

2. Ibid., pp. 24–25.

3. Nikita Khrushchev, *Memoirs of Nikita Khrushchev, Volume 2: Reformer, 1945–64* (Institute of International Studies: Brown University, August 2006), p. 441.

4. PRO, ADM 223, *Quarterly Intelligence Report*, April to June 1956.

5. *ONI Review* (March 1954): p. 83.

6. PRO, ADM 223, *Quarterly Intelligence Report*, January to March 1955, p. 42.

7. Ibid., p. 42.

8. Ibid., p. 30.

9. Ibid., p. 4.

10. Ibid., p. 4.

11. PRO, ADM 223, *Quarterly Intelligence Report*, July to September 1956, pp. 46–50.

12. Ibid., pp. 46–50.

13. PRO, ADM 223, *Quarterly Intelligence Report*, October to December 1955, p. 53.

14. Ibid., p. 61.

15. PRO, ADM 223, *Quarterly Intelligence Report*, April to June 1957, p. 9.

16. PRO, ADM 223, *Quarterly Intelligence Report*, October to December 1955, p. 60.

17. Interview to BBC, *Inside Out*, January 19, 2007.

18. Norman Friedman, *British Destroyers and Frigates: The Second World War and After* (Newbury, UK: Greenhill Books, 2006).

19. Interview to BBC, *Inside Out*, January 19, 2007.

20. PRO, ADM 223, *Quarterly Intelligence Report*, January to March 1957, p. 4.

21. PRO, ADM 223, *Quarterly Intelligence Report*, April to June 1956, pp. 38–48.

22. Ibid.

23. Ibid.

24. Ibid.

25. Ibid., pp. 35–37.

26. Ibid., pp. 38–48.

27. Mario de Archangelis, *Electronic Warfare: From the Battle of Tsushima to the Falklands and Lebanon Conflicts* (Dorset, UK: Blandford Press, 1985).

28. Khrushchev, *Memoirs,* p. 455.

29. PRO, FO 371/122885, Letter from the Foreign Office to the Soviet ambassador in London, May 8, 1956.

30. Why was Commander Crabb diving in the close vicinity of the Soviet cruiser, which was there on a friendly visit? Why and under whose authority was a police officer sent to the hotel at which Commander Crabb was staying? Why did this police officer order that the leaves with the names of Commander Crabb and the man he stayed with be torn from the register? What was the name of that other man and why did the police officer threaten the hotel keeper with action under the Official Secrets Act if he did not allow the register to be altered? PRO, FO 371/122885.

31. PRO, FO 371/122885, Wednesday, May 9, 1956, Statement by the Prime Minister; Sir John Sinclair, the head of MI6 (1953–1956), asked his Foreign Office liaison officer Michael Wright to clear the operation with the Foreign Office. The Foreign Office later claimed that the operation had not been cleared. Sinclair had to resign.

32. PRO, FO 371/122885, from Moscow to Foreign Office, Immediate, DEDIP, Top Secret, departed. 2:22 p.m. May 10, 1956; received. 4:27 p.m. May 10, 1956, following for Sir I. Kirkpatrick.

33. PRO, FO 371/122885, Telegram, departed 10:35 a.m., May 11, 1956, received 11:53 a.m., May 11, 1956.

34. PRO, FO 371/122885, Confidential, from Moscow to Foreign Office, departed. 2:45 P.M., May 16, 1956; received. 4:33 P.M., May 16, 1956.

35. PRO, FO 371/122885, Mr. R. Lambert to the Soviet Embassy, London, July 4, 1956.

36. Interview of Joseph Zverkin by Yigal Serna, quoted at *Divernet* magazine online, www.divernet.com/cgi-bin/articles.pl?id=3072&sc=1040&ac=d&an=3072:How+Buster+Crabb+died.

37. London, *Times*, November 16, 2007.

38. James Rusbridger, *The Intelligence Game: The Illusions and Delusions of Intelligence Espionage* (London: Tauris, 1991), p. 67.

39. "Buster Crabb—the 'Spy' from the Cold War?" Interview on the BBC program *Inside Out*, January 19, 2007.

40. Khrushchev, *Memoirs*, p. 449.

41. SHD/Marine, III BB 2 SEC 112, Captain Poncet, French embassy in Egypt to Captain de Geffrier, Joint Staff 2nd division, and Captain Jouslin, Naval Staff 2nd bureau, Cairo, August 6, 1956.

42. SHD/Marine, 136GG^2 5, Amman papers, Joint Staff 2nd division; Intelligence report from Cairo, August 27, 1956.

43. The figures included seven cruisers, fourteen destroyers, eight (old) destroyers, twelve escorts, three (old) escorts, forty-four long-range submarines, five medium-range submarines, and fifteen short-range submarines. SHD/Marine, Fonds Suez EMG 4, Captain Greffier, chief of the Joint Staff 2nd division, *Possible role for the Soviet Union in Suez,* July 30, 1956.

44. SHD, 136GG^2 4, Papers Amman Telegram from Naval France London 05.16.30/Z to "Marine Paris," September 5, 1956, Top Secret, urgent/operation.

45. SHD, 136GG^2 4, Papers Amman Telegram from Naval France London 12.16.20/Z to "Marine Paris," September 12, 1956, Top Secret, routine.

46. SHD, 136GG^2 5, *Report of Operations in the Middle East,* no. 1011 CCFFO/TS, Vice Admiral P. Barjot, CINC, French Forces in the East to the Minister of National Defense, March 21, 1957, p. 13.

47. SHD, Fonds Suez, CFS, Message EMG/2/S, *Submarines in the Red Sea— Activity at Cheikh Said,* September 18, 1956, Lieutenant Commander Labrousse commanding the navy on the Somali Coast, to the armed forces secretary.

48. SHD, II BB 401 23, Submarine spotted by Lieutenant Guillaume on October 11–12, 1956, no. 112 EM2, Djibouti December 21, 1956, Lieutenant Commander Labrousse commanding the navy on the Somali Coast, to the armed forces secretary.

49. Philippe Masson, *La Crise de Suez (Novembre 1956–Avril 1957)* (Paris: Marine Nationale, Etat-Major général, Service historique, 1966), p. 152.

50. SHD, 136GG^2 5, Papiers Amman, Lettre no. 242A/DN/TS du CA Amman, Attaché Naval à Londres au Général Sir Charles Keightley, Londres, Octobre 8, 1956.

51. SHD, 136GG^2 5, *Report of Operations in the Middle East,* no. 1011 CCFFO/TS, Vice Admiral P. Barjot, CINC, French forces in the Middle East, to Mr. Defense Minister, March 21, 1957, pp. 116–119.

52. PRO, ADM 223, *Quarterly Intelligence Report,* October to December 1956, p. 35.

53. SHD, 136GG^2 5, *Report of Operations in the Middle East,* no. 1011 CCFFO/TS, Vice Admiral P. Barjot, CINC, French forces in the Middle East, to Mr. Defense Minister, March 21, 1957.

54. PRO, ADM 223, *Quarterly Intelligence Report,* October to December 1957, p. 13.

55. U.S. Navy Historical Center, Naval Operational Archives, *Sea-Based Airborne Antisubmarine Warfare 1940–1977,* vol. 1, 1940–1960, February 17, 1978, Prepared for Op-095 under ONR Contract N00014–77-C-0338, 2nd ed., R. F. Cross Associates, Declassified, p. 166.

56. Ibid., p.151.

57. PRO, ADM 223, Quarterly Intelligence Report, October to December 1957, pp. 3–6.

58. Almost regardless of cost, maximum scientific effort has been put into a carefully selected number of fields: nuclear energy and all of its applications; guided missiles of all types, including earth satellites; aircraft and engines; electronic techniques; machine tool design and production; control engineering; high voltage DC transmission; large hydroelectric schemes; the optical industry; and the watch and clock industry.

59. PRO, ADM 223, *Quarterly Intelligence Report,* October to December 1957, pp. 3–6.

60. U.S. Navy Historical Center, Naval Operational Archives, *Sea-Based Airborne Antisubmarine Warfare 1940–1977,* p. 147.

61. Ibid., p. 149.

62. Ibid., p. 151.

63. Ibid., p. 152.

64. PRO, ADM 223, *Quarterly Intelligence Report*, April to June 1957, pp. 14–15.

65. U.S. Navy Historical Center, Naval Operational Archives, *Sea-Based Airborne Antisubmarine Warfare 1940–1977*, p. 152.

66. Ibid., p. 153.

67. Ibid., p. 170.

68. Ibid.

69. Admiral Jerauld Wright, USN.

70. *Washington Post,* October 19, 1958, "Value of Anglo-US Exchange of Nuclear Secrets."

7. Anatomy of Treason, 1958–1964

1. Interview with Lev Vtorygin on Frye Island, Maine, September 25, 1995.

2. Krasov's written account of the Artommanov decection was provided to Peter Huchthausen in 1995.

3. Interview with Lev Vtorygin on Frye Island, Maine, September 25, 1995.

4. Nick Shadrin's real name was Nikolai Fedorovich Artamonov. He become an intelligence celebrity after his defection. Nick was employed as a consultant to ONI and DIA. Accounts of the entire Shadrin episode, from the time he fled his ship near Gdynia, Poland, in June 1959 until his disappearance in Vienna in December 1975 are contained in Henry Hurt, *Nick Shadrin, the Spy Who Never Came Back* (New York: McGraw Hill/Reader's Digest Press, 1981) and William R. Corson and Susan B. Trento, *Widows* (New York: Crown Publishers, 1989).

5. Interview with Lev Vtorigyn, Moscow, July 2005.

6. Robert Warren Herrick wrote two books that were apparently heavily influenced by his close personal contact with the defector Nick Shadrin. Both books drew criticism from "hard line" naval intelligence officers for seeming to contradict the standard view of the growing Soviet navy threat. The books are *Soviet Naval Strategy* (Annapolis, MD: U.S. Naval Institute Press, 1968) and *Soviet Naval Theory and Policy, Gorskov's Inheritance* (Annapolis, MD: U.S. Naval Institute Press, 1988).

7. *New York Times,* June 2, 1967, p. 7.

8. Thierry Wolton, *Le KGB en France* (Paris: Grasset, 1986).

9. The novel *Topaz* by Leo Uris is derived from the confessions of French SDECE agent Philippe Thyraud de Vosjoli, who lived in exile in Mexico allegedly to escape persecution from Soviet moles in the French security apparatus.

10 Interview with French counterintelligence veteran Claude Faure, Paris, May 2008.

11. Michail Boltunov, *Agentuoi GRU Ustanovleno* (Moscow: Russkaya razvedka, 2003), pp. 160–209.

12. Ibid.

13. Ibid., p. 139

14. Ibid.

15. Ibid., pp. 213–214.

16. Ibid.

17. Nigel West, *Venona: The Greatest Secret of the Cold War* (London: HarperCollins Publishers Ltd, 1999).

18. Yevgeny Ivanov and Gennady Sokolov, *The Naked Spy* (Blake Publishing, 1994), p. 230.

19 Ibid.

20. Ibid., p. 245.

21. Ibid., p. 247.

22. Philip Knightley and Caroline Kennedy, *An Affair of State: The Profumo Case and the Framing of Stephen Ward* (New York: Atheneum, 1987), p. 53.

23. Ibid., p. 87.

24. Ibid.

25. Ivanov and Sokolov, *The Naked Spy*, p. 197.

26. Andrew Roth, Obituary of John Profumo, *Guardian* (London), guardian.co.uk, March 10, 2006.

27. Ivanov and Sokolov, *The Naked Spy.*

28. Interview with Lev Vtorigyn, Moscow, June 2007.

8. Cuba, 1962

1. Interview with Lev Vtorigyn, Moscow, July 2005.

2. The book *October Fury* by Peter Huchthausen, published by John Wiley & Sons, 2002, has been printed in the United States, Japan, and Russia. The U.S. Naval Historical Center called the information in the book unique concerning the presence of nuclear-tipped torpedoes aboard the Soviet submarines, and said that the Soviets' guidelines for their rules of engagement in using these nuclear-tipped torpedoes, either launched from submarines or ashore in Cuba, had never been heard of before. Former Secretary of Defense Robert McNamara said in September 2002 that neither he nor President John F. Kennedy had any knowledge that the Soviets had already deployed the nuclear weapons or that they had received permission in advance to launch them if attacked by U.S. forces.

3. NSA, *Top Secret Daunt*, Weekly Comint Economic Briefing, October 5, 1960, www.nsa.gov/public/publi00003.cfm, Cuban Missile Crisis Document Archives.

4. NSA, Secret Kimbo, electrical release, February 1, 1961, dist: 0/D, Spanish-speaking pilot noted in Czechoslovak air activity at Trencin, January 17, 1961, CIA. www.nsa.gov/public/publi00003.cfm, Cuban Missile Crisis Document Archives.

5. Wyman H. Packard, *Century of U.S. Naval Intelligence* (Washington, DC: Naval Historical Center, 1996), p. 88.

6. Ibid.

7. *Inspector General's Bay of Pigs*, text of the CIA Inspector General's 1962 report on the Cuban operation, released to the public on February 21, 1998. Associate Deputy Director for Plans Richard Bissell's rebuttal is available in Peter Kornbluh, ed., *Bay of Pigs Declassified: The Secret CIA Report on the Invasion of Cuba* (New York: New Press, 1998).

8. NSA, chatter between San Antonio de los Banos and Rancho Boyeros, May 23, 1961, in *Secret Sabre*, www.nsa.gov/public/publi00003.cfm, Cuban Missile Crisis Document Archives.

9. NSA, operator conversation between two unknown stations, June 18, 1961, www.nsa.gov/public/publi00003.cfm, Cuban Missile Crisis Document Archives.

10. NSA, May 2, 1962, dry-cargo shipments to and from Cuba in Soviet ships, January 1–March 31, 1962, www.nsa.gov/public/publi00003.cfm, Cuban Missile Crisis Document Archives.

11. NSA, *Secret Sabre*, April 18, 1962, electrical release distribution.

12. NSA, May 17, 1962, electrical release distribution, Cuban air force VHF communications procedure, www.nsa.gov/public/publi00003.cfm, Cuban Missile Crisis Document Archives.

13. This occurred on November 30, 1961.

14. Also attending the session were Brigadier General E. S. Lansdale (OSD), Major James Patchell (OSD), Brigadier General William H. Craig (JCS), Richard Helms (CIA), George McManus (CIA), and another undisclosed CIA representative.

15. CIA, Office of the Secretary of Defense, Memorandum for the Chief of Operations, Cuba Project, February 19, 1962.

16. Of the teams that were already active on the island, the most successful was the one in Pinar del Rio in western Cuba. Its success depended on the maritime resupply of arms and equipment.

17. CIA, Office of the Secretary of Defense, Memorandum for the Special Group (augmented), from Brigadier General Lansdale, Subject: "Review of Operation Mongoose," Top Secret Sensitive, Operation Mongoose, July 25, 1962, CIA electronic library.

18. Alexandre Sheldon-Duplaix's interview with Lev Vtorygin, Moscow, July 2005.

19. Peter Huchthausen's interview with the Russian submariner Lev D. Chernavin in St. Petersburg, 1995. Chernavin was a junior officer aboard a Foxtrot submarine deployed during the October 1962 crisis.

20. NSA, July 19, 1962, Op-922Y Top Secret Dinar Memorandum for the Secretary of the Navy, Subj: "Navy Participation in Increased SIGINT Program for Cuba Ref.: Secdef memo of July 16, 1962, Subj: "Increased SIGINT Program for Cuba," www.nsa.gov/public/publi00003.cfm, Cuban Missile Crisis Document Archives.

21. NSA, P191653Z, FM DIRNSA to CNO, Info CINCLANT, CINCLANTFLT, DIRNAVSECGRULANT, JCS, SSO DIA/NSA Rep CINCLANT, ZEM, Top Secret COMINT Channels, www.nsa.gov/public/publi00003.cfm, Cuban Missile Crisis Document Archives.

22. NSA, Secret Kimbo, electrical release, July 24, 1962, www.nsa.gov/public/publi00003.cfm, Cuban Missile Crisis Document Archives.

23. NSA, Secret Kimbo, electrical release, August 7, 1962, www.nsa.gov/public/publi00003.cfm, Cuban Missile Crisis Document Archives.

24. Ibid.

25. NSA, Secret Kimbo, electrical release, August 17, 1962, www.nsa.gov/public/publi00003.cfm, Cuban Missile Crisis Document Archives.

26. NSA, Secret Kimbo, electrical release, August 24, 1962, www.nsa.gov/public/publi00003.cfm, Cuban Missile Crisis Document Archives.

27. CIA, Memorandum, Subject: "Soviet MRBM in Cuba," October 31, 1962, Document number 3; Mary S. McAuliffe, ed., CIA History Staff, CIA Documents on the Cuban Missile Crisis, October 1992, p. 13.

28. Philippe Thyraud de Vosjoli, *Lamia* (Boston: Little, Brown, 1970).

29. CIA, McCone, "Memorandum on Cuba," August 20, 1962, Document number 5, in Mary S. McAuliffe, ed., CIA History Staff, CIA Documents on the Cuban Missile Crisis, October 1992, p. 19.

30. CIA, McCone, Memorandum for the file, "Discussion in Secretary Rusk's Office at 12 O'clock, August 21, 1962," Document number 6; Memorandum, Subject: "Soviet MRBM in Cuba," October 31, 1962, Document number 3, Mary S. McAuliffe, ed., CIA History Staff, CIA Documents on the Cuban Missile Crisis, October 1992, pp. 21–22, p. 13.

31. CIA, the White House, August 23, 1962, National Security Action Memorandum no. 181, "Presidential Directive on Actions and Studies in Response to New Soviet Bloc Activity in Cuba," to Secretary of State, Secretary of Defense, Attorney General, Acting Director, CIA, General Taylor, August 23, 1962.

32. CIA Memorandum for the Director, Subject: "Action Generated by DCI Cables Concerning Cuban Low-Level Photography and Offensive Weapons," undated, Document number 12, Mary S. McAuliffe, ed., CIA History Staff, CIA Documents on the Cuban Missile Crisis, October 1992, p. 39.

33. Memorandum for Acting Director of Central Intelligence, Subject: "Recent Soviet Military Activities in Cuba," September 3, 1962, Document number 11, Mary S. McAuliffe, ed., CIA History Staff, CIA Documents on the Cuban Missile Crisis, October 1992, pp. 35–37.

34. General Anatoli I. Gribkov and U. Krai Yadernoi Bezdni, *On the Edge of Nuclear Chasm: From the History of the Caribbean Crisis, 1962: Facts, Witnesses, Analysis* (Moscow: Gregori Page, 1998). This book was published in limited numbers specifically for the participants of "The Great Victory of International Socialist Brotherhood." It contains firsthand accounts from Soviet commanders, troops, and sailors. It also includes former top-secret cables, orders, and instructions to the Soviet forces. It is an invaluable source of details of the Soviet side of the episode and clearly portrays the whole crisis as a victory for the Soviet Union.

35. Ibid., pp. 71–72.

36. "Anadyr," Soviet General Staff Archives, file 6, vol. 2, quoted by Gribkov and Bezdni, *On the Edge of Nuclear Chasm*, p. 144.

37. CIA, September 5, 1962, Carter to McCone, Cable, Document number 14, Mary S. McAuliffe, ed., CIA History Staff, CIA Documents on the Cuban Missile Crisis, October 1992, p. 47.

38. USSR Minister of Defense, Marshall of the Soviet Union, R. Malinovskiy, P. P. Chief of the General Staff, Marshall of the Soviet Union, M. Zakharov, September 8, 1962; Source: A. Grikov and W. Smith, *Operation Anadyr: US and Soviet Generals Recount the Cuban Missile Crisis*, translated by Svetlana Savranskaya (Chicago, Tokyo, and Moscow: National Security Archives, 1994).

39. CIA, September 11, 1962, Carter to McCone, Cable, Document number 22, Mary S. McAuliffe, ed., CIA History Staff, CIA Documents on the Cuban Missile Crisis, October 1992, p. 63.

40. Ibid., p. 145.

41. CIA, September 13, 1962, Carter to McCone, Cable, Document number 25, Mary S. McAuliffe, ed., CIA History Staff, CIA Documents on the Cuban Missile Crisis, October 1992, p. 69.

42. Ibid., p. 69.

43. CIA, September 16, 1962, McCone to Carter, Cable, Document number 28, Mary S. McAuliffe, ed., CIA History Staff, CIA Documents on the Cuban Missile Crisis, October 1992, p. 78.

44. CIA, September 17, 1962, Carter to McCone, Cable, Document number 29, Mary S. McAuliffe, ed., CIA History Staff, CIA Documents on the Cuban Missile Crisis, October 1992, p. 80.

45. CIA, September 19, 1962, McCone to Carter, Cable, Document number 32, Mary S. McAuliffe, ed., CIA History Staff, CIA Documents on the Cuban Missile Crisis, October 1992, pp. 88–91.

46. Ibid., p. 91.

47. CIA, September 20, 1962, CIA Information Report, Document number 34, Mary S. McAuliffe, ed., CIA History Staff, CIA Documents on the Cuban Missile Crisis, October 1992, p. 105.

48. CIA, Memorandum for Record, "Minutes of Meeting of the Special Group (Augmented) on Operation Mongoose," October 4, 1962 (the Attorney General; Mr. Johnson; Mr. Gilpatrick; General Taylor; General Landsdale; Mr. McCone; General Carter; Mr. Wilson).

49. Ryurik A. Ketov, "The Cuban Missile Crisis as Seen through a Periscope," *Journal of Strategic Studies*, 28, no. 2 (April 2005): 217–231.

50. CIA, October 16, 1962, CIA Memorandum, "Probable Soviet MRBM Sites in Cuba," Top Secret; Mary S. McAuliffe, ed., CIA History Staff, CIA Documents on the Cuban Missile Crisis, October 1992, pp. 187–192; October 18, 1962, "Joint Evaluation of Soviet Missile Threat in Cuba," Top Secret, Document number 61, Mary S. McAuliffe, ed., CIA History Staff, CIA Documents on the Cuban Missile Crisis, October 1992, pp. 187–192.

51. CIA, Central Intelligence Agency, Memorandum, "USSR/Cuba, Information as of 0600," October 27, 1962.

52. Two days earlier, the Joint Chiefs of Staff had called for 111 ships to conduct the invasion. CIA, Notes taken from transcripts of meetings of the Joint Chiefs of Staff, October–November 1962, dealing with the Cuban missile crisis.

53. Republica de Cuba, Ministerio de las Fuerzas Armadas Revolucionarias, October 24, 1962, Aspectos importantes contenidos en los informes ofrecidos por los jefes militares reunidos el dia 24 de Octubre de 1962 en el EMG con el Cmdte. Fidel Castro [Important aspects contained in the reports presented by the military chiefs during the October 24, 1962 meeting with Commander in Chief Fidel Castro], translated by Gary Goldberg for the Cold War International History Project and the National Security Archives.

54. Havana, October 26, 1962, letter from Castro to Khrushchev, translated by Gary Goldberg for the Cold War International History Project and the National Security Archives, April 2002.

55. Volkogonov Collection, Library of Congress, Manuscript Division, Reel 17, Container 26, translated by Gary Goldberg for the Cold War International History Project and the National Security Archives; Archive of the President of the Russian Federation, Special Declassification, April 2002, translation by Svetlana Savranskaya, the National Security Archives.

56. CIA, Based on the latest low-level reconnaissance, three of the four MRBM sites at San Cristobal and the two sites at Sagua La Grande appeared to be fully operational; Central Intelligence Agency, Memorandum, "USSR/Cuba, Information as of 0600, October 27, 1962.

57. Peter Huchthausen's interview with Marshal Dimitry T. Yazov, Sevastopol, August 9, 1988.

58. Alexandre Sheldon-Duplaix interview with Lev Vtorigyn, Moscow 2005.

59. CIA, Summary record of NSC Executive Committee Meeting N 20, November 5, 1962.

60. Ibid.

61. Telegram from Malinovsky to Pliyev, early November (circa November 5) 1962, Top Secret; Volkogonov Collection, Library of Congress, Manuscript Division, Reel 17,

Container 26, translated by Gary Goldberg for the Cold War International History Project and the National Security Archives; Archive of the President of the Russian Federation, Special Declassification, April 2002, translation by Svetlana Savranskaya, the National Security Archives.

62. The JCS recommended five U-2 flights and fourteen low-level flights (an increase over the current level) to cover ports, the isle of Pines, and seven caves suspected of being weapon-storage sites.

63. Edward J. Marolda, "Cold War to Violent Peace," in W. J. Holland Jr., ed., *The Navy* (Washington, DC: Naval Historical Foundation, 2000).

64. CIA, chairman's talking paper for meeting with the president, November 16, 1962.

65. Havana, telegram, Soviet ambassador for Comrade A. I. Mikoyan, November 22, 1962, translated by Gary Goldberg for the Cold War International History Project and the National Security Archives; Archive of the President of the Russian Federation, Special Declassification, April 2002. Translation by Svetlana Savranskaya, the National Security Archives.

9. Transition to War: Vietnam, 1961–1975

1. Robert J. Hanyok, *Spartans in Darkness: NSA, American Sigint and the Indochina War, 1945–1975*, National Security Agency, Central Security Service, Series VI, Volume 7, Top Secret Comint, declassified by NSA on December 21, 2007, pp. 91–92.

2. Ibid., p. 101.

3. Ibid., p. 176.

4. Ibid., pp.182–183.

5. Edward Marolda, *The U.S. Navy in the Vietnam War: An Illustrated History* (Washington, DC: Brassey's, 2002).

6. Hanyok, *Spartans in Darkness*, p. 179.

7. Wyman H. Packard, *Century of U.S. Naval Intelligence* (Washington, DC: Naval Historical Center, 1996).

8. Hanyok, *Spartans in Darkness*, p. 180.

9. Ibid., p. 184.

10. Ibid., p. 187.

11. Ibid., p. 191.

12. Ibid., pp. 188–191.

13. Ibid., p. 195.

14. Ibid., p.199

15. Ibid.

16. Ibid., p. 198.

17. Ibid., p. 208.

18. NSA Web site, interview with Robert J. Hanyok, October 29, 2004, www.nsa .gov/vietnam.

19. Hanyok, *Spartans in Darkness*, p. 220.

20. Mario de Archangelis, *Electronic Warfare: From the Battle of Tsushima to the Falklands and Lebanon Conflicts* (Dorset, UK: Blandford Press, 1985).

21. Packard, *Century of U.S. Naval Intelligence*, pp. 201–202.

22. De Arcangelis, *Electronic Warfare*, pp. 160–173.

23. Hanyok, *Spartans in Darkness*, p. 104.

24. Ibid., pp. 101–103.

25. Edward J. Marolda, "Cold War to Violent Peace," in W. J. Holland Jr., ed., *The Navy* (Washington, DC: Naval Historical Foundation, 2000).

26. Packard, *Century of U.S. Naval Intelligence*, pp. 115–118.

27. USS *Oxford* Web site, http://members.tripod.com/~USS_OXFORD/.

28. Hanyok, *Spartans in Darkness*, p. 248.

29. Ibid., pp. 302, 335.

30. Glenn E. Helm, Navy Department Library, Naval Historical Center, *Pull Together*, the Newsletter of the Naval Historical Foundation and the Naval Historical Center, vol. 36, no. 1 (Spring/Summer 1997).

31. Hanyok, *Spartans in Darkness*, p. 335.

32. Ibid., pp. 339–340, 370.

33. Ibid., p. 420.

34. U.S. Navy Historical Center, Naval Operational Archives, R. F. Cross Associates, *Sea-based Airborne Antisubmarine Warfare 1940–1977*, Vol. II, 1960–1977, February 17, 1978, Prepared for Op-095 under ONR Contract N00014-77-C-0338, Second Edition, Declassified, p. 61.

35. Interview with Lev Vtorygin, Moscow, July 14, 2007.

10. A Submarine Is Lost and Found, 1968–1974, 1989

1. Interview with Lev Vtorygin on Frye Island, Maine, September 1996.

2. Clyde W. Burleson, *The Jennifer Project* (Englewood Cliffs, NJ: Prentice-Hall, 1977).

3. Letter from Admiral Smith to Kontr Admiral Dygalo, November 1994.

4. S. P. Buran, *Po Sledam Podvodnikh Katastrof* (Moscow: Gildia Masterov Rus, 1992), p. 208.

11. Ocean Surveillance, 1962–1980s

1. Quoted by Ronald J. Kurth, "Gorshkov's Gambit," in *Journal of Strategic Studies* 28, no. 2 (April 2005): pp. 261–280.

2. V. P. Kuzin and V. I. Nikol'skiy, *Voenno-morskoy flot SSSR 1945–1991* (St. Petersburg: Istoricheskoye Morskoye Obshchestvo, 1996), p. 379.

3. Ibid.

4. *ONI Review* (April 1954): p. 145.

5. Public Record Office, ADM 223, *Quarterly Intelligence Report*, April to June 1959, pp. 39–40.

6. D. Miller, "Intelligence and the War at Sea," in W. Kennedy, ed., *The War of Intelligence* (London: Salamander Press, 1984), pp. 174–199.

7. Ibid.

8. Yuri. A. Berkov, *Memoirs*, 1966, www.hotstreams.ru/index.php?option=com_smf&Itemid=27&.

9. Gary Weir and Walther Boyne, *Rising Tide: The Untold Story of the Russian Submarine That Fought the Cold War* (New York: Basic Books, 2003), p. 177.

10. Ibid., pp. 165–169.

11. Wyman H. Packard, *Century of U.S. Naval Intelligence* (Washington, DC: Naval Historical Center, 1996), p. 115.

12. The *Georgetown* had been active off the north coast of Cuba and off the west coast of South America in 1964 and 1965. The *Jamestown* had sailed along the coast of Africa in the summer and the winter of 1964, before being relieved by the *Oxford.* The *Oxford* continued on to the South China Sea and operated there for the remainder of 1965, while the *Pvt. Jose Valdez* conducted similar operations along the Atlantic and Indian Ocean coasts of Africa. In January 1966, the *Jamestown* conducted surveillance operations off the coast of Cambodia and continued to operate in the Western Pacific throughout the year. The *Oxford* was also on active duty in the Southeast Asian area in 1966.

13. Packard, *Century of U.S. Naval Intelligence*, pp. 114–115.

14. Ibid., p. 115.

15. Ibid., p. 116.

16. Ibid., p. 115.

17. Ibid., p. 116.

18. James M. Ennes Jr., *Assault on the "Liberty": The True Story of the Israeli Attack on an American Intelligence Ship* (New York: Random House, 1979). Two survivors of the attack, Jim Ennes and Joe Meadors, maintain a memorial Web site on the USS *Liberty*, www.ussliberty.org/.

19. Harriet Dashiell Schwar and Edward C. Keefer, eds., *Arab-Israeli Crisis and War, 1967*, Vol. XIX in *Foreign Relations of the United States, 1964–1968* (Washington, DC: Department of State, 2004), reviewed by Norman Polmar, U.S. Naval Institute, *Proceedings*, July 2004, p. 84.

20. James Bamford, *Body of Secrets: Anatomy of the Ultra-Secret National Security Agency from the Cold War through the Dawn of a New Century* (New York: Doubleday, 2001). The quotations attributed to Mr. Bamford were taken from a statement he published in the *New Republic.*

21. Jay A. Cristol, *The Liberty Incident: The 1967 Israeli Attack on the U.S. Navy Spy Ship* (Washington, DC: Brassey's, 2002). Jonkers, AFIO WIN 25-02, June 24, 2002, notes that Cristol, "a reputable former naval aviator and esteemed federal judge, spent ten years investigating the incident and concluded that the attack was a tragic mistake by the Israelis." James M. Ennes Jr., *Washington Report on Middle East Affairs*, June/July 2002, comments that Liberty "[s]urvivors see it as a flawed work, packed with evasions and misleading statements. Cristol seems to accept at face value all the arguments that support his case, while he nitpicks, dismisses and ignores entirely the eyewitness reports of survivors and other supporting evidence." Paul Tobin, U.S. Naval Institute *Proceedings*, August 2002, does not agree with Cristol's conclusion that the attack was an unfortunate accident but finds the author's research "rigorous and extensive."

22. E-mail sent to James Bamford, March 3, 2000; Steven Aftergood, "Bamford 'Liberty' Account Repudiated," *Secrecy News*, July 17, 2001, www.fas.org, reported that "[k]ey aspects of . . . Bamford's recent account of the 1967 Israeli attack on the U.S.S. *Liberty* are being disavowed by some of his own sources."

23. "NSA Releases USS Liberty Records," *Secrecy News*, July 9, 2003, www.fas.org. The NSA's June 6, 2007, releases concerning the 1967 Israeli attack on the USS *Liberty* are available at www.nsa.gov/docs/efoia/released/liberty.html; see also text of e-mail sent to James Bamford, March 3, 2000; Jay Cristol, *The Liberty Incident: The 1967 Attack on the U. S. Navy Spy Ship* (Washington, DC: Brassey, 2002).

24. Packard, *Century of U.S. Naval Intelligence*, p. 116.

25. Ibid.

26. Ibid., p. 117.

27. Mitchell B. Lerner, a professor of history at Ohio State University, is the author of *The Pueblo Incident: A Spy Ship and the Failure of American Foreign Policy* (Lawrence: University Press of Kansas, 2002).

28. Richard A. Mobley, *Flash Point North Korea: The Pueblo and EC-121 Crises* (Annapolis, MD: Naval Institute Press, 2003).

29. Packard, *Century of U.S. Naval Intelligence*, p. 118.

30. Interview with a former AGI skipper who did not want to be named, Moscow, July 2006.

31. On April 22, 1968, the *Washington Post* published the comments of Defense Secretary Robert S. McNamara on the value of Soviet spy ships.

32. "MI6 Link to Sunken Trawler, Revealed David Pallister," *Guardian*, September 30, 2000.

33. Office of the Chief of Naval Operations, *Understanding Soviet Naval Developments*, 6th Edition (Washington, DC: Office of the Chief of Naval Operations, Department of the Navy, 1991); Norman Polmar, *Guide to the Soviet Navy*, 5th Edition (Annapolis, MD: Naval Institute Press, 1991).

12. A Naval Intelligence Revolution, 1970s

1. U.S. Navy Historical Center, Naval Operational Archives, R. F. Cross Associates, *Sea-based Airborne Antisubmarine Warfare 1940–1977,* Volume II, 1960–1977, February 17, 1978, prepared for Op-095 under ONR Contract N00014-77-C-0338, Second Edition, Declassified, p. 8.

2. Oleg Kalugin, *The First Directorate: My 32 Years in Intelligence and Espionage against the West* (New York: St. Martin's Press, 1994), p. 84.

3. Robert W. Hunter, *Spy Hunter: Inside the FBI Investigation of the Walker Espionage Case* (Annapolis, MD: Naval Institute Press,1999), p. 122.

4. Ed Offley, *Scorpion Down: Sunk by the Soviets, Buried by the Pentagon: The Untold Story of the USS Scorpion* (New York: Basic Books, 2007), pp. 317–348, gives a full summary of the Walker spy case and the estimates that the information they sold to the Soviets were related to some or all three events: USS *Liberty, Pueblo,* and *Scorpion.*

5. Laura J. Heath, *Analysis of the Systemic Security Weaknesses of the U.S. Navy Fleet Broadcasting System, 1967–1974, as Exploited by CWO John Walker* (master's thesis, Georgia Institute of Technology, 2001), p. 54.

6. Mitchell B. Lerner, *The Pueblo Incident: A Spy Ship and the Failure of American Foreign Policy* (Lawrence: University Press of Kansas, 2002).

7. Ibid.

8. Heath, *Analysis of the Systemic Security Weaknesses of the U.S. Navy Fleet Broadcasting System*, p. 69.

9. Ibid., p. 70.

10. Hunter, *Spy Hunter*, pp. 186–188.

11. Heath, pp. 58–64.

12. Hunter, *Spy Hunter*, p. 141.

13. Victor Cherkashin, *Spy Handler: Memoir of a KGB Officer: The True Story of the Man Who Recruited Robert Hanssen and Aldrich Ames* (New York: Basic Books, 2005), p. 225.

14. John Barron, *Breaking the Ring* (Boston: Houghton Mifflin, 1987), in Heath, p. 83.

15. William Studeman, quoted by Hunter, *Spy Hunter*, pp. 230–234.

16. Ibid., p. 234.

17. Ibid., pp. 241–242.

18. Ibid., p. 232.

19. Wyman H. Packard, *Century of U.S. Naval Intelligence* (Washington, DC: Naval Historical Center, 1996), p. 133. ONI followed up on Nitze's order with its instruction 005430.12 of January 11, 1966, which established within ONI the control and management of the navy's clandestine intelligence collection program.

20. December 7, 1965, from Secretary of the Navy, to: distribution list, CNO (DNI—15 copies); (940—1 copy); Chief BUSANDA; Chief of Naval Material; Chief of Naval Personnel; Chief of Industrial Relations; Comptroller of the Navy; CMC; JAG; CINCLANTFLT; CINCPACFLT; CINCUSNAVEUR; COMNAVFORJAPAN; COMIDEASTFOR; COMCRIBSEAFRON; CINCUSNAVEUR Rep Germany; 2 deleted; copy to DIA; CIA; Subject: Instructions for the Coordination and Control of Navy's Clandestine Intelligence Collection Program; National Security Archives (NSA), www.gwu.edu/~nsarchiv/NSAEBB/NSAEBB46/document1.pdf.

21. NSA, Document number 2, Functions of the 1127th USAF Field Activities Group (AFNIA), "History of the Chief of Staff, Intelligence," July–December 1967, www.gwu.edu/~nsarchiv/NSAEBB/NSAEBB46/document2.pdf.

22. NSA, Document number 3a, "History of Navy Humint—Human Source Intelligence, 1973," www.gwu.edu/~nsarchiv/NSAEBB/NSAEBB46/document3a.pdf.

23. NSA, Document number 3b, "History of Navy Humint—Human Source Intelligence, 1974," www.gwu.edu/~nsarchiv/NSAEBB/NSAEBB46/document3b.pdf.

24. Ibid.

25. Ibid.

26. NSA, Document number 4, From CTF 157 to DNI, December 31, 1975, www.gwu.edu/~nsarchiv/NSAEBB/NSAEBB46/document4.pdf.

27. Christopher A. Ford and David A. Rosenberg, *The Admiral's Advantage: U.S. Navy Operational Intelligence in World War II and the Cold War* (Annapolis, MD: Naval Institute Press, 2005), p. 44.

28. The signatories of the Lausanne Convention were Britain, France, Italy, Japan, Bulgaria, Greece, Romania, the USSR, and Yugoslavia. The United States did not participate in the Montreux Convention, but its provisions are binding on all non–Black Sea naval powers.

29. Ford and Rosenberg, *The Admiral's Advantage*, p. 57.

30. Ibid., p. 63.

31. Sherry Sontag and Christoper Drew, *Blind's Man Bluff: The Untold Story of American Submarine Espionage* (New York: HarperCollins, 1998), pp. 158–224.

32. I. A. Baikov and G. L. Zikov, *Intelligence Operations of the American Submarines* (St. Petersburg, 2002).

33. Cherkashin, *Spy Handler*, p. 143.

34. Baikov and Zikov, *Intelligence Operations*, pp. 190–200.

35. Cherkashin, *Spy Handler*, p. 143.

36. Quoted in Ford and Rosenberg, *The Admiral's Advantage*, p. 73.

37. Interview with Peter Huchthausen and former chief of the Soviet general staff from 1988–1991, General of the Army Moiseev, in Moscow, November 14, 1994.

38. The cause of the explosion and the fire aboard Yankee 1 class SSBN K-219 in October 1986 was believed to have been the result of the same defect, moisture inside the missile silo. The same submarine suffered an earlier casualty of a similar nature in 1973.

39. From Peter Huchthausen's interview in Moscow with Swedish naval attaché Captain Erland Sonnersdedt in August 1987.

40. Nikolai Cherkashin provides a detailed account of the mutiny and includes information from Sablin's family in "The Last Parade," published in *The Log* by Andreevsky Flag, St. Petersburg, 1992.

41. Rubin and YeN-D radar are onboard TU-16 long-range bombers.

42. V. P. Kuzin and V. I. Nikol'skiy, *Voenno-Morskoi flot SSSR 1945–1991: Istoriia sozdaniia poslevoennogo Voenno-Morskogo Flota SSSR i vozmozhnyi oblik flota Rossii* (St. Petersburg: Istoricheskoe Morskoe Obshchestvo, 1996), p. 687.

43. Norman Friedman, *Sea Power and Space* (Annapolis, MD: Naval Institute Press, 2000), pp. 157–162.

44. Victor Suvorov "Fleet Intelligence," in *Inside Soviet Military Intelligence* (London: MacMillan Publishing, 1984).

45. John B.Hattendorf, "The Evolution of the US Navy's Maritime Strategy, 1977–1986," *Naval War College Newport Paper* 19 (Newport: Naval War College Press, 2004), pp. 37–83.

46. Vladimir P. Kuzin and Vladislav I. Nikol'skiy, *Voenno-morskoy flot SSSR 1945–1991* (St. Petersburg: Istoricheskoye Morskoye Obshchestvo, 1996), p. 25.

47. Quoted in Ford and Rosenberg, *The Admiral's Advantage*, p. 84.

48. Ibid., p. 87.

13. War Scare and PSYOPS, 1981–1987

The chapter epigraph is from George F. Kennan, *At a Century's Ending: Reflections 1982–1995* (New York: W. W. Norton, 1996), p. 82, quoted in Benjamin B. Fischer, *A Cold War Conundrum: The 1983 Soviet War Scare* (Langley, VA: CIA, 1997), p. 1.

1. V. P. Kuzin and V. I. Nikol'skiy, *Voenno-morskoy flot SSSR 1945-1991* (St. Petersburg: Istoricheskoye Morskoye Obshchestvo, 1996), p. 25.

2. Ibid., pp. 25–28, 500–509.

3. Ibid., pp. 402–403.

4. Steven T. Usdin, *Engineering Communism* (New Haven, CT: Yale University Press, 2005), pp. 244–247.

5. Interview with Lev Vtorygin, Moscow, July 2007.

6. Ivan M. Kapitanets, *Flot, V Voinas tchetchogo pokolenia* (Moscow: Betche, 2006), pp. 160–161.

7. Interview with Lev Vtorygin, Moscow, July 2007.

8. Kuzin and Nikol'skiy, *Voenno-morskoy flot SSSR 1945–1991*, p. 29.

9. Interview with Lev Vtorygin, Moscow, July 2007.

10. Nikolay Amelko, *V interesas flota I gosudarstva: vostominania admirala* (Moscow: BDTs Press, 2002), pp. 131–144.

11. Interview with Georgi Arbatov, "Why Our Country Would Need Aircraft-Carriers?" *Serving the Motherland*, Soviet TV, June 17, 1990.

12. Information on the *Dixon* and its laser tests published in *Versiya*, December 26, 2000 (provided by Werner Globke).

13. Designated by NATO under the name Squeeze Box.

14. CIA National Estimates NIE 11–15–82.

15. John Hattendorff, "The Evolution of the U.S. Maritime Strategy, 1977–1986," *Newport Paper*, NP19, p. 12.

16. See Seymour Hersh, *The Target Is Destroyed: What Really Happened to Flight 007 and What America Knew about It* (New York: Vintage Books, 1987), p. 221. For recently declassified information on the U.S. overflight program, see "Secrets of the Cold War," *U.S. News and World Report*, March 15, 1993, pp. 30–50.

17. Quoted in Benjamin Fischer, *A Cold War Conundrum: The 1983 Soviet War Scare* (Langley, VA: CIA, 1997).

18. Ibid.

19. Ibid.

20. "New Soviet Bombers Fake Strike against U.S. Navy," *Washington Post*, November 9, 1982.

21. Fischer, *A Cold War Conundrum*.

22. Ibid.

23. Interview with Boris Grigoriev, St. Petersburg, May 2008.

24. Hoffmann Mazhny to Honecker and note on statements by Ogarkov, September 14, 1982, AZN 32643, pp. 117–126, BA-MA. Statement by Kulikov, June 9, 1983, VS, OS, 1987, cj. 75174/4, Central Military Archives, Prague (VÚA).

25. Fischer, *A Cold War Conundrum*.

26. Interview with Lev Vtorygin, New York, May 2005.

27. Interview with Lev Vtorygin, Moscow, June 2007.

28. Ibid.

29. A. Kopakidiy and D. Prokorov, *Imperia GRU, Otcherki istorii rossiskoi voennoi razvedki*, (Moscow: Izdatelstvo Olma Press, 1999), p. 63.

30. Speech of the General Secretary of the CPSU (Gorbachev) at the Sofia Political Consultative Committee Meeting, 22.10.85, Bulgarian Central State Archives, Sofia, Fond 1b, Opis 35, a.e. 1025–85, pp. 1–17. Translation by Vania Petkova and Anya Jouravel, Parallel History Project on Cooperate Security, www.php.isn.ethz.ch.

31. Design 1908—*Akademik Sergey Korolev* (21.250 tons, 17 knots, 79 laboratories with 190 scientists); Design 1909—*Kosmonaut Yuri Gagarin* (53,000 tons, 17 knots, 212 scientists; capable of tracking two space objects at once). See Kuzin and Nikol'skiy, *Voenno-morskoy flot SSSR 1945–1991*, p. 382.

32. 25,000 tons; 22 knots; range: 15,000 miles, autonomy: 120 days. Kuzin and Nikol'skiy, *Voenno-morskoy flot SSSR 1945–1991*, p. 382.

33. Ibid.

34. Serguei Kostine, *Bonjour, Farewell* (Paris: R. Laffont, 1997), p. 131.

35. Interview with Lev Vtorygin. Moscow, June 2007.

36. Kostine, *Bonjour, Farewell*, pp. 142–149.

37. Gus W. Weiss, "The Farewell Dossier: Duping the Soviets," *Studies in Intelligence* 35, no. 9 (1996): 121–128.

38. Ibid.

39. Thomas Reed, *At the Abyss: An Insider's History of the Cold War* (New York: Presidio Press/Ballantine Books, 2004), pp. 268–269.

40. Kostine, *Bonjour, Farewell*, p. 274.

41. Nicolay Ogarkov, press conference of September 9, 1983, on Radio Moscow.

42. Fischer, *A Cold War Conundrum*, note 78.

43. Interview with Captain Lev Vtorygin, Moscow, June 2007.

44. Fischer, *A Cold War Conundrum*, note 80.

45. Ibid., note 81.

46. Three U.S. vessels and three chartered tugs were involved in the search-and-rescue operations: the coast guard cutter *Monro*, the rescue salvage ship USS *Conserver*, and the fleet tug USNS *Narrangansett*, assisted by the *Ocean Bull*, the *Kaiko-Maru 7*, and the *Kaiko-Maru 3*. These vessels towed sideways-scanning sonar capable of locating the black boxes whose signals could last thirty days. They were protected by one cruiser, USS *Sterrett*; two destroyers, USS *Towers* and USS *Elliot*; five frigates, USS *O'Callaghan*, USS *Brooke*, USS *Meyerkord*, USS *Stark*, and USS *Badger*; and two support ships, USNS *Hassayampa* and USS *Wichita*; and by various Japanese and South Korean patrol vessels.

47. *Surface Combatant Forces—7th Fleet Task Force 71 Flight 007 After-Action Report*, dated November 18, 1983.

48. Bert Schlossberg, *Rescue 007: The Untold Story of KAL 007 and Its Survivors* (Princeton, NJ: Xlibris Corp, 2000).

49. *Izvestia*, number 228, October 16, 1992.

50. David F. Winkler, *Cold War at Sea: High Seas Confrontation between the United States and the Soviet Union* (Annapolis, MD: Naval Institute Press, 2000), p. 47.

51. Republican Staff Study of the Committee on Foreign Relations, 1991, in Schlossberg, *Rescue 007*.

52. Interview with Rear Admiral Ivan Ivanov, St. Petersburg, July 2006.

53. Interview with Captain First Rank Lev Vtorygin, Moscow June 2007.

54. Interview with Captain 1st Rank Oleg Malov, Moscow, July 2006.

55. *Izvestia*, May 31, 1991.

56. Ibid.

57. Report by the head of the group, Lieutenant General of Aviation Makarov; Lieutenant General Engineer Tichomirov; Major General Engineer Didenko; Major General of Aviation Stepanv; Major General of Aviation Kovtun; Corresponding Member of Academy of Sciences of the USSR Fedosov to Yuri Andropov, November 28, 1983; Report by Deputy Chief Navigator for Air Force Major General of Aviation Kovtun; Deputy Chief Navigator for VTA Air Force Colonel Polevoi; Chief Navigator-Researcher for LII MAP Ireikin; Head of Navigation Equipment Lab at LII MAP Vlasov; Navigator for Leading Group at TsUMVS MGA Svishev; KGB Representative Korichnev, to Yuri Andropov, November 28, 1983.

58. *Izvestia*, number 228, October 16, 1992.

59. Fischer, *A Cold War Conundrum*, n. 84.

60. Christopher Andrew and Oleg Gordievsky, *KGB: The Inside Story of Its Foreign Operations from Lenin to Gorbachev* (New York: HarperCollins, 1991), p. 583.

61. Vojtech Mastny, "Did East German Spies Prevent a Nuclear War?" in *Stasi Intelligence on NATO, 1969–1989: East German Military Espionage against the West*, edited by Bernd Schaefer and Christian Nuenlist (2003).

62. Don Oberdorfer, *The Turn: From the Cold War to a New Era: The United States and the Soviet Union, 1983–1990* (New York: Poseidon, 1991), p. 67; Fischer, *A Cold War Conundrum*.

63. CIA estimates, "Implications of Recent Soviet Military-Political Activities," May 1984.

64. Interview with Vice Admiral Kviatkovskiy, June 2007.

65. Beth A. Fischer, *The Reagan Reversal Foreign Policy and the End of the Cold War* (Columbia: University of Missouri Press, 2000).

66. Ronald Reagan, *An American Life* (New York: Simon and Schuster, 1990), p. 588.

67. Ibid., p. 257.

68. Kuzin and Nikol'skiy, *Voenno-morskoy flot SSSR 1945–1991*, p. 30.

69. Interview with Lev Vtorygin, Moscow, June 2007.

70. Kapitanets, *Flot, V Voinas tchetchogo pokolenia*, pp. 170–171.

71. Vojtech Mastny, Sven G. Holtsmark, Andreas Wenger, Anna Locher, and Christian Nuenlist, eds., *War Plans and Alliances in the Cold War: Threat Perceptions in the East and West* (New York: Routledge, 2006), pp. 95–117.

72. Interview with Captain Andrzej Makowski, director, Polish Naval Academy (Gdynia), Annapolis, September 2007.

73. Interview with Lev Vtorygin, Moscow, June 2007.

74. Frede P. Jensen, *The Warsaw Pact's Special Target: Planning the Seizure of Denmark*, quoted in Mastny, Holtsmark, Wenger, Locher, and Nuenlist, eds., *War Plans and Alliances in the Cold War,* pp. 95–117.

75. Graham H. Turbiville, "Soviet TVD, Strategic Intelligence and Key Targets in the CONUS (Continental United States)," *Military Review* (January–February 2002).

76. Georgii Georgievich Kosteev, *Neizvestnyi flot: liudi, fakty, problemy,* (Moscow: Ogni, 2004), pp. 344–345.

77. Ibid.

14. Swedish Waters, 1980–1990s

1. Interview with Rear Admiral Emil Svensson, Stockholm, October 2005.

2. Milton Leitenberg, *Soviet Submarine Operations in Swedish Waters (The Washington Papers)* (New York: Praeger, 1987), pp. 33–36.

3. United Press International, October 29, 1981.

4. Mikhail Yakovlev.

5. Leitenberg, *Soviet Submarine Operations in Swedish Waters*, pp. 42–43.

6. *New York Times*, December 24, 1981.

7. Serguei Aprelev, *Pod "chorokh" nachykh "dieselii"* (St. Petersburg: NIKA, 2005), pp. 255–285.

8. E-mail from Rear Admiral Emil Svensson to Alexandre Sheldon-Duplaix.

9. Leitenberg, *Soviet Submarine Operations in Swedish Waters*, pp. 38–39.

10. Ibid., p. 40.

11. Report from the Submarine Commission, Stockholm, Försvarsdepartementet, 1995.

12. Report from the Submarine Commission, Stockholm, Försvarsdepartementet, 2001, English summary, p. 356.

13. UPI, May 6, 1982.

14. UPI, June 6, 1982.

15. UPI, June 9, 1982.

16. The Naval Analysis Group Report for the Hårsfjärden incident (September 27–October 15, 1982) under Rear Admiral Emil Svensson, 1983, Attachment 2, quoted by Ola Tunander, *Some Remarks on the US/UK Submarine Deception in Swedish Waters in the 1980s* (Oslo: International Peace Research Institute), p. 3.

17. *Aftonbladet*, October 1, 1982, the Naval Analysis Group Report, 1983, quoted by Tunander, *Some Remarks on the US/UK Submarine Deception in Swedish Waters in the 1980s*, p. 3.

18. The Naval Analysis Group Report, Stockholm, 1983.

19. UPI, October 5, 1982.

20. UPI, October 6, 1982.

21. AP, October 6, 1982.

22. AP, October 8, 1982.

23. AP, October 9, 1982.

24. AP, October 10, 1982.

25. Special reports about incidents at Mälsten Coastal Defence Base (SRMCDB) signed by Sven-Olof Kviman and Per Andersson (October 11–14, 1982); War Diary of the chief of the Mine Troops and the chief of Mälsten Coastal Defence Base (WDCMT/ CMCDB), Lieutenant Colonel Sven-Olof Kviman (October 6–15, 1982); War Diary of the Chief Naval Base East (WDCNBE), Rear Admiral Christer Kierkegaard, covering the Harsfjarden submarine incident (September 27–October 15 1982). Report from day-to-day decision at the Stockholm Coastal Defence Staff (RSCDS) in Vaxholm under Chief of Staff, Lieutenant Colonel Jan Svenhager (October 5–14, 1982), quoted by Tunander, *Some Remarks on the US/UK Submarine Deception in Swedish Waters in the 1980s*, p. 6.

26. AP, October 11, 1982.

27. AP, October 12, 1982.

28. SRMCDB, WDCMT/CMCDB, WDCNBE, RSCDS, quoted by Tunander, *Some Remarks on the US/UK Submarine Deception in Swedish Waters in the 1980s*, p. 6. See also Tunander, *The Secret War against Sweden: US and British Submarine Deception in the 1980s* (London: Frank Cass, 2004).

29. Tunander, *Some Remarks on the US/UK Submarine Deception in Swedish Waters in the 1980s*, p. 6.

30. Ibid.

31. Two days later, Vice Admiral Stefenson went to Mälsten by helicopter to comfort Kviman and his men, taking the unusual step to forward red roses to Lieutenant Colonel Kviman's wife with a letter commending her husband. The regional coastal defense chief, Brigadier General Lars Hansson later said that he was forced to release a submarine. Ibid., p. 10.

32. Alexandre Sheldon-Duplaix interview with Rear Admiral Emil Svenson and Captain Erland Sonnerstedt, Stockholm, March 17, 2008.

33. AP, October 27, 1982.

34. AP, October 18, 1982.

35. Leitenberg, *Soviet Submarine Operations in Swedish Waters*, pp. 50, 54.

36. Report from the Submarine Commission, Stockholm, Försvarsdepartementet, 1995, pp. 144–146, quoted by Tunander, *Some Remarks on the US/UK Submarine Deception in Swedish Waters in the 1980s*, p. 11.

37. *Time*, May 9, 1983.

38. Leitenberg, *Soviet Submarine Operations in Swedish Waters*, p. 65.

39. AP, "Sweden Sets Off Mines in Search for Submarines North of Capital," May 5, 1983.

40. Leitenberg, *Soviet Submarine Operations in Swedish Waters*, p. 71.

41. Carl Bildt, quoted in Leitenberg, *Soviet Submarine Operations in Swedish Waters*.

42. Reuters, "Sweden Letting Navy Attack Foreign Subs," July 2, 1983.

43. AP, July 20, 1983.

44. Leitenberg, *Soviet Submarine Operations in Swedish Waters*, p. 73.

45. AP, August 25, 1983.

46. AP, July 20, 1983; Leitenberg, *Soviet Submarine Operations in Swedish Waters*, p. 71.

47. R. W. Apple Jr., "Sweden Warns Moscow over Subs and Temporarily Recalls Its Envoy," *New York Times*, April 27, 1983.

48. Serge Schmemann, "Sweden's Charges Denied in Moscow," *New York Times*, April 27, 1983.

49. Leitenberg, *Soviet Submarine Operations in Swedish Waters*, p. 73.

50. Ibid., p. 81.

51. *Krasnaya Zvezda*, February 28, 1984, in Leitenberg, *Soviet Submarine Operations in Swedish Waters*, p. 92.

52. Reuters, "Swedes Hunt 3 Frogmen on Isle Near Navy Base," March 5, 1984.

53. Barnaby J. Feder, "Swedes' Seabed Spy Hunt: No Stone Is Left Unturned," *New York Times*, March 31, 1984.

54. Gordon H. McCormick, *Stranger Than Fiction: Soviet Submarine Operations in Swedish Waters* (Santa Monica, CA: Rand Corporation, 1990), p. 17.

55. AP, "Swedes Fire at a Sub Off the Southeast Coast," April 12, 1985; AP, "Sweden Calls Off Search for Foreign Submarine," April 13, 1985.

56. McCormick, *Stranger Than Fiction*, p. 23.

57. Incursions occurred at Goteborg (March, May), off Karlskrona (April, August), in Gullmarsfjord (June), in Sundsvall (July, November), and near Stockholm (July, November), in McCormick, *Stranger Than Fiction*, p. 17.

58. Contacts were chased in the Aland Sea, near Gotland, Tore (June), south of Hudviksall (July–August), and in the Kalmarsund (July–October), in McCormick, *Stranger Than Fiction*, p. 18; AP, "Swedes Again Cite Evidence of Alien Submarine Activity," January 26, 1986.

59. *Svenska Daglabet*, June 30, 1988, p. 6; McCormick, *Stranger Than Fiction*, p. 21.

60. McCormick, *Stranger Than Fiction*, p. 22.

61. Aprelev, *Pod "chorokh" nachykh "dieselii,"* pp. 255–285.

62. *Dagens Nyheter*, January 29, 1988, in McCormick, *Stranger Than Fiction*, p. 24.

63. Leitenberg, *Soviet Submarine Operations in Swedish Waters*, p. 149.

64. Ibid., p. 154.

65. Report from the Submarine Commission, Stockholm, Försvarsdepartementet, 1995.

66. Anders Mellbourn, '*Efter ubåtsrapporten 1983—Palme tvingad peka ut Sovjet,*' *Dagens Nyheter*, March 6, 1988; Ingvar Carlsson, *Ur skuggan av Olof Palme* (Stockholm: Hjalmarson & Högberg, 1999), p. 75; quoted in Tunander, *Some Remarks on the US/UK Submarine Deception in Swedish Waters in the 1980s*, p. 2.

67. Ibid.

68. *Rapport*, Swedish TV2, March 8, 2000.

69. *Aftonbladet* (October 1, 1982); the Naval Analysis Group report, 1983, Attachment 6, quoted in Tunander, *Some Remarks on the US/UK Submarine Deception in Swedish Waters in the 1980s*, p. 4.

70. Alexandre Sheldon-Duplaix interview with Rear Admiral Emil Svensson, Stockholm, March 17, 2008.

71. Chief Naval Base East, War Diary, (September 27–October 15, 1982), Rear Admiral Christer Kierkegaard, quoted in Tunander, *Some Remarks on the US/UK Submarine Deception in Swedish Waters in the 1980s*, p. 4.

72. The Naval Analysis Group report, 1983, p. 83, in Tunander, *Some Remarks on the US/UK Submarine Deception in Swedish Waters in the 1980s*.

73. Diary of the commander in chief, General Lennart Ljung, 1978–1986 (Stockholm: The Stockholm War Archive), quoted in Tunander, *Some Remarks on the US/UK Submarine Deception in Swedish Waters in the 1980s*, p. 4.

74. WDCMT/CMCDB, October 6–15, 1982, quoted in Tunander, *Some Remarks on the US/UK Submarine Deception in Swedish Waters in the 1980s*, p. 4.

75. *Communication Instructions, Distress and Rescue Procedures*, ACP 135 (E), the Combined Communications-Electronics Board (Australia, Canada, New Zeeland, United Kingdom and United States), March 1996; Chapter 34 Information Concerning Submarines (Canadian National Defence Headquarters), Notice to Mariners, Canadian Hydrographic Service (CHS), Canadian Ministry of Fisheries and Oceans, 2003; *NRL (1998) 75th Anniversary: Awards for Innovation, Celebrating 75 Years of Science and Technology Development for the Navy and the Nation* (Washington, DC: Naval Research Laboratory 1923–1998), www.nrl.navy.mil, quoted in Tunander, *Some Remarks on the US/UK Submarine Deception in Swedish Waters in the 1980s*, p. 47.

76. Ibid.

77. Ola Tunander interview with Bengt Gabrielsson, September 2000, quoted in Tunander, *Some Remarks on the US/UK Submarine Deception in Swedish Waters in the 1980s*.

78. Alexandre Sheldon-Duplaix interview with Rear Admiral Emil Svensson, Stockholm, March 17, 2008.

79. Protocol from the tape recording of underwater sounds in the Mälsten area made by Rolf Andersson from FMV (Swedish Defence Material), October 11–12, 1982; WDCMT; War Diary Chief Naval Base East, Rear Admiral Christer Kierkegaard, September 27–October 15 1982, quoted in Tunander, *Some Remarks on the US/UK Submarine Deception in Swedish Waters in the 1980s*.

80. Tape recordings from Mälsten, 1982 (MUSAC), quoted in Tunander, *Some Remarks on the US/UK Submarine Deception in Swedish Waters in the 1980s*.

81. The Naval Analysis Group Report (1983), Attachment 38.

82. The Naval Analysis Group Report (2001), p. 118.

83. Alexandre Sheldon-Duplaix interview with Per Clason, Stockholm, March 18, 2008.

84. Alexandre Sheldon-Duplaix interview with Rear Admiral Emil Svensson, Stockholm, March 17, 2008.

85. Swedish Government, Ministry of Foreign Affairs, "Record of Investigator's Conversation with Jaruzelski," Memorandum, September 26, 2002, PHP Web site.

86. Alexandre Sheldon-Duplaix interview with Rear Admiral Ivan Ivanov, St. Petersburg, July 2006.

87. Lee Vyborny and Don Davis, *Dark Waters, A Crew Member Reveals the True Story of America's Secret Submarine* (New York: New American Library, 2003), p. 147.

88. E-mail by Ola Tunander to Alexandre Sheldon-Duplaix, June 24, 2008.

89. John McWethy, *World News Tonight*, ABC, March 21, 1984.

90. Gary Stubblefield, *Inside the US Navy SEALs* (Osceola, WI: Motorbooks International, 1995), p. 144.

91. Ibid., p. 134.

92. Ola Tunander interview with Einar Ansteensen, December 1999, in Tunander, *Some Remarks on the US/UK Submarine Deception in Swedish Waters in the 1980s*, p. 16.

93. Ola Tunander interview with James Schlesinger, June 1993, in Tunander, *Some Remarks on the US/UK Submarine Deception in Swedish Waters in the 1980s*, p. 16.

94. E-mail from Ola Tunander to Alexandre Sheldon-Duplaix. June 24, 2008.

95. John P. Craven, *The Silent War—the Cold War Battle beneath the Sea* (New York: Simon & Schuster, 2001), p. 137.

96. Sherry Sontag and Christopher Drew, *Blind Man's Bluff: The Untold Story of American Submarine Espionage* (New York: Public Affairs, 1998), p. 356; see also USS *Seawolf* Web site, www.seawolf-ssn575.com/ssn575/uss_seawolf_history_continued.htm.

97. *Striptease*, Swedish TV2, April 11, 2000.

98. *Rapport*, Swedish TV1, November 21, 2001.

99. Tunander, *Some Remarks on the US/UK Submarine Deception in Swedish Waters in the 1980s*, p. 14.

100. Jim Ring, *We Come Unseen—the Untold Story of Britain's Cold War Submariners* (London: John Murray, 2001), p. 133; Tunander, *The Secret War against Sweden*, p. 14.

101. Tunander, *The Secret War against Sweden*, p. 116.

102. Göran Stütz, *Opinion 87—en opinionsundersökning om svenska folkets inställning till några samhälls- och försvarsfrågor hösten 1987* (Stockholm: Styrelsen för psykologiskt försvar, 1987), p. 64, quoted in Ola Tunander, *The Secret War against Sweden*, p. 20. These figures refer to a study by the Swedish Board of Psychological Defence (Stutz, 1987), p. 64.

103. Ibid.

104. Alexandre Sheldon-Duplaix interview with Rear Admiral Emil Svensson, Stockholm, March 17, 2008.

105. Alexandre Sheldon-Duplaix interview with Lev Vtorygin, Moscow, June 2007.

15. Spies in Uniform, 1980s

1. Brian Moynahan, *The Claws of the Bear: The History of the Soviet Armed Forces 1917 through Present* (London: Hutchinson, 1989), pp. 386–387.

2. The British, the French, and the Americans had military missions in Berlin, based on bilateral agreements with the commander of the Western Group of Soviet Forces. The British mission was the first and had a total of thirty touring officers. The French military mission suffered the loss of an NCO, Adjutant Chief Philippe Marriotti, when one of its tour vehicles was deliberately rammed near Halle, East Germany, by a Soviet army KRAZ-214 heavy truck in 1983. (Interview with Wing Commander Colin Campbell, RAF, and Colonel Jean Paul Huet, July 27, 1994.)

16. Soviet, U.S., . . . or Others? 1940s–1980s

1. On May 12, 1990, the official Soviet military journal *Krasnaia Zvezda* published an article by General Colonel A. Simonov on a UFO observed above the Baikonour space center; see also *Krasnaia Zvezda,* October 12 1991.

2. *New York Times*, February 28, 1960.

3. Jean Claude Bourret's interview of Defense Minister Robert Galley, February 21, 1974 (Radio Channel France Inter); see also two French government publications: Capitaine Kervendal and Charles Garreau, "Sur les traces des soucoupes volantes," *Revue d'études et d'informations de la Gendarmerie Nationale*, 87, 1st trimester, 1971; Gaston Alexis, "L'armée de l'air face aux OVNI," *Armées d'Aujourd'hui*, April 1976.

4. Lord Hill-Norton, foreword to *Above Top Secret: The Worldwide UFO Cover-Up*, by Timothy Good (London: Sidwick and Jackson, 1996).

5. Earlier, on February 16, 1987, Mikhail Gorbachev had confirmed to Russian journalists that Ronald Reagan had called his attention on this issue during the Geneva Summit on November 18–20, 1986. The U.S. president publicly addressed this topic on at least two more occasions on December 4, 1985 at Fallston High School, Maryland, and on May 4, 1988, at the National Strategy Forum in Chicago.

6. Reginald V. Jones, *Most Secret War, British Scientific Intelligence, 1939–1945* (London: Hamilton, 1978), p. 510.

7. Telegram from the U.S. Embassy to the Department of State, Stockholm, July, 11, 1946. Quoted in Good, *Above Top Secret*.

8. Telegrams from the U.S. Embassy to the Department of State, Stockholm, July 11, 1946, and August 29, 1946. Quoted in Good, *Above Top Secret*.

9. Jones, *Most Secret War*, p. 510.

10. Ibid.

11. *New York Times*, quoted in Good, *Above Top Secret*, p. xxxiii.

12. Air Force Base of Wright Field Dayton was later renamed Wright Patterson Air Base. On March 14, 1949, the newly created United States Air Force issued a secret listing of all reported sightings of UFOs in the United States and abroad since July 1947; NARA NND 917033, Record Group 38, RG 38, ONI Exhibits File 1942–58, Box 6.

13. Secret listing of all reported sightings of UFOs in the United States and abroad since July 1947; NARA NND 917033, Record Group 38, RG 38, ONI Exhibits File 1942–58, Box 6; Richard M. Dolan, *UFOs and the National Security State: Chronology of a Cover-Up* 1941–1973 (Charlottesville, VA: Hampton Roads Pub., 2002), pp. 395–397.

14. National Investigations Committee on Aerial Phenomena: "Pacing of Navy Missile," www.nicap.org/wsands2mc.htm; Robert McLaughlin, "How Scientists Tracked a Flying Saucer," www.nicap.org/true-mc.htm; memorandum, "Oct/1949 IG Special Inquiry into LA Times Article," www.nicap.org/nmexico/wsands490614docs.htm.

15. Ibid.

16. Air Material Command Opinion Concerning "Flying Discs" to Commanding General Army Air Force, Washington, D.C. Attention: Brig. General George Schulgen, September 23, 1947, NARA, RG 18, Records of the Army Air Force, AAG OOO General C.

17. Stanton Friedman, *Crash at Corona: The U.S. Military Retrieval and Cover-Up of a UFO* (New York: Paragon House, 1992), pp. 47–48.

18. Jeffrey M. Dorwart, *Conflict of Duty: The U.S. Navy's Intelligence Dilemma, 1919–1945* (Annapolis, MD: Naval Institute Press, 1983), p. 147.

19. The Parallel Universe of T. Townsend Brown, www.ttbrown.com.

20. Christopher Andrew, *Her Majesty's Secret Service: The Making of the British Intelligence Community* (New York: Penguin Books, 1987), pp. 465–467.

21. The Parallel Universe of T. Townsend Brown, www.ttbrown.com.

22. Margareth Cheney, *Tesla, Man Out of Time* (New York: Simon & Schuster, 2001), pp. 331–342.

23. Mario de Arcangelis, *Electronic Warfare: From the Battle of Tsushima to the Falklands and Lebanon Conflicts* (Dorset, UK: Blandford Press, 1985), p. 22.

24. The Parallel Universe of T. Townsend Brown, www.ttbrown.com.

25. Friedman, *Crash at Corona*, pp. 39–40.

26. Philipp Ziegler, *Mountbatten, the Official Biography* (London: Collins, 1985), p. 494.

27. Fleet Logistic Air Wing, Atlantic/Continental Air Transport Squadron One U.S. Naval Air Station Patuxent River, Maryland, February 10, 1951, Memorandum Report to Commanding Officer, Air Transport Squadron One, "Subj: Report of Unusual Sighting on Flight 125/9," February 1951, NICAP.

28. Donald Keyhoe, *Aliens from Space: The Real Story of Unidentified Flying Objects* (Garden City, NY: Doubleday, 1973), pp. 65–66.

29. Ibid.

30. Commander Edward P. Stafford, U.S. Navy (Retired), "Cosmic Curiosity," *Naval Institute Naval History*, October 2004.

31. "Operation Mainbrace Sightings, 1952, www.ufocasebook.com/operationmain-brace1952.html.

32. Ibid.

33. At the time of the incident, conditions were as follows: "Date—November 14, 1949, Time 1830 GMT, Position 26'47.5', 56°-51' E. Wind NW'ly force 1. Sea calm with slight surface ripples; no swell. Air 75° (Fahr.), sea 83°. Visibility very good. A clear, bright night with no moon. Vessels course 157° T. Speed through the water 11.6 knots. Actual speed over the bottom, approximately 9 knots due to strong head current. (Very strong streams are encountered in this area.) At no time were any unusual deviations of the magnetic compass observed."; J. R. Bodler, "An Unexplained Phenomenon of the Sea," United States Naval Institute *Proceedings* (January 1952): 66–67.

34. *Air Intelligence Digest* 7, no. 12 (December 1954), the Computer UFO Network, www.cufon.org.

35. NARA RG 38, Office of Naval Intelligence, informal letter to USNA Moscow, January 13, 1955, the Computer UFO Network, www.cufon.org.

36. The Parallel Universe of T. Townsend Brown, www.ttbrown.com.

37. Under CINPACFLT instruction 3820.3 and CINLANTFLT instruction 03360.2C; Antonio F. Rullan, "Blue Book UFO Reports at Sea by Ships, Analysis of the Blue Book Ship Database," December 10, 2002, Martinez, CA.

38. Rullan, "Blue Book UFO Reports at Sea by Ships, Analysis of the Blue Book Ship Database."

39. Wyman H. Packard, *Century of U.S. Naval Intelligence* (Washington, DC: Naval Historical Center, 1996), p. 51.

40. The first MERINT report that reached Project Blue Book was sent on November 5, 1957, by the SS *Hampton Roads*, navigating south southwest of New Orleans. The first report to Project Blue Book by a U.S. Navy ship was on April 7, 1955.

41. Chester Grusinski, *UFO Sightings from the U.S.S Franklin D. Roosevelt* (Dublin, OH: MidOhio Research Associates, 1995); *Ohio UFO Notebook*, no.9 (1995): pp. 21–23.

42. Interview with Vice Admiral Kviatkovskiy, Moscow, June 2007.

43. Herman Koltchin, *Fenomen NLO, Vgliad Iz Rossii* (St. Petersburg: Liss, 1997), p. 103.

44. Ibid., p. 104.

45. Ibid.

46. Nikolai Shigin, unpublished paper on "Kvakiers," 2006, p. 2.

47. Ibid.

48. Koltchin, pp. 344–347.

49. Ibid.

50. Ibid.

51. Alexandre Sheldon-Duplaix interview with retired captain first rank Evguenii Litvinov, Department of Geography, Academy of Sciences of the Russian Federation, St. Petersburg, July 2007.

52. Koltchin, pp. 40, 104.

53. Shigin, p. 2.

54. Alexandre Sheldon-Duplaix interview with retired captain first rank Evguenii Litvinov, Anomaly Committee, Department of Geography, Academy of Sciences of the Russian Federation, St. Petersburg, July 2007.

55. Jim Kopf, "U.S. Aircraft Carrier Stopped by UFO," NICAP files, www.nicap.org/jfk.htm.

56. Koltchin, p. 76, 77.

57. Ibid., pp. 76, 77.

58. Ibid., p. 109.

59. DIA Defense Information Evaluation Report, IR No. 6846013976, September 22, 1976, by Major Roland B. Evans, USAF, Military Capability Analyst. A three-page Department of Defense message was obtained by Charles Huffer in 1977 under the Freedom of Information Act. After midnight on September 19, 1976, two successive Iranian air force F-4 Phantom interceptors attempted to catch a radar-visual brightly lit UFO reported above Teheran. As the pursuit continued, two brightly lit objects emerged from the first object. One pilot attempted to fire a missile, but his weapons control panel went off and he lost all communications.

60. Two-Dimensional Asymmetrical Thrust Capacitor, Protection, patent no. 6,317,310; reference no. TOP8–80.

61. National Investigations Committee on Aerial Phenomena (NICAP), www.nicap.org.

62. Ibid., p. 40.

63. Interview by V. Prasdichev, "Phantoms of the Depths," REN TV, Russia, 2005.

17. The Legacy of the Soviet Navy, 1989 and After

1. The Igor Boechin article "Investigating a Catastrophe: Sarcophagus at Bear Island," in the Russian magazine *Technika-Molodezhi*, contains precise details surrounding the conditions at the site of the sunken *Komsomolets.* Boechin claims that the area is one of the most active in underwater growth and marine life.

2. William Broad, *New York Times*, September 9, 1994.

3. From Dmitri A. Romanov's *Tragedy of Submarine* Komsomolets: *Arguments of the Constructor* (St. Petersburg: Assotsiatsiia izdatelei Sankt-Peterburga, 1993), p. 252.

4. Peter Huchthausen interview with the constructor Yuri Kormilitsin, September 25, 1995, St. Petersburg.

5. Peter Huchthausen interview with Dmitri A. Romanov, St. Petersburg, September 1995.

6. Nikolai Mormul's book *La Dramatique Histoire de Sous-Marins Nucléaires Soviétiques*, with Lev Giltsov and Leonid Ossipenko (Paris: Robert Laffont, 1992) and S. P. Buran's *Po Sledam Podvodnikh Katastrof* (Moscow: Gilda Masterov Rus, 1992) contain tables of submarine accidents citing details of the casualties and the men lost. Not all losses were recorded in the two books. Additional material regarding the losses appears in the vastly more detailed Greenpeace work by Joshua Handler titled *Soviet /Russian Submarine Accidents: 1956–1994*. The *Yablokov Report* also contains details that do not appear in any of the previous sources.

7. Mormul's book was first published in France in 1992. When it was finally printed in Russia in 1996, the Russian version—*Atomnaya Podvonaya Epopeya, Podvigii, Neydatchii, Katastrofi* (Moscow: Izdatelstvo Borgess, 1994)—had been stripped of much of his disclosures about dumping and all of the data regarding how the Soviet navy leadership had repressed him personally.

8. Mormul's Russian-language edition of *Atomnaya Podvonaya Epopeya*, chap. 9, pp. 291–310, gives a detailed account of what is known and not known about nuclear waste dumping during the Soviet years.

9. Peter A. Huchthausen, Igor Kurdin, and R. Alan White, *Hostile Waters* (New York: St. Martin's Press, 1997).

10. Handler, *Soviet Russian Submarine Accidents*, p. 7.

11. Deep Sea Radiological Environmental Monitoring reports on the sites of USS *Thresher* and USS *Scorpion*, 1993.

12. *Bellona Report*, chapter 1.3.1, "Economic Conditions," p. 15.

13. Ibid., p. 18.

14. *Nezavivisimaya Gazeta*, April 22, 1995, interview with commander in chief of the Northern Fleet Oleg A. Yerofeev.

15. Vladimir Kiselyov of the Institute for the Problems of the Safe Development of Atomic Energy, www.bellona.org/subjects/1140451820.2.

16. "Andreyeva Bay—Time to Avert a Catastrophe, Anna Kireeva, May 23, 2008, translated by Charles Digges, www.bellona.org/subjects/1140451820.2.

17. P. J. Capelotti, *Our Man in the Crimea: Commander Hugo Koehler and the Russian Civil War* (Columbia: University of South Carolina Press, 1991), p. 12.

18. A former naval intelligence officer and attaché in Sweden, retired navy captain Edmund Pope wrote an account of his misadventures in Russia: Edmund Pope and Tom Shachtman, *Torpedoed: An American Businessman's True Story of Secrets, Betrayal, Imprisonment in Russia, and the Battle to Set Him Free* (Boston: Little, Brown, 2001).

PHOTO CREDITS

INDEX

Page numbers in *italics* indicate illustrations.